CAMBRIDGE TRACTS IN MATHEMATICS

---

202 Canonical Ramsey Theory on Polish Spaces

CAMBRIDGE TRACTS IN MATHEMATICS

A complete list of books in the series can be found at www.cambridge.org/mathematics. Recent titles include the following:

169. Quantum Stochastic Processes and Noncommutative Geometry. By K. B. SINHA and D. GOSWAMI
170. Polynomials and Vanishing Cycles. By M. TIBĂR
171. Orbifolds and Stringy Topology. By A. ADEM, J. LEIDA, and Y. RUAN
172. Rigid Cohomology. By B. LE STUM
173. Enumeration of Finite Groups. By S. R. BLACKBURN, P. M. NEUMANN, and G. VENKATARAMAN
174. Forcing Idealized. By J. ZAPLETAL
175. The Large Sieve and its Applications. By E. KOWALSKI
176. The Monster Group and Majorana Involutions. By A. A. IVANOV
177. A Higher-Dimensional Sieve Method. By H. G. DIAMOND, H. HALBERSTAM, and W. F. GALWAY
178. Analysis in Positive Characteristic. By A. N. KOCHUBEI
179. Dynamics of Linear Operators. By F. BAYART and É. MATHERON
180. Synthetic Geometry of Manifolds. By A. KOCK
181. Totally Positive Matrices. By A. PINKUS
182. Nonlinear Markov Processes and Kinetic Equations. By V. N. KOLOKOLTSOV
183. Period Domains over Finite and p-adic Fields. By J.-F. DAT, S. ORLIK, and M. RAPOPORT
184. Algebraic Theories. By J. ADÁMEK, J. ROSICKÝ, and E. M. VITALE
185. Rigidity in Higher Rank Abelian Group Actions I: Introduction and Cocycle Problem. By A. KATOK and V. NIŢICĂ
186. Dimensions, Embeddings, and Attractors. By J. C. ROBINSON
187. Convexity: An Analytic Viewpoint. By B. SIMON
188. Modern Approaches to the Invariant Subspace Problem. By I. CHALENDAR and J. R. PARTINGTON
189. Nonlinear Perron–Frobenius Theory. By B. LEMMENS and R. NUSSBAUM
190. Jordan Structures in Geometry and Analysis. By C.-H. CHU
191. Malliavin Calculus for Lévy Processes and Infinite-Dimensional Brownian Motion. By H. OSSWALD
192. Normal Approximations with Malliavin Calculus. By I. NOURDIN and G. PECCATI
193. Distribution Modulo One and Diophantine Approximation. By Y. BUGEAUD
194. Mathematics of Two-Dimensional Turbulence. By S. KUKSIN and A. SHIRIKYAN
195. A Universal Construction for Groups Acting Freely on Real Trees. By I. CHISWELL and T. MÜLLER
196. The Theory of Hardy's Z-Function. By A. IVIĆ
197. Induced Representations of Locally Compact Groups. By E. KANIUTH and K. F. TAYLOR
198. Topics in Critical Point Theory. By K. PERERA and M. SCHECHTER
199. Combinatorics of Minuscule Representations. By R. M. GREEN
200. Singularities of the Minimal Model Program. By J. KOLLÁR
201. Coherence in Three-Dimensional Category Theory. By N. GURSKI
202. Canonical Ramsey Theory on Polish Spaces. By V. KANOVEI, M. SABOK and J. ZAPLETAL
203. A Primer on the Dirichlet Space. By O. EL-FALLAH, K. KELLAY, J. MASHREGHI and T. RANSFORD

Canonical Ramsey Theory on Polish Spaces

VLADIMIR KANOVEI

Institute for Information Transmission Problems of the Russian Academy of Sciences

and

Moscow State University of Railway Engineering

MARCIN SABOK

Institute of Mathematics of the Polish Academy of Sciences

and

Uniwersytet Wrocławski

JINDŘICH ZAPLETAL

University of Florida

and

Institute of Mathematics, Academy of Sciences, Czech Republic

CAMBRIDGE
UNIVERSITY PRESS

CAMBRIDGE
UNIVERSITY PRESS

University Printing House, Cambridge CB2 8BS, United Kingdom

Published in the United States of America by Cambridge University Press, New York

Cambridge University Press is part of the University of Cambridge.

It furthers the University's mission by disseminating knowledge in the pursuit of education, learning, and research at the highest international levels of excellence.

www.cambridge.org
Information on this title: www.cambridge.org/9781107026858

First published 2013

Printed in the United Kingdom by CPI Group Ltd, Croydon CR0 4YY

A catalogue record for this publication is available from the British Library

ISBN 978-1-107-02685-8 Hardback

Contents

	Preface	*page* vii
1	**Introduction**	1
	1.1 Motivation	1
	1.2 Basic concepts	3
	1.3 Outline of results	7
	1.4 Navigation	10
	1.5 Notation	12
2	**Background facts**	14
	2.1 Descriptive set theory	14
	2.2 Invariant descriptive set theory	16
	2.3 Forcing	18
	2.4 Absoluteness and interpretations	23
	2.5 Generic ultrapowers	31
	2.6 Idealized forcing	33
	2.7 Concentration of measure	42
	2.8 Katětov order and coding functions	42
3	**Analytic equivalence relations and models of set theory**	45
	3.1 The model $V[x]_E$	45
	3.2 The model $V[[x]]_E$	54
	3.3 The poset P_I^E	56
4	**Classes of equivalence relations**	62
	4.1 Smooth equivalence relations	62
	4.2 Countable equivalence relations	63
	4.3 Equivalence relations classifiable by countable structures	68
	4.4 Hypersmooth equivalence relations	76
	4.5 Analytic vs. Borel equivalence relations	78

5 Games and the Silver property 80
5.1 Integer games connected with σ-ideals 80
5.2 Determinacy conclusions 86
5.3 The selection property 90
5.4 A Silver-type dichotomy for a σ-ideal 97

6 The game ideals 100
6.1 σ-ideals σ-generated by closed sets 100
6.2 Porosity ideals 120
6.3 Laver forcing 134
6.4 Fat tree forcings 136
6.5 The Lebesgue null ideal 139

7 Benchmark equivalence relations 147
7.1 The E_0 ideal 147
7.2 The E_1 ideal 150
7.3 The E_2 ideal 172

8 Ramsey-type ideals 187
8.1 Halpern–Läuchli cubes 187
8.2 Silver cubes 190
8.3 Ellentuck cubes 196
8.4 Milliken cubes 202
8.5 Matet cubes 207

9 Product-type ideals 218
9.1 Finite products and the covering property 218
9.2 Finite products of the E_0 ideal 222
9.3 Infinite products of perfect sets 232
9.4 Products of superperfect sets 240

10 The countable support iteration ideals 244
10.1 Iteration preliminaries 245
10.2 The weak Sacks property 247
10.3 A game reformulation 248
10.4 The canonization result 255
10.5 The iteration of Sacks forcing 260
10.6 Anticanonization results 261

References 264
Index 268

Preface

We wrote this book to lay the foundation to an area of descriptive set theory parallel to the canonical Ramsey theory in finite combinatorics. The book develops the theory of analytic and Borel equivalence relations, a subject that has received a great deal of attention over the past two decades, and connects it with the abstract methods of forcing and generic extensions.

The subject grew quickly from near nonexistence in 2010 to a large body of interconnected results of considerable sophistication by the end of 2012. An entirely new landscape was created, with a number of satisfactory general results as well as many avenues open for further investigation.

The target audience consists of graduate students and researchers with basic experience in descriptive set theory and forcing. We hope that the book will help them to sharpen the understanding of Borel and analytic equivalence relations.

We would like to thank Bohuslav Balcar, Clinton Conley, Benjamin Miller, Christian Rosendal and Sławek Solecki for many helpful comments and suggestions. We are also indebted to Adrian Mathias for allowing us to include Theorem 8.17 in this book.

It is our pleasure to acknowledge the financial support from several sources that made possible our cooperation during the period 2010–2012. Vladimir Kanovei was partially supported through the RFBR grant 01-13-000006. Marcin Sabok was supported through the MNiSW (Polish Ministry of Science and Higher Education) grant N201418939, the Foundation for Polish Science, the Czech–Polish cooperation AIP project MEB051006 and INFTY, the ESF research networking programme. Jindřich Zapletal was partially supported through the following grants: AIP project MEB051006, AIP project MEB060909, Institutional Research Plan No. AV0Z10190503 and grant IAA100190902 of the Grant Agency of the Academy of Sciences of the Czech Republic, NSF grants DMS 1161078 and 0801114, and INFTY, the ESF research networking programme. Last, but not least, the authors would like to thank the Fields Institute for Research in Mathematical Sciences for its support and hospitality during the Thematic Program on Forcing and its Applications in the fall of 2012, during which this book was completed.

1
Introduction

1.1 Motivation

The Ramsey theory starts with a classical result:

Fact 1.1 *For every partition of pairs of natural numbers into two classes there is a* homogeneous *infinite set: a set $a \subset \omega$ such that all pairs of natural numbers from a belong to the same class.*

It is not difficult to generalize this result for partitions into any finite number of classes. An attempt to generalize further, for partitions into infinitely many classes, hits an obvious snag: every pair of natural numbers could fall into its own class, and then certainly no infinite homogeneous set can exist for such a partition. Still, there seems to be a certain measure of regularity in partitions of pairs even into infinitely many classes. This is the beginning of canonical Ramsey theory.

Fact 1.2 (Erdős–Rado (Erdős and Rado 1950)) *For every equivalence relation E on pairs of natural numbers there is an infinite* homogeneous *set: a set $a \subset \omega$ on which one of the following happens:*

(i) $p \mathrel{E} q \leftrightarrow p = q$ *for all pairs* $p, q \in [a]^2$;
(ii) $p \mathrel{E} q \leftrightarrow \min(p) = \min(q)$ *for all pairs* $p, q \in [a]^2$;
(iii) $p \mathrel{E} q \leftrightarrow \max(p) = \max(q)$ *for all pairs* $p, q \in [a]^2$;
(iv) $p \mathrel{E} q$ *for all pairs* $p, q \in [a]^2$.

In other words, there are four equivalence relations on pairs of natural numbers such that any other equivalence can be *canonized*: made equal to one of the four equivalences on the set $[a]^2$, where $a \subset \omega$ is judiciously chosen infinite set. It is not difficult to see that the list of the four primal equivalence relations is irredundant: it cannot be shortened for the purposes of this theorem. It is also not difficult to see that the usual Ramsey theorems follow from the canonical version.

Further generalizations of these results can be sought in several directions. An exceptionally fruitful direction considers partitions and equivalences of substructures of a given finite or countable structure, such as in Nešetřil (2005). Another direction seeks to find homogeneous sets of larger cardinalities. In set theory with the axiom of choice, the search for uncountable homogeneous sets of arbitrary partitions leads to large cardinal axioms (Kanamori 1994), and this is one of the central concerns of modern set theory. A different approach will seek homogeneous sets for partitions that have a certain measure of regularity, typically expressed in terms of their descriptive set theoretic complexity in the context of Polish spaces (Todorcevic 2010). This is the path this book takes. Consider the following classical results:

Fact 1.3 (Mycielski 1964) *If R is a meager relation on a Polish space X, then there is a perfect set consisting of pairwise non-R-related elements.*

Fact 1.4 (Silver 1970) *For every partition $[\omega]^{\aleph_0} = B_0 \cup B_1$ into an analytic and coanalytic piece, one of the pieces contains a set of the form $[a]^{\aleph_0}$, where $a \subset \omega$ is some infinite set.*

Here, the space $[\omega]^{\aleph_0}$ of all infinite subsets of natural numbers is considered with the usual Polish topology which makes it homeomorphic to the space of irrational numbers. This is the most influential example of a Ramsey theorem on a Polish space. It deals with Borel partitions only as the Axiom of Choice can be easily used to construct a partition with no homogeneous set of the requested kind.

Are there any canonical Ramsey theorems on Polish spaces concerning sets on which Borel equivalence relations can be canonized? A classical example of such a theorem is the Silver dichotomy:

Fact 1.5 (Silver 1980) *If E is a coanalytic equivalence relation on a Polish space X, then either there is a perfect set consisting of pairwise E-related elements, or the space X decomposes into countably many E-equivalence classes.*

As a consequence, a coanalytic equivalence relation must have perfect set of pairwise inequivalent elements, or a perfect set of pairwise equivalent elements. If one wishes to obtain sets on which the equivalence relation is simple that are larger than just perfect, the situation becomes more complicated. Another classical result starts with an identification of Borel equivalence relations E_γ on the space $[\omega]^{\aleph_0}$ for every function $\gamma : [\omega]^{<\aleph_0} \to 2$ (the exact statement and definitions are stated in Section 8.3) and then proves the following:

Fact 1.6 (Prömel–Voigt, Mathias (Prömel and Voigt 1985; Mathias 1977)) *If $f : [\omega]^{\aleph_0} \to 2^\omega$ is a Borel function then there is γ and an infinite set $a \subset \omega$ such that for all infinite sets $b, c \subset a$, $f(b) = f(c) \leftrightarrow b\ E_\gamma\ c$.*

Thus, this theorem deals with *smooth* equivalence relations on the space $[\omega]^{\aleph_0}$, i. e., those equivalences E for which there is a Borel function $f : [\omega]^{\aleph_0} \to 2^\omega$ such that $b\ E\ c \leftrightarrow f(b) = f(c)$, and shows that such equivalence relations can be canonized to a prescribed form on a Ramsey cube. Other similar results can be found in the work of Otmar Spinas (Spinas 2001a,b; Klein and Spinas 2005). In the realm of nonsmooth equivalence relations, we have, for example:

Fact 1.7 (Connes–Feldman–Weiss (Connes *et al.* 1981)) *Suppose that X is a Polish space with an amenable countable Borel equivalence relation E on it and a quasi-invariant Borel probability measure μ. Then there is a Borel set $B \subset X$ of μ-mass 1 on which E is an orbit equivalence relation of a Borel action of $\mathbb{Z}$.*

The principal aim of this book is to expand this line of research in two directions. First, we consider canonization properties of equivalence relations more complicated than smooth in the sense of the Borel reducibility complexity rating of equivalence relations (Kanovei 2008; Gao 2009). It turns out that distinct complexity classes of equivalence relations possess various canonization properties, and most of the interest and difficulty lies in the nonsmooth cases. Second, we consider the task of canonizing the equivalence relations on Borel sets which are large from the point of view of various σ-ideals. Again, it turns out that σ-ideals commonly used in mathematical analysis greatly differ in their canonization properties, and a close relationship with their forcing properties, as described in Zapletal (2008), appears. Canonization theorems can then be applied to obtain ergodicity results for classical Borel equivalence relations such as E_2 and F_2.

1.2 Basic concepts

The central canonization notion of the book is the following:

Definition 1.8 A σ-ideal I on a Polish space X has *total canonization* for a class of equivalence relations if for every Borel I-positive set $B \subset X$ and every equivalence relation E on B in this class there is a Borel I-positive set $C \subset B$ consisting only of pairwise E-inequivalent elements or only of pairwise

E-equivalent elements. Total canonization (without the class of equivalence relations mentioned) means the total canonization for the class of all analytic equivalence relations.

The classes of equivalence relations considered in this book are nearly always closed under Borel reducibility; the broadest class would be that of all analytic equivalence relations. The total canonization is closely related to an ostensibly stronger notion:

Definition 1.9 A σ-ideal I on a Polish space X has the *Silver property* for a class of equivalence relations if for every Borel I-positive set $B \subset X$ and every equivalence relation E on B in this class, either there is a Borel I-positive set $C \subset B$ consisting of pairwise E-inequivalent elements, or B decomposes into a union of countably many E-equivalence classes and an I-small set. The Silver property (without the class of equivalence relations mentioned) means the Silver property for the class of all Borel equivalence relations.

Thus, the usual Silver theorem can be restated as the Silver property for coanalytic equivalence relations for the σ-ideal of countable sets. Unlike the total canonization, the Silver property introduces a true dichotomy, as the two options presented cannot coexist for σ-ideals containing all singletons: the Borel I-positive set C from the first option would have to have a positive intersection with one of the countably many equivalence classes from the second option, which is of course impossible. The Silver dichotomy also has consequences for undefinable sets. If there is *any* I-positive set $C \subset B$ consisting of pairwise E-inequivalent elements, then there must be a *Borel* such set, simply because the second option of the dichotomy is excluded by the same argument as above.

The Silver property certainly implies total canonization within the same class of equivalence relations: either there is a Borel I-positive set $B \subset C$ consisting of pairwise inequivalent elements, or one of the equivalence classes from the second option of the dichotomy must be I-positive and provides the second option of the total canonization. On the other hand, total canonization does not imply the Silver property as shown in Section 8.1. In many common cases, the various uniformization or game theoretic properties of the σ-ideal I in question can be used to crank up the total canonization to the Silver property. This procedure is described in Section 5.4 and it is the only way used to argue for the Silver property in this book.

There are a number of tricks used to obtain total canonization for various restricted classes of equivalence relations used in this book. However, the total canonization for all analytic equivalence relations is invariably proved via the following notion originating in Ramsey theory:

Definition 1.10 A σ-ideal I on a Polish space X has the *free set property* if for every I-positive analytic set $B \subset X$ and every analytic set $D \subset B \times B$ with all vertical sections in the ideal I there is a Borel I-positive set $C \subset B$ such that $(C \times C) \cap D \subseteq \mathrm{id}$.

The free set property immediately implies the total canonization for analytic equivalence relations. If E is an analytic equivalence relation on an I-positive Borel set $B \subset X$, then either there is an I-positive E-equivalence class, which immediately yields the second option of total canonization, or, the relation $E \subset B \times B$ has I-small vertical sections, and the E-free I-positive set postulated by the free set property yields the first option of total canonization. It is not at all clear how one would argue for the opposite implication though, and we are coming to the first open question of this book.

Question 1.11 Is there a σ-ideal on a Polish space that has total canonization for analytic equivalence relations, but not the free set property?

The free set property is typically verified through fusion arguments as in Theorem 6.8, or through some version of the mutual generics property. This is the first place in this book where forcing makes explicit appearance through the following notion:

Definition 1.12 (Zapletal 2008) If I is a σ-ideal on a Polish space X, then the symbol P_I denotes the partial order of I-positive Borel subsets of X, ordered by inclusion.

Definition 1.13 A σ-ideal I on a Polish space X has the *mutual generics property* if for every Borel I-positive set B and every countable elementary submodel M of a large enough structure containing I and B there is a Borel I-positive subset $C \subset B$ such that its points are pairwise generic for the product forcing $P_I \times P_I$.

The mutual generics property is a strengthening of properness of the poset P_I, as the characterization of properness (Fact 2.50) shows. It implies the free set property by Proposition 2.57, but it is not implied by it by Theorem 8.1. It is somewhat ad hoc in that in many cases, the arguments demand that the product forcing is replaced by various reduced product forcings, see Sections 7.1 or 8.1. The resulting tools are very powerful and flexible, and where we cannot find them, we spend some effort proving that no version of mutual generics property can hold. The concept used to rule out all its versions is of independent interest:

Definition 1.14 A σ-ideal I on a Polish space X has a *square coding function* if there are a Borel I-positive set $B \subset X$ and a Borel function $f : B \times B \to 2^\omega$ such that for every analytic I-positive set $C \subset B$, we have $f''(C \times C) = 2^\omega$. The σ-ideal I has a *rectangular coding function* if there are Borel I-positive sets $B_0, B_1 \subset X$ and a Borel function $g : B_0 \times B_1 \to 2^\omega$ such that, for every analytic I-positive sets $C_0 \subset B_0$ and $C_1 \subset B_1$, we have $g''(C_0 \times C_1) = 2^\omega$.

Clearly, if a σ-ideal has a square coding function then no I-positive Borel set below the critical set B (which is always equal to the whole space in this book) can consist of points pairwise generic over a given countable model M, since some pairs in this set code via the function f, for example a transitive structure isomorphic to M. We will show that many σ-ideals have a square or rectangular coding function (Theorem 6.11 or 6.22) and that coding functions can be abstractly obtained from the failure of canonization of equivalence relations in many cases (Corollaries 7.35, 7.51, and 10.30).

In the cases where total canonization is unavailable, we can still provide a number of good canonization results. In the spirit of traditional Ramsey-theoretic notation, we introduce an arrow to record them. This is the central definition of the book:

Definition 1.15 Let $\mathbf{E}$, $\mathbf{F}$ be two classes of equivalence relations, and let I be a σ-ideal on a Polish space X. $\mathbf{E} \to_I \mathbf{F}$ denotes the statement that for every I-positive Borel set $B \subset X$ and an equivalence $E \in \mathbf{E}$ on B there is a Borel I-positive set $C \subset B$ such that $E \restriction C \in \mathbf{F}$.

The strongest anti-canonization results we can achieve will be cast in terms of a *spectrum* of a σ-ideal. Note that it has no obvious counterpart in the canonical Ramsey theory on finite structures.

Definition 1.16 An analytic equivalence relation E is in the *spectrum* of a σ-ideal I on a Polish space X if there is an I-positive Borel set $B \subset X$ and an equivalence relation F on B which is Borel bireducible with E, and also for every I-positive Borel set $C \subset B$, $F \restriction C$ remains bireducible with E.

For example, the equivalence relations such as E_0 and F_2 are in the spectrum of the meager ideal, and E_{K_σ} is in the spectrum of the ideals associated with Silver forcing and Laver forcing. From the point of view of Ramsey theory, finding a nontrivial equivalence relation in the spectrum amounts to a strong negative result. However, results of this sort may be quite precious in themselves and have further applications.

1.3 Outline of results

For a good number of σ-ideals, we prove the strongest canonization results possible. As a motivational example, we include:

Theorem 1.17 (Corollary 6.16) *The σ-ideal on ω^ω σ-generated by compact sets has the Silver property. Restated, for every Borel equivalence relation E on ω^ω, exactly one of the following holds:*

(i) there is a closed set $C \subset \omega^\omega$ homeomorphic to ω^ω consisting of pairwise inequivalent points;
(ii) ω^ω is covered by countably many E-classes and countably many compact sets.

One satisfactory general theorem in this direction proved in this book states the following:

Theorem 1.18 (Theorem 6.8 and Corollary 6.9) *Let I be a σ-ideal on a compact space, σ-generated by a coanalytic family of compact sets. If I is calibrated, then it has the free set property, total canonization for analytic equivalence relations, and the Silver property.*

The class of calibrated σ-ideals introduced by Kechris and Louveau (1989) contains many σ-ideals commonly studied in abstract analysis: the σ-ideal on the unit interval σ-generated by closed sets of measure zero, the σ-ideal on the unit circle σ-generated by closed sets of uniqueness, or (up to a technical detail) the σ-ideal on the Hilbert cube σ-generated by compact sets of finite dimension.

In most cases, such a strong canonization result either is not available or we do not know how to prove it. Situations where the total canonization fails because there are clearly identifiable obstacles to it are of great interest. Therefore, we strive to canonize up to the known obstacles:

Theorem 1.19 (Theorem 9.3) *Let $n \in \omega$, let $\{I_i : i \in n\}$ be σ-ideals on respective compact spaces $\{X_i : i \in n\}$, σ-generated by coanalytic collection of compact sets and such that every I_i-positive analytic set has an I_i-positive compact subset. If $\{B_i : i \in n\}$ are Borel I_i-positive sets and E is an analytic equivalence relation on $\prod_i B_i$, there is a set $a \subset n$ and Borel I-positive sets $\{C_i \subset B_i : i \in n\}$ such that $E \restriction \prod_i C_i = \mathrm{id}_a$, where id_a is the equality on indices in the set a. In other words, writing I for the σ-ideal of those Borel subsets of $\prod_i X_i$ that do not contain a product of I_i-positive Borel sets, we have*

$$\text{analytic} \to_I \{\mathrm{id}_a : a \subset n\}.$$

Theorem 1.20 (Theorem 7.1) *Let I be the σ-ideal on 2^ω generated by Borel sets on which the equivalence relation E_0 is smooth. If B is a Borel I-positive set and E is an analytic equivalence relation, then there is a Borel I-positive set $C \subset B$ such that either C consists of pairwise inequivalent elements, or C consists of pairwise equivalent elements, or $E \restriction C = E_0$. In other words,*

$$\text{analytic} \to_I \{\text{id}, \text{ev}, E_0\}.$$

Theorem 1.21 (Corollary 7.7) *Let I be the σ-ideal on $(2^\omega)^\omega$ generated by Borel sets on which the equivalence relation E_1 is Borel reducible to E_0. If B is a Borel I-positive set and E is an a hypersmooth equivalence relation on B, then there is a Borel I-positive set $C \subset B$ such that either C consists of pairwise E-inequivalent elements, or C consists of pairwise E-equivalent elements, or $E \restriction C = E_1$. In other words,*

$$\text{hypersmooth} \to_I \{\text{id}, \text{ev}, E_1\}.$$

Theorem 1.22 (Theorem 7.36) *Let I be the σ-ideal on 2^ω generated by Borel sets on which the equivalence relation E_2 is essentially countable. If B is a Borel I-positive set and E is an an equivalence relation on B Borel reducible to $=_J$ for an F_σ P-ideal J on ω, then there is a Borel I-positive set $C \subset B$ such that either C consists of pairwise E-inequivalent elements, or C consists of pairwise E-equivalent elements, or $E \restriction C = E_2$. In particular,*

$$\ell_p \text{ equivalences} \to_I \{\text{id}, \text{ev}, E_2\}.$$

In a similar vein, we apply the canonization techniques to achieve a number of ergodicity results for certain classical Borel equivalence relations.

Definition 1.23 If E, F are equivalence relations on Polish spaces X and Y, we say that E is *F-generically ergodic* if for every Borel homomorphism from E to F there is a comeager subset of X that is mapped into a single F-equivalence class. Similarly, if μ is a Borel probability measure on X, we say that E is *μ, F-ergodic* if for every homomorphism from E to F, there is a subset of X of full μ-mass that is mapped into a single F-equivalence class.

Hjorth and Kechris (Kanovei 2008, theorem 13.5.3) showed that E_2 is F-generically ergodic for every equivalence relation F classifiable by countable structures. We replace their turbulence techniques with Ramsey theory and prove a similar result.

Theorem 1.24 (Theorem 6.66) *Let E be an equivalence relation on 2^ω; the space 2^ω is equipped with the usual Borel probability measure μ. If $E_2 \subseteq E$*

and E is classifiable by countable structures then E has a co-null class. Restated, E_2 is μ, F-ergodic for every equivalence relation F classifiable by countable structures.

Theorem 1.25 (Theorems 6.24 and 6.67) *Let E be an analytic equivalence relation on $(2^\omega)^\omega$; the space $(2^\omega)^\omega$ is equipped with the usual product topology and the product Borel probability measure μ. If $F_2 \subseteq E$ and F_2 is not Borel reducible to E, then E has a comeager class as well as a co-null class. Restated, for every analytic equivalence relation F exactly one of the following holds:*

(i) $F_2 \leq_B F$;
(ii) F_2 is F-generically ergodic and F_2 is μ, F-ergodic.

The weakest canonization results we obtain reduce the Borel reducibility complexity of equivalence relations to a specified class. It should be noted that no parallel to results of this kind exists in the realm of finite or countable canonization results.

Theorem 1.26 (Theorems 9.26 and 9.27) *Let E be an equivalence relation on $(2^\omega)^\omega$ which is classifiable by countable structures or Borel reducible to equality modulo an analytic P-ideal on ω. Then there are nonempty perfect sets $\langle P_n : n \in \omega \rangle$ such that $E \restriction \prod_n P_n$ is smooth. In other words, writing I for the σ-ideal of Borel subsets of $(2^\omega)^\omega$ which do not contain an infinite product of nonempty perfect sets,*

$$\text{classifiable by countable structures} \rightarrow_I \text{smooth},$$
$$(\leq_B =_J) \rightarrow_I \text{smooth}$$

whenever J is an analytic P-ideal on ω.

Theorem 1.27 (Theorem 8.17, originally by Mathias) *Let E be an essentially countable Borel equivalence relation on $[\omega]^{\aleph_0}$. Then there is an infinite set $a \subset \omega$ such that $E \restriction [a]^{\aleph_0}$ is Borel reducible to E_0.*

There is a number of anticanonization results. Again, the classification of analytic equivalence relations via their Borel reducibility complexity allows us to prove theorems that have no immediate counterpart in the finite or countable realm:

Theorem 1.28 (Corollary 6.55) *Whenever I is a σ-ideal such that P_I is proper and adds a dominating real, then E_{K_σ} is in the spectrum of I.*

Many results in the book either connect canonization properties of σ-ideals with the forcing properties of the associated quotient forcing P_I of Borel I-positive sets ordered by inclusion as studied in Zapletal (2008), or use that connection in their proofs. This allows us to tap a wealth of information amassed about such forcings by people otherwise not interested in descriptive set theory. Of special note is the connection between analytic equivalence relations and intermediate forcing extensions, explained in Chapter 3. It leads to the following general results:

Definition 1.29 Let E be an equivalence relation on a Polish space X, let I be a σ-ideal on X, and let $B \subset X$ be a set. We say that $E \restriction B$ is *I-ergodic* (or just *ergodic* if the σ-ideal is clear from the context) if for all Borel I-positive sets $C, D \subset B$ there are E-equivalent points $x \in C$, $y \in D$. The relation $E \restriction B$ is *nontrivially I-ergodic* any two Borel I-positive sets $C, D \subset B$ contain points $x \in C$ and $y \in D$ which are E-related and also points $x' \in C$ and $y' \in D$ that are E-unrelated.

Theorem 1.30 *Let I be a σ-ideal on a Polish space X such that the quotient poset P_I is proper.*

(i) (Theorem 3.1) *If P_I adds a minimal forcing extension, then for every I-positive Borel set $B \subset X$ and every analytic equivalence relation E on B there is a Borel I-positive $C \subset B$ such that $E \restriction C = \mathrm{id}$ or $E \restriction C$ is ergodic.*

(ii) (Corollary 4.10) *If P_I is nowhere c.c.c. (countable chain condition) and adds a minimal forcing extension then it has total canonization for equivalence relations classifiable by countable structures.*

(iii) (Theorem 4.9) *If P_I adds only finitely many real degrees then every equivalence relation classifiable by countable structures simplifies to an essentially countable equivalence relation on a Borel I-positive set.*

1.4 Navigation

Chapter 2 contains background material which at first glance has nothing to do with the canonization of equivalence relations. Most results are standard and stated without proofs, although there are exceptions such as the proof of canonical interpretation of $\mathbf{\Pi}^1_1$ on $\mathbf{\Sigma}^1_1$ classes of analytic sets in generic extensions (Theorem 2.44). Section 2.6 describes the basic treatment of quotient posets as explored in Zapletal (2008). For a σ-ideal I on a Polish space X we write P_I for the poset of Borel I-positive subsets of X ordered by inclusion, and outline its basic forcing properties.

Chapter 3 explains the main tool underlying much of the current of thought contained in this book: the correspondence between analytic equivalence relations and models of set theory. It enables proofs of canonization theorems essentially by a discussion of the possibilities for intermediate forcing extensions of the P_I models; such possibilities have been in many cases explored independently by mathematicians interested purely in forcing.

Chapter 4 discusses several classes of equivalence relations and their impact on the structure of forcing extensions. Thus, the smooth equivalence relations (Section 4.1) correspond to models of the form $V[x]$, where x is a real. The countable Borel equivalence relations (Section 4.2) are associated with a very specific intermediate extension containing no new reals. The hypersmooth equivalence relations (Section 4.4) are mirrored in infinite descending sequences of real degrees. The precise evaluation of these relationships helps us to obtain several remarkable general theorems; for example, if the quotient poset P_I is proper, nowhere c.c.c., and adds a minimal forcing extension then the σ-ideal I has total canonization for equivalence relations classifiable by countable structures (Corollary 4.10).

The following chapters contain the main contributions of this book. They proceed from one class of σ-ideals to another, with the intention of proving the strongest canonization theorem possible. The task at hand is enormous and a great number of natural σ-ideals remains untouched. We have arranged the material in such a way that these chapters are mostly independent of each other, and the occassional cross-references are inessential.

Chapter 5 isolates a broad class of σ-ideals that have integer games associated to them. This class includes every naturally definable σ-ideal generated by Borel sets of bounded Borel rank known to the authors. These σ-ideals share a number of abstract uniformization and complexity properties; most importantly for our concerns, the games can be used to upgrade total canonization results to Silver-style dichotomies for the σ-ideals, as discussed in Sections 5.1 and 5.4.

Chapter 6 then deals with many particular cases of game σ-ideals, including the meager ideal (Section 6.1.3) and the Lebesgue null ideal (Section 6.5).

Chapter 7 considers the equivalence relations E_0, E_1, and E_2 and their associated σ-ideals and proves strong canonization theorems for them. The arguments typically use Gandy–Harrington-type forcings. Chapter 8 deals with several σ-ideals associated with such classical partition theorems as the Halpern–Läuchli, Silver, or Milliken theorems.

Chapter 9 considers the σ-ideals obtained as products of finitely or countably many σ-ideals. We show that in the case of a finite product of σ-ideals generated by closed sets, with the covering property, the best canonization theorem possible in fact holds – every analytic equivalence relation simplifies to equality on a fixed set of coordinates in the product. A similar strong canonization theorem is achieved in the case of a finite product of the E_0 ideal, with the help of the Shelah–Zapletal partition theorem on products of finite sets with submeasures. As soon as the quotient forcings of the σ-ideals in the product are not bounding, such strong canonization theorems are impossible to achieve. In Section 9.4, we consider the simplest possible case corresponding to the two-dimensional product of Miller forcing and provide examples of equivalence relations that cannot be canonized there. The infinite products are much more challenging. In Section 9.3, we deal only with the infinite products of perfect sets, and prove a canonization theorem up to the class of smooth equivalence relations.

Finally, Chapter 10 looks at the σ-ideals obtained in the course of countable support iterations, or in other words the σ-ideals obtained as transfinite Fubini products. We show that if the iterands satisfy the weak Sacks property, then every equivalence relation reducible to E_{K_σ} in the iteration simplifies to an equivalence relation on one of the iterands (Theorem 10.1).

1.5 Notation

We employ the set theoretic standard notation as used in Jech (2002). If T is a tree then $[T]$ denotes the set of all of its infinite branches. If $x \in 2^\omega$ and $m \in \omega$ then $x \text{ flip } m$ is the element of 2^ω that differs from x exactly and only at its m-th entry. If $h \in 2^{<\omega}$ is a finite binary sequence then $x \text{ rew } h$ is the sequence in 2^ω obtained from x by rewriting its initial segment of corresponding length with h. If σ is a winning strategy for one of the players in an integer game and $y \in \omega^\omega$ is a sequence of moves of the other player, the symbol $\sigma * y$ denotes the sequence that the strategy σ produces in response to y.

The identity equivalence relation on any underlying set is id, and ev is the equivalence relation making every two points of the underlying set equivalent. If E is an equivalence relation on a set X and $x \in X$ then $[x]_E$ denotes the equivalence class of x. If $B \subset X$ is a set then $[B]_E$ denotes the *saturation* of the set B, the set $\{x \in X : \exists y \in B \; x \; E \; y\}$. If E, F are equivalence relations on respective Polish spaces X, Y, we write $E \leq_B F$ if E is *Borel reducible* to F, in other words if there is a Borel function $f : X \to Y$ such that $x_0 \; F \; x_1 \leftrightarrow f(x_0) \; E \; f(x_1)$. If $f : X \to Y$ is a Borel function between two Polish spaces and F is an equivalence relation on the space Y, then $f^{-1}F$ is the *pullback* of F, the equivalence relation E on the space X defined by $x_0 \; E \; x_1 \leftrightarrow f(x_0) \; F \; f(x_1)$; so $E \leq_B F$.

If P is a forcing notion, τ is a P-name, and $G \subset P$ is a generic filter, then τ/G is the valuation of the name τ according to the generic filter G. If P has an obvious "canonical" generic object x (typically, an element of a Polish space), we will abuse this convention to write also τ/x for the valuation. If M is a model of set theory and x is a set, we say that x is M-generic if x belongs to a (set) generic extension of M. If κ is a cardinal then H_κ denotes the collection of sets whose transitive closure has cardinality $< \kappa$. Several theorems in the book use a rather weak large cardinal assumption stated as "ω_1 is inaccessible to the reals," which is to say that for every $x \in 2^\omega$, the ordinal ω_1 as computed in the model $L[x]$, the class of sets constructible from x, is countable.

Theorem is used for a self-standing statement ready for applications outside of this book. Claim is an intermediate result within a proof of a theorem. Fact is a result that has been obtained elsewhere and will not be proved in this book; this does not in any way intimate that it is an unimportant, easy, or peripheral result.

2
Background facts

2.1 Descriptive set theory

Familiarity with basic concepts of descriptive set theory is assumed throughout. Kechris (1994) serves as a standard reference.

Definition 2.1 A collection I of subsets of a Polish space X is $\mathbf{\Pi}^1_1$ *on* $\mathbf{\Sigma}^1_1$ if for every analytic set $A \subset 2^\omega \times X$ the set $\{y \in 2^\omega : \{x \in X : \langle y, x \rangle \in A\} \in I\}$ is coanalytic.

Fact 2.2 (Kechris 1994, theorem 35.10) *If I is $\mathbf{\Pi}^1_1$ on $\mathbf{\Sigma}^1_1$ then every analytic set in I has a Borel superset in I.*

Fact 2.3 (Luzin–Novikov (Kechris 1994, theorem 18.10)) *If $A \subset X \times Y$ is a Borel set with countable vertical sections, then A is the countable union of graphs of Borel functions from X to Y.*

Fact 2.4 (Kechris 1994, theorem 18.6) *Let X, Y be Polish spaces. If $A \subset X \times Y$ is a Borel set then the set $B = \{x : A_x$ is not meager$\}$ is Borel, and the set $A \cap (B \times Y)$ has Borel uniformization.*

Fact 2.5 (Arsenin–Kunugui (Kechris 1994, theorem 18.18)) *Let X, Y be Polish spaces. If $A \subset X \times Y$ is a Borel set with K_σ vertical sections, then A has a Borel uniformization.*

Fact 2.6 (Jankov–von Neumann (Kechris 1994, theorem 18.1)) *Let X, Y be Polish spaces. If $A \subset X \times Y$ is an analytic set then it has a $\sigma(\mathbf{\Sigma}^1_1)$-measurable uniformization. In particular, the uniformization is Baire measurable.*

Fact 2.7 (Luzin–Suslin (Kechris 1994, theorem 15.1)) *A Borel one-to-one image of a Borel set is Borel.*

In several instances, we will need the tools of effective descriptive set theory.

Fact 2.8

(i) (Kanovei 2008, corollary 2.8.5) *The set* $\{(x, y) \in 2^\omega \times 2^\omega : y \in \Delta^1_1(x)\}$ *is* Π^1_1.

(ii) *Every* $\Sigma^1_1(x)$ *subset of* 2^ω *with a non-*$\Delta^1_1(x)$*-element is uncountable.*

(iii) (Harrington *et al.* 1988, lemma 1.2) (Effective reflection) *If* I *is a* $\Pi^1_1(z)$ *on* $\Sigma^1_1(z)$ *collection of analytic sets and* $A \in I$ *is* $\Sigma^1_1(z)$*, then* A *has a* $\Delta^1_1(z)$ *superset in* I.

(iv) (Kanovei 2008, theorem 2.8.2) (Coding of Borel sets) *There are a* Σ^1_1 *set* $\mathrm{An} \subset 2^\omega \times \omega \times 2^\omega$*, a* Π^1_1 *set* $\mathrm{Coan} \subset 2^\omega \times \omega \times 2^\omega$ *and a* Π^1_1 *set* $\mathrm{Code} \subset 2^\omega \times \omega$ *such that for every* $\langle z, e\rangle \in \mathrm{Code}$ *it is the case that* $\mathrm{Coan}_{ze} = \mathrm{An}_{ze}$*, and for every* $\Delta^1_1(z)$ *set* $B \subset 2^\omega$ *there is* $e \in \omega$ *such that* $\langle z, e\rangle \in \mathrm{Code}$ *and* $B = \mathrm{Coan}_{ze} = \mathrm{An}_{ze}$.

(v) *If* I *is a* $\Pi^1_1(z)$ *on* $\Sigma^1_1(z)$ *collection of analytic sets, then the union of all* $\Sigma^1_1(z)$ *sets in* I *is* $\Pi^1_1(z)$.

Proof We provide the argument for the last item. Write $B = \bigcup(\Sigma^1_1(z) \cap I)$. By *(iii)*, $B = \bigcup(\Delta^1_1(z) \cap I)$. The coding in *(iv)* tells us that $y \in B$ if there is $e \in \omega$ such that $\langle z, e\rangle \in \mathrm{Code}$, $\mathrm{An}_{ze} \in I$, and $y \in \mathrm{Coan}_{ze}$, which is a $\Pi^1_1(z)$ statement as I is $\Pi^1_1(z)$ on $\Sigma^1_1(z)$. □

Let X be a recursively presented Polish space. The Gandy–Harrington forcing is the poset of nonempty Σ^1_1 subsets of X, ordered by inclusion. The poset has obvious relativizations for every real parameter.

Fact 2.9 *If* G *is a Gandy–Harrington generic filter, then* $\bigcap G = \{x_{\mathrm{gen}}\}$ *for some point* $x_{\mathrm{gen}} \in X$*, and* $G = \{A \in \Sigma^1_1 : x_{\mathrm{gen}} \in A\}$.

As an interesting application of the Gandy–Harrington forcing, we will prove the following:

Proposition 2.10 *Let* $A \subset (2^\omega)^\omega$ *be a* Σ^1_1 *set containing a sequence* x *such that* $x(n) \notin \Delta^1_1$ *for every number* $n \in \omega$*. Then for every countable set* $a \subset 2^\omega$ *there is a sequence* $x' \in A$ *such that* $\mathrm{rng}(x') \cap a = 0$.

Proof Let $A' = \{x \in A : \forall n\ x(n) \notin \Delta^1_1\}$; this is a nonempty Σ^1_1 set. Let M be a countable elementary submodel of a large enough structure containing the set a, and let G be a Gandy–Harrington-generic filter over M containing A', with the associated generic point $x \in A'$. We claim that $\mathrm{rng}(x) \cap a = 0$.

To verify the equality, for every point $y \in 2^\omega$ and every number $n \in \omega$, consider the set $D_{yn} = \{B \subset A' : B \text{ is } \Sigma^1_1 \text{ and for every } z \in B, z(n) \neq y\}$. We will prove that the set D_{yn} is open dense in the Gandy–Harrington forcing below A'. Let $B' \subset A'$ be a nonempty Σ^1_1 set. Consider its projection into the n-th coordinate; this is a nonempty Σ^1_1 subset of 2^ω containing only non-Δ^1_1-elements; therefore, it is uncountable and it must contain some element distinct from y, separated from y by some basic open neighborhood O. Then $B = \{x \in B' : x(n) \notin O\}$ is the desired nonempty Σ^1_1 set in D_{yn} below B'.

Now, by the genericity of the filter G, for every $y \in a$ and $n \in \omega$, the filter G contains some set $B \in D_{yn}$. Since $x \in B$, it must be the case that $x(n) \neq y$. In other words, $\text{rng}(x) \cap a = 0$ as required. □

2.2 Invariant descriptive set theory

If E, F are Borel or analytic equivalence relations on Polish spaces X, Y then $E \leq_B F$ (E is *Borel reducible* to F) denotes the fact that there is a Borel function $f : X \to Y$ such that $x_0 \; E \; x_1 \leftrightarrow f(x_0) \; F \; f(x_1)$. Bireducibility is an equivalence relation on the class of all Borel equivalence relations. $\leq_B$ turns into a complicated ordering on the bireducibility classes. There are several equivalences occupying an important position in this ordering:

Definition 2.11

(i) id is the identity equivalence relation.
(ii) ev, or the *full equivalence relation*, is the equivalence relation that makes all elements of its domain equivalent.
(iii) E_0 is the equivalence on 2^ω defined by $x \; E_0 \; y \leftrightarrow$ the symmetric difference $x \Delta y$ is finite.
(iv) E_1 is the equivalence relation on $(2^\omega)^\omega$ defined by $\vec{x} \; E_1 \; \vec{y} \leftrightarrow \{n \in \omega : \vec{x}(n) \neq \vec{y}(n)\}$ is finite.
(v) E_2 is the equivalence relation on 2^ω defined by $x \; E_2 \; y \leftrightarrow \Sigma\{1/(n+1) : x(n) \neq y(n)\} < \infty$. F_2 is the equivalence relation on $(2^\omega)^\omega$ defined by $\vec{x} \; F_2 \; \vec{y} \leftrightarrow \text{rng}(\vec{x}) = \text{rng}(\vec{y})$.
(vi) E_{K_σ} is the equivalence on $\Pi_n(n+1)$ defined by $x \; E_{K_\sigma} \; y \leftrightarrow \exists m \forall n \; |x(n) - y(n)| < m$.
(vii) If J is an ideal on ω then $=_J$ is the equivalence on $\mathcal{P}(\omega)$ defined by $x =_J y$ if $x \Delta y \in J$.

Definition 2.12 An equivalence E on a Polish space X is:

(i) *smooth* if it is Borel reducible to the identity;
(ii) *hypersmooth* if it is an increasing union of a countable sequence of smooth equivalence relations;
(iii) *countable* if it is Borel and all of its equivalence classes are countable;
(iv) *essentially countable* if it is Borel reducible to a countable Borel equivalence relation;
(v) *classifiable by countable structures* if there is a countable relational language L such that E is Borel reducible to the equivalence relation of isomorphism of models of L with universe ω;

(vi) *an orbit equivalence relation* if there is a continuous action of a Polish group G on X such that for all $x, y \in X$, $x \ E \ y$ if and only if there is $g \in G$ such that $g \cdot x = y$.

Fact 2.13 (Feldman–Moore (Feldman and Moore 1977; Gao 2009, theorem 7.1.4)) *For every countable Borel equivalence relation E on a Polish space X there is a countable group G and a Borel action of G on X whose orbit equivalence relation is equal to E.*

Fact 2.14 (Silver dichotomy (Silver 1980; Kanovei 2008, section 10.1)) *For every coanalytic equivalence relation E on a Polish space X, either X is covered by countably many equivalence classes or there is a perfect set of mutually E-inequivalent points.*

Corollary 2.15 *For every coanalytic equivalence relation E on a Polish space X and every analytic set $A \subset X$, either A intersects only countably many E-equivalence classes, or it contains a perfect set of pairwise E-inequivalent elements.*

Proof Let $f : \omega^\omega \to X$ be a continuous function such that $\text{rng}(f) = A$, and let $F = f^{-1}E$; thus, F is a coanalytic equivalence relation on ω^ω and the Silver dichotomy can be applied to it. The corollary follows immediately. □

Fact 2.16 (Glimm–Effros dichotomy (Harrington *et al.* 1990; Kanovei 2008, section 10.4)) *For every Borel equivalence relation E on a Polish space, either $E \leq_B \text{id}$ or $E_0 \leq_B E$.*

Fact 2.17 (Hjorth (Hjorth 2000)) *If $E \leq_B E_2$ is a Borel equivalence relation, then either E is essentially countable, or $E_2 \leq_B E$.*

Fact 2.18 (Kechris–Louveau (Kechris and Louveau 1997; Kanovei 2008, section 11.3)) *For every Borel equivalence relation $E \leq_B E_1$ on a Polish space, either $E \leq_B E_0$ or $E_1 \leq_B E$.*

Fact 2.19 (Rosendal (Rosendal 2005; Kanovei 2008, section 6.6)) *Every K_σ equivalence relation on a Polish space is Borel reducible to E_{K_σ}.*

Proposition 2.20 *There is an F_σ-ideal I on a countable set such that $=_I$ is bireducible with E_{K_σ}.*

Proof Let $\text{dom}(I)$ be the countable set of all pairs $\langle n, m\rangle$ such that $n \in \omega$ and $m \leq n$ and let $a \in I$ if there is a number $k \in \omega$ such that for every number $n \in \omega$, there are at most k many numbers m with $\langle n, m\rangle \in a$. This ideal is clearly F_σ and so $=_I \leq_B E_{K_\sigma}$ by the above fact. On the other hand, $E_{K_\sigma} \leq =_I$, as the reduction $f : \Pi_n(n+1) \to \mathrm{dom}(I)$ defined by $f(x) = \{\langle n, m\rangle : m \leq x(n)\}$ shows. □

Fact 2.21 (Rosendal (Rosendal 2005; Kanovei 2008, chapter 18)) *For every Borel equivalence relation E there is a Borel ideal I on ω such that $E \leq_B =_I$.*

Several results in this book can be restated in the language of ergodic theory, using the following notion:

Definition 2.22 Let E, F be analytic equivalence relations on respective Polish spaces X, Y. We say that E is *F-generically ergodic* if every Borel homomorphism from E to F stabilizes on a comeager subset of X. Similarly, if μ is a Borel probability measure on X, we say that E is *F, μ-ergodic* if every Borel homomorphism from E to F stabilizes on a set of full μ-mass.

2.3 Forcing

Familiarity with basic forcing concepts is assumed throughout the book. If $\langle P, \leq\rangle$ is a partial ordering and $Q \subset P$ then Q is said to be *regular* in P if every maximal antichain of Q is also a maximal antichain in P; restated, for every element $p \in P$ there is a *pseudoprojection* of p into Q, a condition $q \in Q$ such that every strengthening of q in Q is still compatible with p.

Regarding extensions of models of set theory, we will use the standard nomenclature of Jech (2002, theorem 13.22 ff). If V is a transitive model of ZF set theory containing all ordinals and a (perhaps $a \notin V$) is a set, then $V(a)$ is the smallest transitive model of ZF containing V as a subset and a as an element, and $V[a]$ is the smallest model W of ZF containing V as a subset and such that $a \cap W \in W$. The two models are formed by transfinite induction in parallel to the $L(a)$ and $L[a]$ constructions, except at successor stages of the induction, all subsets of the constructed part which belong to V are added. If a is in addition a generic filter over V, the models $V[a]$ and $V(a)$ coincide and there is an alternative description of them as a forcing extension: $V(a) = V[a] = \{\tau/a : \tau \text{ is a name in } V\}$.

In a number of places, we will use weakly homogeneous forcings.

Definition 2.23 A poset P is *weakly homogeneous* if for every pair of elements $p, q \in P$ there is an automorphism $\pi : P \to P$ such that the elements $p, \pi(q) \in P$ have a lower bound.

Fact 2.24 (Kanamori 1994, proposition 10.19) *If P is a weakly homogeneous notion of forcing, ϕ a formula of the language of set theory, and v a finite sequence of sets, then the statement $\phi(\check{v})$ of the P-forcing language is decided by the largest condition in P.*

One weakly homogeneous poset we will use all the time is the *collapse*. Whenever κ is a cardinal, $\mathrm{Coll}(\omega, \kappa)$ is the poset of finite partial functions from ω to κ ordered by reverse extension. The following fact sums up the homogeneity properties of the collapse orders.

Fact 2.25 (Jech 2002, corollary 26.10) *Let κ be a cardinal, P a poset of size $< \kappa$, and $G \subset \mathrm{Coll}(\omega, \kappa)$ be a generic filter. In $V[G]$, for every V-generic filter $H \subset P$ there is a filter $K \subset \mathrm{Coll}(\omega, \kappa)$ generic over $V[H]$ such that $V[G] = V[H][K]$.*

In several places, we need to invoke a forcing factorization theorem, and the following is the exact statement that we will use:

Fact 2.26 (Forcing factorization theorem) *If P, Q are posets and τ is a P-name for a Q-generic filter, then there is a condition $q \in Q$ such that in a large collapse extension, for every filter $H \subset Q$ generic over V containing the condition q there is a filter $G \subset P$ generic over V such that $H = \tau/G$.*

Proof Let κ be a cardinal larger than $|\mathcal{P}(P)|$. Consider the statement ϕ of the Q-forcing language saying "$\mathrm{Coll}(\omega, \kappa) \Vdash$there is a V-generic filter $G \subset \check{P}$ such that $\dot{H} = \check{\tau}/G$"; here, $\dot{H}$ is the Q-name for the generic filter on Q. Note that Fact 2.25 implies that the largest condition in $\mathrm{Coll}(\omega, \kappa)$ decides the statement on the right of the forcing relation, as all its parameters are in the model $V[H]$.

First argue that there must be a condition $q \in Q$ that forces ϕ. If this fails then $Q \Vdash \neg\phi$. Let $G \subset P$ be a filter generic over V, and let $K \subset \mathrm{Coll}(\omega, \kappa)$ be a filter generic over $V[G]$. The usual characterization of $\mathrm{Coll}(\omega, \kappa)$ as in Zapletal (2008, lemma 26.7) shows that the model $V[G][K]$ is a $\mathrm{Coll}(\omega, \kappa)$-extension of V. Let $H = \tau/G$. Now, Fact 2.25 shows that $V[G][K]$ is a $\mathrm{Coll}(\omega, \kappa)$-extension of the model $V[H]$. By the forcing theorem and the failure of ϕ, there should be no V-generic filter on P in the model $V[G][K]$ that gives H. However, there is such a filter there, namely G; a contradiction.

Thus, fix a condition $q \in Q$ forcing ϕ; we claim that q works as required. Let $K \subset \mathrm{Coll}(\omega, \kappa)$ be a filter generic over V, and let $H \subset Q$ be a V-generic filter in $V[K]$, containing the condition q. By Fact 2.26, $V[K]$ is a $\mathrm{Coll}(\omega, \kappa)$ extension of $V[H]$, and the forcing theorem applied to ϕ and H shows that there is a filter $G \subset P$ generic over V in the model $V[K]$ such that $H = \tau/G$. □

Corollary 2.27 *If Q is a nowhere c.c.c. forcing then $Q \Vdash$ the Q-forcing extension is not a c.c.c. extension of the ground model.*

Proof If this failed then by the forcing theorem, there would be a c.c.c. forcing P and a P-name τ such that $P \Vdash \tau \subset \check{Q}$ is a generic filter over V. Let $q \in Q$ be a condition as in Fact 2.26, and let $A \subset Q$ be an uncountable antichain below q. For every $r \in A$ there must be $p_r \in P$ forcing $\check{r} \in \tau$: in a large collapse extension, there is a filter $H \subset Q$ generic over V containing r, by the choice of the condition q there is a filter $G \subset P$ generic over V in that large extension such that $H = \tau/G$, and by the forcing theorem applied to $V[G]$, the filter G must contain some condition p_r forcing $\check{r} \in \tau$. Since τ is forced to be a filter, the conditions $\{p_r : r \in A\}$ are pairwise incompatible, contradicting the c.c.c. of P. □

The following definition sums up the common forcing properties used in this book:

Definition 2.28 A forcing P is:

(i) *proper* if for every condition $p \in P$ and every countable elementary submodel M of large enough structure containing P, p there is a *master* condition $q \leq p$ which forces the generic filter to meet all dense subsets in M in conditions in M;
(ii) *nowhere c.c.c.* if there is an uncountable antichain below every condition;
(iii) *bounding* if every function in ω^ω in its extension there is a ground model function with larger value on every entry.
(iv) The forcing *preserves outer Lebesgue measure* if any ground model set of reals has the same outer measure whether evaluated in the ground model or in the extension.
(v) The forcing *preserves Baire category* if any ground model set of reals is meager in the ground model if and only if it is meager in the extension.
(vi) The forcing has the *Laver property* if for every point $x \in \omega^\omega$ in the extension that is dominated by a ground model point, there is a function $g : \omega \to [\omega]^{<\aleph_0}$ in the ground model such that for every $n \in \omega$, $x(n) \in g(n)$ and $|g(n)| = n + 1$.

An *independent real* is a set $a \subset \omega$ in a generic extension such that no infinite set $b \subset \omega$ in the ground model is either a subset of or disjoint from a. A *dominating real* is a function $x \in \omega^\omega$ in a generic extension such that every function $y \in \omega^\omega$ is modulo finite dominated by x.

The phrase "countable elementary submodel of a large enough structure" appears in essentially every paper dealing with proper forcing. The large enough structure should be interpreted as the structure with universe H_κ, the collection of sets of hereditary cardinality $< \kappa$ for some cardinal κ such that H_κ contains every object named in the proof before the phrase occurs and its powerset. The structure has a binary relation $\in$ and a designated unary predicate for every object named in the proof before the phrase occurs.

If M is a model of some portion of set theory and $P \in M$ is a poset, we will say that a filter $G \subset P$ is P-generic over M if for every dense set $D \subset P$ in M there is a condition $p \in D \cap M \cap G$. If the poset P adds a "canonical" object nominally different from a generic filter (such as some point $\dot{x}_{\text{gen}}$ in a Polish space from which the generic filter is recoverable), we will extend this usage to say that x is P-generic over M if the filter on P recovered from x is. In contradistinction, the phrase "a is M-generic" will mean that a belongs to some generic extension of the transitive isomorph of M; it will be used only if M is wellfounded.

The descriptive set theoretic part of the interplay between forcing a countable elementary submodels is encapsulated in the following uniformization theorem.

Theorem 2.29 *Let E be an equivalence relation on a Polish space X Borel reducible to an orbit equivalence relation, let P be a partial order, τ a P-name for an element of X, and let M be a countable elementary submodel of a large structure containing E, τ, P. Then:*

(i) the set $B = \{x \in X : \exists h \subset P \cap M$ a filter generic over M such that $\tau/h \mathrel{E} x\}$ is Borel;
(ii) there is a Borel function $h : B \to \mathcal{P}(P \cap M)$ such that for every $x \in B$, $h(x)$ is a filter generic over M such that $\tau/h(x) \mathrel{E} x$.

Proof We will first choose several objects that do not take into account the model M. Let G be a Polish group acting continuously on a Polish space Y, and let $f : X \to Y$ be a Borel reduction of E to the orbit equivalence relation of the action. Let Q be the Cohen forcing for G' – the poset of nonempty open subsets of G ordered by inclusion, adding a point $\dot{g}_{\text{gen}} \in G$. Let R be the complete Boolean algebra generated by the $P \times Q$-name $\sigma = \dot{g}_{\text{gen}} \cdot \dot{f}(\tau)$ for an element of the space Y. Let κ be a regular cardinal larger than $|\mathcal{P}(P \times Q)|$. For the purposes of the following claim, treat σ as a $\text{Coll}(\omega, \kappa) \times R$-name.

Claim 2.30 *The poset $\text{Coll}(\omega, \kappa) \times R$ forces that there is a filter $h \subset P$ generic over V such that $f(\tau/h)$ is orbit equivalent to σ.*

Proof Let $\dot{S}$ be the R-name for the remainder forcing $(P \times Q)/R$ as in Jech (2002, exercise 16.3) or Bell (2005, problem 6.37). Whenever $h_C \subset$

$\mathrm{Coll}(\omega, \kappa)$, $h_R \subset R$ is a pair of mutually generic filters on $\mathrm{Coll}(\omega, \kappa)$ and R over V, the set $\mathcal{P}(\dot{S}/h_R) \cap V[h_R]$ is countable in $V[h_C, h_R]$ and so there is a filter $h_S \subset \dot{S}/h_R$ generic over $V[h_R]$. This yields a pair of filters $h_P \subset P$, $h_Q \subset Q$ in the model $V[h_C, h_R]$ mutually generic over V extending the filter h_R. The filter $h = h_C$ is as required in the claim. Indeed, $\dot{g}_{\mathrm{gen}}/h_Q \cdot f(\tau/h_C) = \sigma/h_R$ by the definitions. □

Finally, let χ be a $\mathrm{Coll}(\omega, \kappa) \times R$-name for the filter P-generic over V obtained in Claim 2.30. By the elementarity of the model M, it is possible to find all the objects G, Y, f, σ, χ so that they belong to the model M. We are finally ready for the descriptive set theoretic part of the proof.

Say that a point $y \in Y$ is *R-generic over M* if there is a filter $h \subset R \cap M$ generic over M such that $y = \sigma/h$.

Claim 2.31 *The set $A \subset Y$ is Borel.*

Proof Consider the set $\mathcal{P}(R \cap M)$ with the usual topology; this is a Polish space. The set F of filters R-generic over M is its G_δ subset. The map $h \mapsto \sigma/h$ from F to Y is continuous by the forcing theorem, and it is one-to-one as the name σ generates the algebra R. The set A is equal to the range of this function, and as a one-to-one continuous image of a Borel set it is Borel by Fact 2.7. □

Claim 2.32 *For every $x \in X$, $x \in B$ if and only if for comeagerly many elements $g \in G$ we have $g \cdot f(x) \in A$.*

Proof For the left-to-right direction, suppose that $x \in B$. Then there is a filter $h \subset P$ generic over M such that $\tau/h \; E \; x$. Find an element $g_0 \in G$ such that $g_0 \cdot f(\tau/h) = f(x)$. The set $C = \{k \in G : k \text{ is } Q\text{-generic over } M[h]\}$ is comeager in G, and so is its shift Cg_0^{-1}. It will be enough to show that for each $g \in Cg_0^{-1}$, $g \cdot f(x) \in A$ holds. To verify the membership, find $k \in C$ such that $g = kg_0^{-1}$, note that $g \cdot f(x) = kg_0^{-1} f(x) = k \cdot f(\tau/h)$, and note that the last expression confirms the membership in the set A since the pair h, k is mutually generic over M.

For the right-to-left direction, suppose that for some $g \in G$, $g \cdot f(x) \in A$. Then, there is a pair $h_P \subset P$, $h_Q \subset Q$ of filters mutually generic over M such that $\sigma/h_P \times h_Q = g \cdot f(x)$. The filter h_P witnesses that $x \in B$. □

It follows that the set B is Borel, since it is obtained by the application of the Vaught transform to the Borel set A (Kechris 1994, theorem 16.1 or Gao 2009, theorem 3.2.7). For the construction of the uniformizing function h, consider the set $D = \{\langle x, g, h_C\rangle : \text{(a) } g \cdot f(x) \in A \text{ and (b) } h_C \subset \mathrm{Coll}(\omega, \kappa) \text{ is generic}$

over the model $M[g \cdot f(x)]\}$. Treat D as a subset $X \times G \times F$, where F is the space of filters on $\text{Coll}(\omega, \kappa) \cap M$ equipped with the topology inherited from $\mathcal{P}(\text{Coll}(\omega, \kappa)) \cap M$. Note that $X \times G \times F$ is Polish and D is Borel in it. For every point $x \in B$, the vertical section D_x is comeager in the space $G \times F$ by the Kuratowski–Ulam theorem (Kechris 1994): there are comeagerly many elements $g \in G$ satisfying (a) by Claim 2.32, and to each such g there are comeagerly many filters satisfying (b). By comeager uniformization (Fact 2.4), there is a Borel function $k : B \to G \times F$ uniformizing the Borel set D, with component functions k_0, k_1. Let $h : B \to \mathcal{P}(P \cap M)$ be the function assigning to each $x \in B$ the point $\chi/(k_0(x) \cdot f(x), k_1(x))$. This Borel function works as required. □

2.4 Absoluteness and interpretations

The main proof tool in this book is the comparison of various models of set theory. We must address the issue of translation of various concepts of descriptive set theory between such models, and show that the translations are to some degree faithful. While the resulting discussion is widely percieved as trivial, it is formally necessary for all of our proofs, and there are several instances of actual nontrivial mathematical content.

Fact 2.33 (Mostowski (Jech 2002, theorem 25.4)) *Let ϕ be a $\Sigma^1_1(a)$ formula with one free variable and a parameter $a \in \omega^\omega$. Then for every $x \in \omega^\omega$, the truth value of $\phi(x)$ is the same in all transitive models of set theory containing a, x.*

Further assumptions on the wellfounded models yield more absoluteness, as in the following classical fact:

Fact 2.34 (Shoenfield (Jech 2002, theorem 25.20)) *Let ϕ a $\Pi^1_2(a)$ formula with one free variable and a parameter $a \in \omega^\omega$. Then for every $x \in \omega^\omega$, the truth value of $\phi(x)$ is the same in all transitive models of set theory containing a, x and all countable ordinals, in particular it is the same in all forcing extensions.*

The Shoenfield absoluteness is used in essentially every argument in this book, either explicitly or tacitly through the use of canonical interpretation of various concepts of descriptive set theory. The issue of canonical interpretations is not addressed anywhere in the literature, and in this section we spend some time making the (rather obvious) definitions explicit. Let V be a ground model, and $V[G]$ some generic extension of V. The following definitions and claims take place in the extension $V[G]$.

Definition 2.35 An *interpretation* of a V-complete metric space $\langle X, d\rangle \in V$ is a map $\phi : X \to Y$ in $V[G]$ where $\langle Y, e\rangle$ is a $V[G]$-complete metric space, $\text{rng}(\phi) \subset Y$ is dense and ϕ is an isometry between $\langle X, d\rangle$ and $\langle \text{rng}(\phi), e\rangle$.

It is quite clear that an interpretation of a complete metric space exists (as some completion of $\langle X, d\rangle$ in the model $V[G]$) and it is essentially unique. Whenever $\phi : X \to Y$ and $\psi : X \to Z$ are two interpretations of a complete metric space then there is a unique isometry $\chi : Y \to Z$ such that $\chi \circ \phi = \psi$. This immediately follows from the fact that $\psi \circ \phi^{-1}$ is an isometry of dense subsets of Y and Z, and such an isometry extends uniquely to an isometry of the complete metric spaces in question.

It is not difficult to verify that with any usual definable metrics, the identity maps between $(2^\omega)^V$ and $(2^\omega)^{V[G]}$, between $(\omega^\omega)^V$ and $(\omega^\omega)^{V[G]}$, $\mathbb{R}^V$ and $\mathbb{R}^{V[G]}$, the corresponding versions of the Hilbert cube are all interpretations of complete metric spaces.

Definition 2.36 An *interpretation* of a Polish space $X \in V$ is a map $\phi : X \to Y$ to a Polish space $Y \in V[G]$ such that for some V-complete metric $d \in V$ on X and some complete metric e on Y metrizing the respective Polish topologies, ϕ is an interpretation of $\langle X, d\rangle$ in $\langle Y, e\rangle$.

Here, we must address the question of uniqueness more carefully in order to show the independence on the choice of the complete metric on X.

Claim 2.37 *Whenever $\phi_0 : X \to Y_0$ and $\phi_1 : X \to Y_1$ are two interpretations of a Polish space then there is a unique homeomorphism $\chi : Y_0 \to Y_1$ such that $\chi \circ \phi_0 = \phi_1$.*

For the proof, we will use the following notion. If $\langle X, d\rangle$ is a metric space and s, t are two ω-sequences of its points, we will say that s, t are *d-close* if the sequence u given by $u(2n) = s(n)$ and $u(2n+1) = t(n)$ is d-Cauchy.

Proof Let d_0, e_0 be the complete metrics on X and Y_0 showing that ϕ_0 is an interpretation. Let d_1, e_1 be the complete metrics on X and Y_1 showing that ϕ_1 is an interpretation. Let $a \subset X$ be a countable dense set in V. Consider the following sentence θ: for every d_0-Cauchy sequence s of elements of a there is an d_1-Cauchy sequence t which is d_0-close to s, and if t, t' are two such d_1-Cauchy sequences, then they are d_1-close. The sentence θ certainly holds in V since the metrics d_0 and d_1 yield the same completion X of the set a. Since θ is $\mathbf{\Pi}^1_2$, by the Shoenfield absoluteness it holds in $V[G]$ as well.

Now let $\chi : Y_0 \to Y_1$ be defined as follows: whenever $y \in Y_0$ is a point, there must be a d_0-Cauchy sequence s of points in a such that $y = \lim \phi_0 \circ s$. Use the sentence θ to find a d_1-Cauchy sequence t in a which is d_0-close to s, and define $\chi(y) = \lim \phi_1 \circ t$. It is routine to check that χ is a well-defined homeomorphism.

The uniqueness of χ follows from the fact that on the dense set $\phi_0'' X \subset Y_0$, χ must be equal to $\phi_1 \circ \phi_0^{-1}$. $\square$

The next step is the definition of canonical interpretations of analytic subsets of Polish spaces. For this, if $\phi : X \to Y$ is an interpretation of a Polish space X and $C \subset X$ is a V-closed set in V, write $C^\phi =$ the closure of $\phi'' C$ in the space Y. We will also make use of interpretations of products. If $\phi_0 : X_0 \to Y_0$ and $\phi_1 : X_1 \to Y_1$ are interpretations of Polish spaces, then the map $\phi_0 \times \phi_1 : X_0 \times X_1 \to Y_0 \times Y_1$ defined by $\phi_0 \times \phi_1(x_0, x_1) = \langle \phi_0(x_0), \phi_1(x_1) \rangle$ is easily seen to be an interpretation of the product.

Definition 2.38 Suppose that in the ground model, X_0, X_1 are Polish spaces and $A \subset X_0$ is an analytic set, $A = p(C)$ for some closed set of the product $X_0 \times X_1$. If in the extension $\phi_0 : X_0 \to Y_0$ and $\phi_1 : X_1 \to Y_1$ are interpretations, we define the *interpretation* A^{ϕ_0} of A to be the set $p(C^{\phi_0 \times \phi_1})$.

The following two claims show that the definition of the interpretation of A does not depend on the choice of C or ϕ_1, and that the interpretation of A is contained in all interpretations of X in the same way.

Claim 2.39 *Suppose that in the ground model, X_0, X_1, X_2 are Polish spaces, and $C \subset X_0 \times X_1$ and $D \subset X_0 \times X_2$ are closed sets such that $p(C) = p(D)$ holds in V. Whenever ϕ_0, ϕ_1, ϕ_2 are interpretations, then $p(C^{\phi_0 \times \phi_1}) = p(D^{\phi_0 \times \phi_2})$.*

Proof Let d_0, d_1, d_2 be complete metrics on X_0, X_1, X_2 in V and e_0, e_1, e_2 be complete metrics on Y_0, Y_1, Y_2 showing that ϕ_0, ϕ_1, ϕ_2 are interpretations. Let $a \subset C$ and $b \subset D$ be countable dense sets in V. For a sequence s of points in C, write $\bar{s}$ for the sequence of the first coordinates of points in s; similarly for sequences of points in D. Consider the following sentence θ: whenever s is a $d_0 \times d_1$-Cauchy sequence in a then there is a $d_0 \times d_2$-Cauchy sequence in b such that $\bar{t}$ is d_0-close to $\bar{s}$. Since $p(C) = p(D)$ holds in V, θ holds in V. Now, since the sentence θ is $\mathbf{\Pi}^1_2$, it also holds in $V[G]$ by the Shoenfield absoluteness.

Now move to $V[G]$. We will show that $p(C^{\phi_0 \times \phi_1}) \subseteq p(D^{\phi_0 \times \phi_2})$; the opposite inclusion follows by symmetry. Suppose that $y \in Y_0$ belongs to $p(C^{\phi_0 \times \phi_1})$. This means that there must be a $d_0 \times d_1$-Cauchy sequence s of points in a such that $\phi_0 \circ \bar{s}$ converges to y. Applying θ, find a $d_0 \times d_2$-Cauchy sequence t of points in b such that $\bar{s}$ is d_0-close to $\bar{t}$. Then y is the first coordinate of the limit of t, showing that $p(D^{\phi_0 \times \phi_2})$ contains y as required. □

Claim 2.40 *Let X be a Polish space and A an analytic subset of X in the ground model. In the generic extension:*

(i) if $\phi : X \to Y$ is an interpretation of X then $\phi'' A = A^\phi \cap \phi'' X$;
(ii) if $\phi_0 : X \to Y_0$ and $\phi_1 : X \to Y_1$ are interpretations of X and $\chi : Y_0 \to Y_1$ is the unique commuting homeomorphism, then $\chi'' A^{\phi_0} = A^{\phi_1}$.

Interpretations of analytic sets preserve inclusion and commute with countable unions and intersections in the ground model.

Proof The elementary arguments using the Mostowski and Shoenfield absoluteness are left to the reader. □

We will now record several claims that show that various operations on Polish spaces in the ground model have faithful interpretations in the generic extension.

Claim 2.41 (Interpretation of hyperspaces) *Suppose that X is a Polish space in the ground model, and $\phi_0 : X \to Y_0$ is its interpretation. Then in the generic extension:*

(i) (existence) *there is an interpretation $\psi_0 : (K(X))^V \to K(Y_0)$ such that for every compact set $K \in K(X)$, $\psi_0(K) = K^{\phi_0}$;*
(ii) (uniqueness) *if $\phi_1 : X \to Y_1$ and $\psi_1 : (K(X))^V \to K(Y_1)$ are interpretations as in (i) and $\chi : Y_0 \to Y_1$, $\theta : K(Y_0) \to K(Y_1)$ are the unique commuting homeomorphisms, then for every compact set $K \subset Y_0$, $\theta(K) = \chi'' K$.*

Proof The interpretation ψ_0 is defined by the demand that for every $K \in (K(X))^V$, $\psi_0(K) = K^{\phi_0}$. We must first show that the range of ψ_0 consists of compact subsets of Y. Let d, e be the complete metrics on X, Y witnessing that ϕ_0 is an interpretation. As K is compact, it is totally bounded in d, so $\phi_0'' K$ is totally bounded in e and its closure in Y is compact as desired.

To show that ψ_0 is an interpretation, let d' and e' be the Hausdorff metrics on $(K(X))^V$ and $K(Y_0)$ derived from d, e. It is immediate that ψ_0 is an isometry with respect to d' and e'. Also, $\text{rng}(\psi_0)$ is dense in $K(Y_0)$ – it contains the set $\text{rng}(\phi_0)^{<\aleph_0}$, which is dense in $K(Y_0)$ since $\text{rng}(\phi_0)$ is dense in Y.

For (*ii*), note that the map $\theta' : K(Y_0) \to K(Y_1)$ defined by $\theta'(K) = \chi'' K$ is a homeomorphism. By (*i*), the homeomorphism θ agrees with θ' on the set $\operatorname{rng}(\phi_0)^{<\aleph_0}$, which is dense in $K(Y_0)$ as the set $\operatorname{rng}(\phi_0)$ is dense in Y_0. It follows that $\theta' = \theta$ as desired. □

Claim 2.42 (Interpretation of spaces of probability measures) *Suppose that X is a Polish space in the ground model, and $\phi_0 : X \to Y_0$ is its interpretation. Then in the generic extension:*

(i) (existence) *there is an interpretation $\psi_0 : (P(X))^V \to P(Y_0)$ such that for every measure $\mu \in V$ and every open set $O \in V$, $\psi_0(\mu)(O^\phi) = \mu(O)$;*
(ii) (uniqueness) *if $\phi_1 : X \to Y_1$ and $\psi_1 : (P(X_1))^V \to P(Y_1)$ are different interpretations as in (i), if $\chi : Y_0 \to Y_1$ and $\theta : P(Y_0) \to P(Y_1)$ are the unique commuting homeomorphisms, then for every measure μ on Y_0, $\theta(\mu)$ is the push-forward of μ by χ.*

Proof Let $\mu \in (P(X))^V$, the value $\psi_0(\mu)$ is the unique $\nu \in P(Y_0)$ such that the following statement $(\star)$ holds: for every $O \subset X$ open in V, $\nu(O^{\phi_0}) = \mu(O)$. The existence and uniqueness of such ν must be verified using the Shoenfield absoluteness.

First, if ν satisfying $(\star)$ exists, it must be unique. Any Borel probability measure is determined by its values on open sets, and every open subset $O \subset Y$ is an increasing union $\bigcup_n O_n^{\phi_0}$ for some open sets $O_n \in V$. Then, $(\star)$ together with the countable additivity determines the value $\nu(O)$ as $\sup_n \mu(O_n)$.

Second, we must verify that ν satisfying $(\star)$ actually exists. Let $\mathcal{B}$ be the countable algebra in V generated by the basic open subsets of X. Let $\mathcal{B}^\phi = \{B^\phi : B \in \mathcal{B}\}$. This is an algebra of subsets of Y since interpretations preserve Boolean operations on Borel sets by Claim 2.40. Let $\bar{\nu}$ be the function on $\mathcal{B}^\phi$ defined by $\bar{\nu}(B^\phi) = \mu(B)$. We will show that $\bar{\nu}$ is a σ-additive measure on $\mathcal{B}^\phi$. Once this is done, the Carathéodory extension theorem shows that $\bar{\nu}$ has a unique extension ν to the σ-algebra generated by $\mathcal{B}^\phi$, which is the σ-algebra of all Borel subsets of Y. The Borel probability measure ν will then satisfy $(\star)$.

For every set $B \in \mathcal{B}$ consider the following sentence θ: every sequence $\langle B_n : n \in \omega \rangle$ of sets in $\mathcal{B}$ either does not consist of pairwise disjoint sets, or its union is not B, or $\Sigma_n \mu(B_n) = \mu(B)$. The sentence θ holds in V by the σ-additivity of μ. It is not difficult to restate θ as a $\mathbf{\Pi}^1_2$ sentence, and therefore will hold in $V[G]$ as well by the Shoenfield absoluteness. It follows that $\bar{\nu}$ is a σ-additive measure in $V[G]$ as required. □

In the remainder of the book, we will exploit the existence and essential uniqueness of interpretations in generic extensions in the following standard way:

Convention 2.43 If X is a Polish space in the ground model and we pass to a generic extension, we will use the same letter X to denote (some) interpretation of X in the extension, while the symbol X^V will denote the version of X in the ground model. We will always assume that the identity map between X^V and X is the interpretation, so that $X^V \subset X$. Similarly, for an analytic set $A \subset X$ in the ground model, we will use the same letter A to denote the interpretation of A in the extension, while the symbol A^V is reserved for the ground model version of the set A.

Similar usage may prevail for the hyperspaces, spaces of Borel probability measures, and the like. While this constitutes a minor abuse of notation, the previous remarks should clarify any confusion.

While the existence and uniqueness of interpretations above was an entirely routine matter, we will also need the more involved interpretation of definable σ-ideals on Polish spaces. Here, the tools from effective descriptive theory must be used. This nontrivial result is certainly missing in section 3.8 and theorem 5.1.9 in Zapletal (2008).

Theorem 2.44 *Every $\mathbf{\Pi}^1_1$ on $\mathbf{\Sigma}^1_1$ class of analytic sets has a canonical $\mathbf{\Pi}^1_1$ on $\mathbf{\Sigma}^1_1$ interpretation in every generic extension. If the class is a σ-ideal of analytic sets in the ground model, then its canonical interpretation in the generic extension is again a σ-ideal of analytic sets.*

Proof Let I be a $\mathbf{\Pi}^1_1$ on $\mathbf{\Sigma}^1_1$ collection of analytic sets on a Polish space X. For simplicity we assume that the space X is recursively presented; otherwise, replace it with its homeomorph, a G_δ-subset of the recursively presented space $[0, 1]^\omega$, and insert the code for the G_δ subset in appropriate places in the argument.

Let $A \subset 2^\omega \times X$ be a good universal Σ^1_1 set in the sense of Kanovei (2008, section 2.6). The set $C = \{y \in 2^\omega : A_y \in I\}$ is coanalytic; therefore, both A and C have a canonical interpretation in any forcing extension. The collection $\{A_y : y \in C\}$, as interpreted in the generic extension, is a natural candidate for the canonical interpretation of I in the generic extension. It is necessary to check that in the generic extension, this is a $\mathbf{\Pi}^1_1$ on $\mathbf{\Sigma}^1_1$ collection of analytic sets, and the interpretation does not depend on the choice of the good universal analytic set A. These features will both follow very quickly as soon as we check that, in the generic extension, the set C still defines a collection of analytic sets, in the sense that the sentence $\phi =$"$\forall y, z \in 2^\omega\ (A_y = A_z) \to (y \in C \leftrightarrow z \in C)$" holds in the generic extension. The sentence ϕ holds in the ground model; the challenge resides in the fact that ϕ appears to be more complex than $\mathbf{\Pi}^1_2$, and therefore it is impossible to use Shoenfield absoluteness to transfer it to the generic extension. We need to resort to a patch that includes the use of effective descriptive set theory. Assume for simplicity that the set C is lightface Π^1_1; otherwise insert the parameter used in its definition in the appropriate spots in the argument below.

Fix a standard parametrization of Borel subsets of X such as in Kanovei (2008, theorem 2.8.2). That is, there are Π^1_1 sets $D \subset 2^\omega$ and $F \subset 2^\omega \times X$ such that for every $y \in D$ the equality $A_z = F_z$ holds, and for every parameter $z \in 2^\omega$ and every $\Delta^1_1(z)$ set $G \subset X$ there is $y \in \Delta^1_1(z)$ such that $F_z = G$. Moreover, this property of the sets A, D, and F is provable in ZFC.

Claim 2.45 *For every $z \in C$ there is $y \in \Delta^1_1(z)$ such that $y \in D \cap C$ and $A_z \subseteq A_y$. Moreover, this is a Π^1_2 statement.*

Proof The first sentence is just the effective reflection, Fact 2.8(*iii*). For the evaluation of the complexity, note that $y \in D \wedge A_z \subseteq A_y$ is a Π^1_1 statement about z and y: for $y \in D$, $A_z \subseteq A_y$ is equivalent to $\forall x\ x \notin A_z \vee x \in F_y$. The complexity evaluation then follows from the well-known fact that the quantifier $\exists y \in \Delta^1_1(z)$ translates into a universal quantifier (Kanovei 2008, corollary 2.8.6) □

Claim 2.46 *For every $z \in 2^\omega$ and every $y \in D \cap C$, if $A_z \subseteq A_y$ then $z \in C$. Moreover, this is a Π^1_2 statement.*

Proof The first sentence just says that the collection I is closed under taking an (analytic) subset. This statement has an easy folklore proof: assume that $P \subset R \subset X$ are analytic sets and $R \in I$. Choose an analytic non-Borel set $S \subset 2^\omega$ and consider the analytic set $T = (S \times R) \cup (2^\omega \times P) \subset 2^\omega \times X$. The set $U = \{y \in 2^\omega : T_y \in I\}$ is coanalytic as I is $\mathbf{\Pi}^1_1$ on $\mathbf{\Sigma}^1_1$. The vertical sections T_y for $y \in S$ are equal to R, so $S \subseteq U$. Since the set S is not coanalytic, there must be a point $y \in U \setminus S$. For such a point y, $T_y = P$ and we conclude that $P \in I$ as well.

For the complexity computation, recall from the previous proof that $y \in D \wedge A_z \subseteq A_y$ is a Π^1_1 statement about y and z, and rewrite the first sentence of the claim as $\forall z \forall y\ (y \notin D \cap C) \vee \neg(y \in D \wedge A_z \subseteq A_y) \vee (z \in C)$. The formula to the right of the universal quantifiers is a disjunction of Π^1_1 and Σ^1_1 formulas, therefore Π^1_2, and the complexity estimate follows. □

The sentence ϕ formally follows from the conjunction of the two statements in the claims: if $z_0, z_1 \in 2^\omega$ are points such that $A_{z_0} \subseteq A_{z_1}$ and $z_1 \in C$, then the first claim provides a point $y \in D \cap C$ such that $A_{z_1} \subseteq A_y$, and the second claim applied to y_0, z then shows that $z_0 \in C$. The statements in the claims are Π^1_2 and therefore will hold in any forcing extension by Shoenfield absoluteness; so even ϕ will hold in every forcing extension.

It follows immediately that the canonical interpretation does not depend on the choice of the analytic set $A \subset 2^\omega \times X$. Let $A^* \subset 2^\omega \times X$ be an arbitrary analytic set and $C^* = \{y \in 2^\omega : A^*_y \in I\}$. We will show that in every forcing extension, for every $y \in 2^\omega$, y belongs to (the interpretation of) C^* if and only if A^*_y belongs to the interpretation of I described above. Since the set A is good universal, there is a continuous function $f : 2^\omega \to 2^\omega$ such that for every $y \in 2^\omega$, $A^*_y = A_{f(y)}$; moreover, $V \models y \in 2^\omega\ y \in C^* \leftrightarrow f(y) \in C$ by the definition of the set C^*. This sentence is $\boldsymbol{\Pi}^1_2$ in the parameter f, and therefore persists to the generic extension. Thus, even in the generic extension $C^* = \{y \in 2^\omega : A^*_y \in I\}$.

The (canonical interpretation of the) collection I remains $\boldsymbol{\Pi}^1_1$ on $\boldsymbol{\Sigma}^1_1$ in every generic extension, again due to the choice of the good universal set $A \subset 2^\omega \times X$. In the extension, if $A^* \subset 2^\omega \times X$ is an analytic set, then there must be a continuous function $f : 2^\omega \to 2^\omega$ such that for every $y \in 2^\omega$, $A^*_y = A_{f(y)}$. It follows that the set $C^* = \{y \in 2^\omega : A^*_y \in I\}$ is coanalytic, since it is simply the f-preimage of the (interpretation of) the set C.

Finally, if I is a σ-ideal of analytic sets in the ground model, then this is easily restated as a Π^1_2 statement using the good universal set A, and therefore it persists to all forcing extensions. □

We exploit the previous theorem in the following way. If I is a $\boldsymbol{\Pi}^1_1$ on $\boldsymbol{\Sigma}^1_1$ σ-ideal, we use the same letter I to denote its canonical interpretation in every generic extension.

As a final remark, in this book we are interested in many properties of $\boldsymbol{\Pi}^1_1$ on $\boldsymbol{\Sigma}^1_1$ σ-ideals, most of which seem to be impossible to restate as Π^1_2 formulas, and their absoluteness into forcing extension remains open to question. For example, in Sabok (2012a), it is proved that the statement "I is σ-generated by closed sets" is absolute for such ideals. The following remains open:

Question 2.47 Let I be a $\boldsymbol{\Pi}^1_1$ on $\boldsymbol{\Sigma}^1_1$ σ-ideal on a Polish space.

(i) If I is c.c.c., is (the canonical interpretation of) I c.c.c. in every forcing extension?
(ii) If the quotient forcing P_I is proper, is it again proper in every forcing extension?
(iii) If the quotient forcing P_I is $< \omega_1$-proper, is it again $< \omega_1$-proper in every forcing extension?

For example, the assumptions of Theorem 5.21 have to be stated in a rather awkward way, since we do not have a general positive answer to the last item.

2.5 Generic ultrapowers

In several arguments throughout the book, we will exploit the technique of (illfounded) generic ultrapower outlined in Jech (2002, section 35). Let X be an arbitrary set and I be an ideal on it, and consider the quotient poset $R_I = \mathcal{P}(X)/I$. This is entirely different from, and should not be confused with, the quotient posets of Borel sets on Polish spaces. Whenever $G \subset R_I$ is a generic filter, one can form in the extension $V[G]$ the *generic ultrapower*, the model N of equivalence classes of ground model functions with domain X, where f is equivalent to g if $\{x \in X : f(x) = g(x)\} \in G$, and $[f] \in^N [g]$ if $\{x \in X : f(x) \in g(x)\} \in G$. A modified proof of the Łoś theorem shows that $j : V \to N$ defined by $j(a) = [f_a]$, where f_a is the constant function with the single value a, is an elementary embedding. It is necessary to comment on the wellfoundedness of the model N. Even though at some point in its cumulative hierarchy the model N may not be wellfounded, we will always identify the wellfounded initial part of N with its transitive isomorph.

Claim 2.48 *If I is a σ-ideal then $j \restriction H_{\aleph_1} = \mathrm{id}$. Also, in such a case $j(\omega) = \omega$, $j(2^\omega) = 2^\omega \cap N$ and $j(\omega^\omega) = \omega^\omega \cap N$.*

Proof We will first argue that $\omega^N = j''\omega$. To verify the equality, return to the ground model and assume that $S \in R_I$ is a condition forcing $[f] \in \omega^N$. We may thin the set S if necessary so that $S = \mathrm{dom}(f)$ and $f''S \subset \omega$. By the σ-additivity of the σ-ideal I, there must be a number $n \in \omega$ such that the set $T \subset S, T = \{x \in S : f(x) = n\}$ is I-positive. Such a condition T forces $[f] = j(n)$. A density argument then shows that $\omega^N = j''\omega$.

It follows immediately that ω^N is wellfounded and therefore identified with ω, and for every $a \subset \omega$, $j(a) = a$. Also, $j(2^\omega) = 2^\omega \cap N$ as $j(2^\omega)$ consists of all binary $j(\omega) = \omega$-sequences, and similarly for ω^ω.

Now, if $b \in H_{\aleph_1}$ is a set, then it is a subset of some transitive set $c \in H_{\aleph_1}$, and the structure $(b, c, \in)$ is isomorphic to some $(b', \omega, \in')$ where $\in'$ is a binary relation on ω and $b' \subset \omega$. Now, $j(b', \omega, \in') = (b', \omega, \in')$ by the previous paragraph, so $\in' = j(\in')$ is isomorphic to the $\in$-relation in $j(c)$ in the model N by the elementarity of j. In other words, c is $\in$-isomorphic to $j(c)$, $j(c)$ is in the wellfounded part of the model N and $c = j(c)$, since c and $j(c)$ are $\in$-isomorphic transitive sets and therefore equal. By the elementarity of j again,

$j(b)$ is the image of $b' = j(b')$ under the unique isomorphism of $\in\restriction c$ and $\in'$, therefore $j(b) = b$ as desired. □

We will use the ultrapower technique exclusively through the following theorem. Recall the language of interpretations of Polish spaces in generic extensions introduced in Section 2.4.

Theorem 2.49 *For every uncountable cardinal λ, in some generic extension $V[G]$ there is a generic ultrapower $j : V \to N$ such that $j''H_\lambda$ exists in N and it is countable there. Moreover, every Polish space X in V has an interpretation $\phi : X \to Y$ in $V[G]$ such that:*

(i) for every analytic subset $A \subset X$ in the ground model, $j(A) \subset A^\phi \cap j(X)$;
(ii) for every Borel set B in the ground model, $j(B) = B^\phi \cap j(X)$.

Note that both properties of the embedding j will survive into any further generic extension of the ground model. It may happen that there is an analytic set A such that $j(A) \neq A^\phi \cap j(X)$. For example, if there is an illfounded countable ordinal $\beta \in N$ and A is the set of illfounded linear orders on ω, then N contains some linear ordering x on ω which is isomorphic to β, and then $x \in A^\phi$ (as β is illfounded) and $x \notin j(A)$ (as $N \models x$ is isomorphic to an ordinal). In such a case, the model N may be incorrect about such items as codes for Borel sets, etc.

Proof The argument is entirely standard. Let I be the σ-ideal on nonstationary subsets of $[H_\lambda]^{\aleph_0}$, and let R_I be the quotient poset. Let $G \subset R_I$ be a generic filter over V and form the generic ultrapower $j : V \to N$. We claim that this embedding works.

The first sentence of the theorem follows from the normality of the nonstationary ideal on $[H_\lambda]^{\aleph_0}$. Let i be the identity function on $[H_\lambda]^{\aleph_0}$. Consider the equivalence class $[i] \in N$, and observe that $[i] = j''H_\lambda$. For the right-to-left inclusion, if $a \in H_\lambda$ then for a club subset of $b \in [H_\lambda]^{\aleph_0}$ it is the case that $a \in b$; in other words, the values of function f_a constantly equal to a belong to the values of the identity function everywhere but on a nonstationary set. By the Łoś theorem, $N \models j(a) = [f_a] \in [\mathrm{id}]$. For the left-to-right inclusion, return to the ground model. If S is a condition forcing $N \models [f] \in [i]$ for some function f, then for all but nonstationarily many $b \in S$ it is the case that $f(b) \in b$, and the normality of the nonstationary ideal yields a stationary subset $T \subset S$ and $a \in H_\lambda$ such that $\forall b \in T\ f(b) = a$. Clearly, $T \Vdash [f] = j(a)$, and a density argument concludes the proof of the left-to-right inclusion. Finally, $N \models j''H_\lambda$ is countable by the Łoś theorem again, since the values of the function i are all countable sets.

Let X be any Polish space in V, and choose a complete metric $d \in V$ for it. Note first that $\langle j(X), j(d) \rangle$ is indeed a metric space in $V[G]$ by Claim 2.48, since the $j(d)$-distances are real numbers in $V[G]$. Let Y be any completion of $j(X)$ in $V[G]$, and let $\phi : X^V \to Y$ be the map defined by $\phi(x) = j(x)$. This is certainly an isometry by the elementarity of j; we will show that its range is dense in Y. Choose a countable dense set $a \subset X^V$ in the ground model. Then $j(a) = j''a$ holds, since $\omega^N = \omega$. The set $j(a) = j''a$ is dense in $j(X)$ by the elementarity of j, and so it is dense in Y as well. Thus, ϕ is indeed an interpretation; we claim that it has properties (*i*) and (*ii*).

As the first step towards (*i*), suppose that $C \subset X$ is a closed set in V. We claim that $j(C) \subset C^\phi$ holds. To verify the inclusion, note that by the definition of interpretation, C^ϕ is the closure of $j''C$ in Y. Just as in the previous paragraph choose a countable dense set $a \subset C$ in V, observe that $j(a) = j''a$ and note that the set $j''a = j(a) \subset j(C)$ is dense by the elementarity of j. Thus, the closure of $j''C$ in Y (which includes $j''a$) must contain $j(C)$.

To prove (*i*), suppose that $A \subset X$ is an analytic set in V, a projection of some closed set $C \subset X \times \omega^\omega$. Write ψ for the identity map from $(\omega^\omega)^V$ to $(\omega^\omega)^{V[G]}$; this is an interpretation of the Baire space in $V[G]$. Applying the previous paragraph to the space $X \times \omega^\omega$, its interpretation $\phi \times \psi$ and the set C, we see that $j(C) \subset j(X) \times (\omega^\omega)^N$ is a subset of the interpretation $C^{\phi \times \psi} \subset Y \times \omega^\omega$. Now $j(A) = p(j(C))$ by the elementarity of j, and $A^\phi = p(C^{\phi \times \psi})$ by the definition of the interpretation. Thus, $j(A) \subset A^\phi$ as desired.

For (*ii*), if $B \subset X$ is a Borel set in the ground model, then both sets $B, X \setminus B$ are analytic. The sets $j(B)$ and $j(X \setminus B)$ are complementary subsets of $j(X)$ by the elementarity of j, and they are subsets of the respective interpretations B^ϕ and $(X \setminus B)^\phi$ in the space Y by (*i*). These interpretations are also complementary in the space Y, since interpretation preserves intersections and unions of analytic sets (Claim 2.40). This immediately implies that $j(B) = B^\phi \cap j(X)$. □

2.6 Idealized forcing

The techniques of this book depend heavily on the theory of idealized forcing as developed in Zapletal (2008). Let X be a Polish space and I be a σ-ideal on it. The symbol P_I stands for the partial ordering of all Borel I-positive subsets of X ordered by inclusion; we will alternately refer to it as the *quotient forcing* of the σ-ideal I. Its separative quotient is the σ-complete Boolean algebra of Borel sets modulo I. The forcing extension is given by a unique point $x_{gen} \in X$ such that the generic filter consists of exactly those Borel sets in the ground model containing x_{gen} as an element.

If I is a σ-ideal on a Polish space X, M is an elementary submodel of a large enough structure with $X, I \in M$, and $x \in X$ is a point, we will say that x is P_I-generic over M if for every open dense set $D \subset P_I$, $D \in M$, there is $B \in D \cap M$ with $x \in B$. The set $g \subset P_I \cap M$ consists of those $B \in P_I \cap M$ for which $x \in B$ is an M-generic filter on $P_I \cap M$. We then consider $M[x]$ as the transitive model obtained from the transitive isomorph of M by adjoining the filter g. If $\tau \in M$ is a P_I-name then τ/x denotes the evaluation of the transitive collapse image of τ according to the filter g. Similar terminology will be applied in the case that M is instead some transitive model of set theory and $M \models I$ is a σ-ideal on a Polish space X.

Fact 2.50 (Zapletal 2008, proposition 2.2.2) *Let I be a σ-ideal on a Polish space X. The following are equivalent:*

(i) P_I is proper;
(ii) for every Borel I-positive set $B \subset X$ and every countable elementary submodel M of a large structure with $I \in M$, the set $C \subset B$ of all points in B that are P-generic over the model M is I-positive.

There are many consequences of this characterization:

Fact 2.51 *Let I be a σ-ideal on a Polish space X such that the quotient P_I is proper. Then:*

(i) (Zapletal 2008, proposition 2.3.1) $V[x_{\text{gen}}] \cap 2^\omega = \{f(\dot{x}_{\text{gen}}) : f$ *is a Borel function coded in the ground model*$\}$;
(ii) (Zapletal 2008, proposition 2.3.2) *for every Borel set $B \in P_I$ and every analytic set $A \subset B \times \omega^\omega$ with nonempty vertical sections, there is a Borel I-positive set C and a Borel function $f : C \to \omega^\omega$ such that $f \subset A$;*
(iii) (Zapletal 2008, proposition 2.3.3) *for every Borel set $B \in P_I$ and every Borel function $f : B \to \omega^\omega$ there is a Borel I-positive set $C \subset B$ such that $f''C$ is Borel.*

The following simple fact is used throughout the book:

Fact 2.52 (Zapletal 2008, theorem 3.3.2) *Suppose that I is a σ-ideal on a Polish space X such that the quotient poset P_I is proper. The following are equivalent:*

(i) P_I is bounding;
(ii) every Borel I-positive set has a compact I-positive subset, and every Borel function on an I-positive domain is continuous on a compact I-positive set.

The following properties deal with σ-ideals without any reference to forcing; nevertheless, they are particularly useful when idealized forcing is considered.

Definition 2.53 (Zapletal 2008, definition 3.1.1) A σ-ideal I on a Polish space X has the *continuous reading of names* if for every Borel I-positive set $B \subset X$ and every Borel function $f : B \to 2^\omega$ there is a Borel I-positive set $C \subset B$ such that $f \restriction C$ is continuous.

Definition 2.54 (Zapletal 2008, definition 3.2.1) A pair of σ-ideals I, J on respective Polish spaces X, Y has the *Fubini property* if whenever $B_0 \subset X$ and $B_1 \subset Y$ are Borel I- and J-positive sets respectively, then the product $B_0 \times B_1$ cannot be covered by the union of two Borel subsets of $X \times Y$, one with all vertical sections in J and the other with all horizontal sections in I.

Definition 2.55 (Zapletal 2008, definition 3.11.1) A pair of σ-ideals I, J on respective Polish spaces X, Y has the *rectangular Ramsey property* if whenever $B_0 \subset X$ and $B_1 \subset Y$ are Borel I- and J-positive sets respectively and the product $B_0 \times B_1$ is covered by the union $\bigcup_n D_n$ of countably many Borel subsets of $X \times Y$, then there are Borel I- and J-positive sets $C_0 \subset B_0$ and $C_1 \subset B_1$ respectively such that the product $C_0 \times C_1$ is a subset of one of the sets forming the union. The definition naturally extends to finite or countable collections of σ-ideals.

The main purpose of the rectangular Ramsey property is the computation of the σ-ideal associated with forcing products. Let $\{I_n : n \in \omega\}$ be σ-ideals on respective Polish spaces X_n for $n \in \omega$. The *box product ideal* $\prod_n I_n$ is the collection of all Borel sets $B \subset \prod_n X_n$ containing a subset of the form $\prod_n B_n$, where each $B_n \subset X_n$ is a Borel I_n-positive subset. If the rectangular Ramsey property holds, then the box product K is in fact a σ-ideal and the product of the posets $\prod_n P_{I_n}$ with full support is naturally isomorphic to a dense subset of the quotient P_K. Note what the properness criterion (Fact 2.50) says in case of two ideals: if the rectangular Ramsey property for I, J holds and the poset $P_I \times P_J$ is proper, then for every countable elementary submodel M of a large enough structure, there are Borel I- and J-positive sets B and C respectively such that whenever $x \in B$ and $y \in C$ then the pair $\langle x, y \rangle$ is $P_I \times P_J$-generic over M. The converse implication is also true. The main tool for the verification of the rectangular Ramsey property in this book is the following:

Fact 2.56 (Zapletal 2008, theorem 5.2.6) *If I_n for $n \in \omega$ are $\mathbf{\Pi}^1_1$ on $\mathbf{\Sigma}^1_1$ σ-ideals on Polish spaces such that the quotient posets P_{I_n} are all proper, bounding, and preserve Baire category, then the collection $\langle I_n : n \in \omega \rangle$ has* *the rectangular Ramsey property. The box product ideal* $K = \prod_n I_n$ *is* $\mathbf{\Pi}^1_1$ *on* $\mathbf{\Sigma}^1_1$*, and the quotient poset* P_K *is proper, bounding, and preserves Baire category.*

In a number of choiceless contexts, the rectangular Ramsey property may hold even for partitions into arbitrary sets (Bagaria and DiPrisco 2009).

The mutual genericity property is considerable strengthening of properness, and it is a variation of the rectangular Ramsey property. Here, we record the following:

Proposition 2.57 *Let* I *be a* σ*-ideal on a Polish space* X*. The mutual genericity property implies the free set property.*

Proof Suppose that $B \subset X$ is a Borel I-positive set and $D \subset B \times B$ is an analytic sets whose vertical sections are in the σ-ideal I. Let M be a countable elementary submodel of a large enough structure containing B, I, and use the mutual genericity property to find a Borel I-positive set $C \subset B$ consisting of points pairwise $P_I \times P_I$-generic over M. Thinning out the set C if necessary, we may assume that its intersections with sets in the model M are either I-positive or empty. We claim that $D \cap (C \times C) \subset \text{id}$, verifying the free set property.

Suppose for contradiction that $x, y \in C$ are distinct points such that $\langle x, y\rangle \in D$. Since these points are generic for the product $P_I \times P_I$ over the model M, there must be a condition $\langle B_x, B_y\rangle \in (P_I \times P_I) \cap M$ which forces the generic pair to belong to $\dot{D}$, and moreover $x \in B_x$, $y \in B_y$ and $B_x \cap B_y = 0$. Since the intersection $B_y \cap C$ is nonempty, it is in fact I-positive. Whenever $z \in B_y \cap C$ is a point, then the $P_I \times P_I$-generic filter over M determined by the pair x, z contains the condition $\langle B_x, B_y\rangle$. Therefore by the forcing theorem $M[x, z] \models \langle x, z\rangle \in D$ and by analytic absolutenes between the model $M[x, y]$ and V, $\langle x, z\rangle \in D$. Thus, the vertical section D_x contains the I-positive set $B_y \cap C$ and so is itself I-positive, contradicting the assumed properties of the set D. □

Definable c.c.c. forcings form a very special class that has been investigated repeatedly, and we will need several general properties of this class. We will first show that every such poset adds a real, and then analyze the possibilities for such reals. We make the $\mathbf{\Pi}^1_1$ on $\mathbf{\Sigma}^1_1$ assumption on our σ-ideals so as to remain within ZFC framework; if one is willing to make large cardinal assumptions, the conclusions will hold true of a much larger class of σ-ideals than just $\mathbf{\Pi}^1_1$ on $\mathbf{\Sigma}^1_1$.

Theorem 2.58 *Let I be a $\mathbf{\Pi}^1_1$ on $\mathbf{\Sigma}^1_1$ σ-ideal on a Polish space X such that the quotient forcing P_I is proper. Every c.c.c. intermediate extension of the P_I extension is either equal to the ground model or contains a new real.*

In other words, the quotient does not add new branches through Suslin trees and Suslin algebras. Note that the intermediate c.c.c. extension may not be given by a single real; posets such as adding $\aleph_1$ Cohen reals with finite support may embed to P_I under suitable circumstances.

Proof Suppose for contradiction that the intermediate extension is effected by a c.c.c. forcing Q, a regular subforcing of P_I which generates a nontrivial forcing extension and does not add reals. Let M be a countable elementary submodel of a large enough structure containing Q and I. Since no real is added by Q, it forces that the intersection of its generic filter $\dot{G}$ with the model M belongs to the ground model. Note also that this intersection must be Q-generic over M: every antichain of Q is countable, therefore if it is in M, then it is a subset of M and so Q forces the filter $\dot{G}$ to contain an element of such an antichain which is also in M. Fix a maximal antichain of conditions in the poset Q deciding the value of $M \cap \dot{G}$. The antichain is countable, and so we can list all possible values for the intersection as $\{g_n : n \in \omega\}$. Note that as the forcing Q generates a nontrivial forcing extension, none of the filters $\{g_n : n \in \omega\}$ belong to the model M.

Let $\kappa = |\mathcal{P}(P_I)|$, let $h \subset \mathrm{Coll}(\omega, \kappa) \cap M$ be a filter in V, generic over all the models $M[g_n]$; such a filter exists as the models contain together only countably many antichains of the poset $\mathrm{Coll}(\omega, \kappa) \cap M$. By the product forcing theorem, for every $n \in \omega$ the intersection $M[h] \cap M[g_n]$ is equal to M, in particular $g_n \notin M[h]$. The Borel set B of P_I-generic reals over M is equal to $\bigcap\{\bigcup(D \cap M) : D \in M \text{ is an open dense subset of } P_I\}$; therefore, it has a Borel code in $M[h]$ as the model $M[h]$ can see that the unions and intersections forming it are countable. The set B is I-positive in V by properness of P_I and Fact 2.50, this positivity is an analytic statement by the complexity assumption on I, and therefore it transfers to the wellfounded model $M[h]$. Let $f : B \to \mathcal{P}(Q \cap M)$ be the function defined by $f(x) = \{q \in Q \cap M : \exists C \in P_I \cap M\ x \in C \wedge C \text{ forces } q \text{ to belong to the } Q\text{-generic filter}\}$. This is a Borel function with a code in the model $M[h]$. There are now two cases:

Case 1 Either, the model $M[h] \models f$ is I-to-one. The statement to the right of the $\models$ sign is coanalytic since I is $\mathbf{\Pi}^1_1$ on $\mathbf{\Sigma}^1_1$, therefore transfers to the model V, and there B forces that the intersection of the Q-generic filter with M is not in the ground model, contradicting the assumptions.

Case 2 Or, $M[h] \models \exists g \subset Q \cap M\ f^{-1}\{g\} \notin I$. In this case, let $C = f^{-1}\{g\}$ and observe that $C \subset B$ is a Borel I-positive set and $C \Vdash \check{g}$ is the intersection of the Q-generic filter with M. Since $g \in M[h]$ and $M[h]$ contains none of the filters g_n for $n \in \omega$, this contradicts the choice of the list $\{g_n : n \in \omega\}$. □

For the following fact, recall the notion of a *Maharam algebra* (Jech and Balcar 2006). A function ϕ from a σ-algebra B to the reals is a *continuous submeasure* if for all $a, b \in B$, $\phi(a \vee b) \leq \phi(a) + \phi(b)$, $a \leq b$ implies $\phi(a) \leq \phi(b)$, $\phi(0) = 0$, and ϕ is continuous in decreasing sequences: if $\langle a_i : i \in \omega \rangle$ is a decreasing sequence in B then $\phi(\bigwedge_i a_i) = \inf_i \phi(a_i)$. ϕ is *strictly positive* if $\phi(a) = 0$ implies $a = 0$. A Maharam algebra is a σ-algebra carrying a strictly positive continuous submeasure.

Theorem 2.59 *Let I be a c.c.c. $\mathbf{\Pi}^1_1$ on $\mathbf{\Sigma}^1_1$ σ-ideal on a Polish space containing all singletons and not the whole space. Then:*

(i) if P_I is bounding, then it is a Maharam algebra;
(ii) if P_I is not bounding, then it adds a Cohen real.

Proof For (*i*), if P_I is bounding and I is $\mathbf{\Pi}^1_1$ on $\mathbf{\Sigma}^1_1$, then Player II has a winning strategy in the bounding game (Zapletal 2008, theorem 3.10.7). Fremlin (Jech and Balcar 2006, theorem 7.5) proved that the winning strategy yields a strictly positive continuous submeasure ϕ on the Polish space such that $I = \{A : \phi(A) = 0\}$.

The proof of (*ii*) is a simplification of an argument of Shelah (Bartoszynski and Judah 1995, theorem 3.6.47). We first isolate a Mycielski style claim of independent interest. Whenever $k : [\omega]^{\aleph_0} \to [\omega]^{\aleph_0}$ is a function and $a \in [\omega]^{\aleph_0}$ is such that $k(a) \subset a$, write $\bar{k}(a) = \bigcup\{$all intervals (n, m) such that $n \in k(a)$ and m is the least element of a larger than $n\}$.

Claim 2.60 *Whenever $k : [\omega]^{\aleph_0} \to [\omega]^{\aleph_0}$ is a Borel function such that $k(a) \subset a$ for all $a \in [\omega]^{\aleph_0}$, there is a perfect set $P \subset [\omega]^{\aleph_0}$ such that for every $a \neq b \in P$, the set $\bar{k}(a) \cup \bar{k}(b)$ contains all but finitely many natural numbers.*

Proof Let $\pi : \omega \to 2^{<\omega}$ be a one-to-one extension-preserving enumeration. By induction on $i \in \omega$ build nonempty finite sets s_i and infinite sets $c_i \subset \omega$ so that:

I1. $\max(s_i) < \min(s_{i+1})$, $s_{i+1} \subset c_i$, $c_{i+1} \subset c_i$;
I2. for every $a \in [c_i]^{\aleph_0}$ and every $j < i$, writing $u = \bigcup\{s_k : \pi(k) \subseteq \pi(j)\}$ we have $\max(u) \in k(u \cup a)$.

To perform the induction step, suppose that s_j, c_j for $j < i$ have been found. Write $v = \bigcup\{s_k : \pi(k) \subsetneq \pi(i)\}$ and by induction on $l \in \omega$ build increasing numbers m_l and infinite sets $d_l \subset \omega$ so that:

J1. $d_0 = c_{i-1}, d_{l+1} \subset d_l, m_l \in d_l$;
J2. one of the following holds:

(a) for every $a \in [d_{l+1}]^{\aleph_0}$, we have $m_l \in k(v \cup \{m'_l : l' \leq l\} \cup a)$;
(b) for every $a \in [d_{l+1}]^{\aleph_0}$, we have $m_l \notin k(v \cup \{m'_l : l' \leq l\} \cup a)$.

The induction on l uses the Galvin–Prikry theorem (Fact 8.14). At some finite stage clause J2 (a) must be satisfied; otherwise, the value $k(v \cup \{m_l : l \in \omega\})$ would have to be a subset of v and therefore finite. If clause J2 (a) is satisfied at l, the set $s_i = \{m_{l'} : l' \leq l\}$ and $c_i = d_{l+1}$ will complete the induction step i.

Once the induction on i is performed, for every $z \in 2^\omega$ let $a_z = \bigcup_i s_{\pi^{-1}(z \restriction i)}$ and observe that the set $\{a_z : z \in 2^\omega\}$ works as desired. To check this, let $y \neq z \in 2^\omega$, let $t = \pi(i)$ be their longest common initial segment, and let $n > \max(s_i)$. There are two cases:

Case 1 Suppose first that n belongs to the convex closure of the set s_j for some j such that $\pi(j) \subset y$. Writing j_0, j_1 respectively for the largest number below j and smallest number greater than j such that $\pi(j_0), \pi(j_1) \subset z$, we see that $\max(s_{j_0}) \in k(a_z)$ by I2, and $\max(s_{j_0}) < n \leq \max(s_j) < \min(s_{j_1}) = \min(a_z \setminus \max(s_{j_0}) + 1)$ by I1, and so $n \in \bar{k}(a_z)$.

Case 2 If Case 1 fails, write j_0, j_1 respectively for the numbers such that $\pi(j_0), \pi(j_1)$ are successive initial segments of y such that $\max(s_{j_0}) < n < \min(s_{j_1})$. Again, use I2 to argue that $\max(s_{j_0}) \in k(a_y)$ and conclude that $n \in \bar{k}(a_y)$. □

Now, suppose that I is a $\mathbf{\Pi}^1_1$ on $\mathbf{\Sigma}^1_1$ c.c.c. σ-ideal such that the quotient P_I adds an unbounded real. Use this assumption and Fact 2.51 to find a Borel I-positive set $B \subset X$ and a Borel function $f : B \to \omega^\omega$ such that $B \Vdash \dot{f}(\dot{x}_{\text{gen}})$ is unbounded. This means that for every compact set $K \subset \omega^\omega$ the preimage $f^{-1}K$ is in the ideal I. Now let $\chi : \omega \to 2^{<\omega}$ be an enumeration with infinite repetitions, and for every $a \in [\omega]^{\aleph_0}$ let $h_a : \omega^{\leq\omega} \to 2^{\leq\omega}$ be the function such that $h_a(x)$ is the concatenation of the binary strings $t_0, t_1, \ldots$ where $t_i = \chi(|a \cap x(i)|)$ for all $i \in \text{dom}(x)$. We claim that there is a set $a \in [\omega]^{\aleph_0}$ such that the function $h_a \circ f : B \to 2^\omega$ represents a Cohen real; i.e., the $h_a \circ f$-preimages of closed nowhere dense sets are in I.

Suppose for contradiction that this fails. The set $A = \{\langle a, K\rangle : a \in [\omega]^{\aleph_0}, K \in K(2^\omega)$ is nowhere dense and $f^{-1}h_a^{-1}K \notin I\}$ is analytic as the

σ-ideal I is $\mathbf{\Pi}^1_1$ on $\mathbf{\Sigma}^1_1$, and all its vertical sections are nonempty by the assumptions. By the uniformization for the Ramsey null ideal (Fact 2.51 together with the properness of Mathias forcing), there is an infinite set $c \subset \omega$ and a Borel function $g : [c]^{\aleph_0} \to K(2^\omega)$ such that the graph of g is a subset of A.

Let $\{t_i : i \in \omega\}$ be a fixed enumeration of $\omega^{<\omega}$. For every $a \in [c]^{\aleph_0}$ and every $i \in \omega$ let $n_i(a)$ be the least number larger than $n_{i-1}(a)$ such that for every $j < i$, $[h_a(t_j)^\frown \chi(n_i(a))] \cap g(a) = 0$. Such a number must exist as the set $g(a)$ is nowhere dense. Let $k(a) = \{n_i(a)\text{-th element of } a : i \in \omega\}$. Use Claim 2.60 to find a perfect set $P \subset [c]^{\aleph_0}$ such that for $a \neq b \in P$, the set $\bar{k}(a) \cup \bar{k}(b)$ contains all but finitely many natural numbers. We will show that the set $\{(h_a \circ f)^{-1} g(a) : a \in P\}$ forms an antichain in P_I; this will contradict the assumed c.c.c. of P_I.

To this end, fix $a \neq b \in P$; we will show that the set $C = h_a^{-1} g(a) \cap h_b^{-1} g(b) \subset \omega^\omega$ is compact. The set C is certainly closed. To see that it is compact, choose a string $t \in \omega^{<\omega}$; say $t = t_i$ for some $i \in \omega$. Whenever $n \in \omega$ is greater than the $n_i(a)$-th element of a and belongs to $\bar{k}(a)$, then $[t^\frown n] \cap C = 0$ holds by the definition of k; the same conclusion follows if n is greater than the $n_i(b)$-th element of b and belongs to $\bar{k}(b)$. Since the set $\bar{k}(a) \cup \bar{k}(b) \subset \omega$ is co-finite, this means that for every $t \in \omega^{<\omega}$, for all but finitely many $n \in \omega$ we have $[t^\frown n] \cap C = 0$. Thus, the set C is compact. Now, the initial choice of the function f means that the set $f^{-1}C$ must be in I; in other words, the conditions $(h_a \circ f)^{-1} g(a)$ and $(h_b \circ f)^{-1} g(b)$ are incompatible in P_I as desired. □

Corollary 2.61 *Let I, J be c.c.c. $\mathbf{\Pi}^1_1$ on $\mathbf{\Sigma}^1_1$ σ-ideals on Polish spaces containing all singletons and not the whole space. The pair I, J does not have the rectangular Ramsey property.*

Proof Theorem 2.59 shows that we only have to consider the meager ideal and the null ideals for the various continuous submeasures. As shown in Farah and Zapletal (2004), pairs of these σ-ideals even fail to have the Fubini property unless they are two null ideals obtained from Borel probability measures or two meager ideals obtained from some Polish topologies. In these two cases, the rectangular property fails again. The simplest way to verify the failure is to consider the equivalence E_0 on 2^ω and the usual Borel probability measure or topology on 2^ω. Apply the Steinhaus or Pettis theorem to see that there cannot be a product of two Borel positive sets which is either disjoint from E_0 or a subset of E_0. □

Corollary 2.62 *Let I be a c.c.c. $\mathbf{\Pi}^1_1$ on $\mathbf{\Sigma}^1_1$ σ-ideal on a Polish space containing all singletons and not the whole space. The quotient forcing adds an independent real.*

Proof Maharam algebras add independent reals by Balcar *et al.* (2005, lemma 6.1), and a Cohen real is an independent real in its usual presentation. The rest is handled by Theorem 2.59. □

Corollary 2.63 *Suppose that I is a $\mathbf{\Pi}^1_1$ on $\mathbf{\Sigma}^1_1$ σ-ideal on a Polish space such that the quotient poset P_I is proper and:*

(i) either P_I adds no independent real;
1. or[(ii)] I has the rectangular Ramsey property.

Then every c.c.c. intermediate extension of the P_I extension is equal to the ground model.

Proof Suppose for contradiction that $B \in P_I$ forces that W is a nontrivial c.c.c. extension. By Theorem 2.58, it contains a new real τ. Thinning B down if necessary, we can find a Borel function $f : B \to 2^\omega$ so that $B \Vdash \tau = \dot{f}(\dot{x}_{\text{gen}})$. Let J be the σ-ideal on 2^ω consisting of sets whose f-preimage is in I. J is clearly $\mathbf{\Pi}^1_1$ on $\mathbf{\Sigma}^1_1$, by the nontriviality assumption it contains all singletons, and the poset P_J must be c.c.c. We will now show that the properties in (*i*) or (*ii*) would have to be inherited by J, contradicting Corollary 2.62 or 2.61 respectively.

For (*i*), note that $B \Vdash \tau$ is P_J-generic over V: if $A \subset P_J$ is a maximal antichain, then it is countable, the set $\bigcup A$ is Borel and J-small and so $\{x \in B : f(x) \notin \bigcup A\} \in I$. Thus, as P_I adds no independent reals, it must be the case that P_J adds no independent reals.

For (*ii*), suppose that $D_0, D_1 \subset 2^\omega$ are J-positive Borel sets and $D_0 \times D_1 = \bigcup_n A_n$ is a union of countably many Borel sets; we will find Borel J-positive subsets $D_0', D_1' \subset D_0, D_1$ such that $D_0' \times D_1' \subset A_n$ for some $n \in \omega$. Let $B_0 = f^{-1}D_0$ and $B_1 = f^{-1}D_1$; these are Borel I-positive subsets of B. Let $A_n' = f^{-1}A_n$ for every n; so $B_0 \times B_1 = \bigcup_n A_n'$. Use the rectangular Ramsey property of I to find Borel I-positive sets $B_0' \subset B_0$, $B_1' \subset B_1$ such that $B_0' \times B_1' \subset A_n'$ for some fixed n, use Fact 2.51(*iii*) to thin them down if necessary so that their images are Borel, and let $D_0' = f''B_0'$ and $D_1' = f''B_1'$. These are J-positive Borel sets witnessing the rectangular Ramsey property of J. □

2.7 Concentration of measure

In several places, we will make good use of the concentration of measure phenomenon on simple finite metric spaces. The following is the simplest context:

Definition 2.64 A *Hamming cube* of dimension n is the set 2^n equipped with the normalized counting measure and the normalized Hamming distance, i. e., $d(x, y) = |\{m \in n : x(m) \neq y(m)\}|/n$.

Hamming cubes are a special case of product metric spaces with product measures. The interaction of the metric and the measure in them is governed by the following calculation:

Fact 2.65 (Pestov 2006, theorem 4.2.5) *Let $X = \prod_{i \in n} M_i$ be a product of finite metric spaces, each with metric d_i of diameter a_i and a probability measure μ_i. Equip X with the product measure μ and the ℓ_1-sum d of the metrics d_i. For every $\varepsilon > 0$ and every set of μ-mass at least $1/2$, its d-ε-neighborhood has μ-mass at least $1 - 2\exp(-\varepsilon^2/16 \sum_i a_i^2)$.*

Corollary 2.66 *Assume that $c, d \subset X$ are sets, both of of μ-mass greater than $2\exp(-\varepsilon^2/16 \sum_i a_i^2)$. Then there are points $u \in c$ and $v \in d$ of d-distance at most 2ε.*

Proof Let $\bar{c}, \bar{d} \subset X$ be the ε-neighborhoods of c, d respectively. We will show that both $\bar{c}, \bar{d}$ have μ-mass greater than $1/2$. It then follows immediately that the intersection $\bar{c} \cap \bar{d}$ is nonempty, containing some point w, which must be at $\leq \varepsilon$-distance from points $u \in c$ and $v \in d$. The triangle inequality shows that u, v are as required.

To see that $\mu(\bar{c}) > 1/2$, suppose for contradiction that this fails and so $\mu(X \setminus \bar{c}) \geq 1/2$. Fact 2.65 shows that the ε-neighborhood of $X \setminus \bar{c}$ has μ-mass greater than $1 - \mu(c)$, and so it must intersect the set c. This contradicts the definition of the set $\bar{c}$. The case of $\bar{d}$ is the same. □

2.8 Katětov order and coding functions

In a number of places, to construct rectangular (or square) coding functions we produce Borel functions f such that images of rectangles (or squares) with I-positive analytic sides have nonempty interior. This is an equivalent job, as the following proposition shows:

Proposition 2.67 *Suppose that I is a σ-ideal on a Polish space X and $g : X \times X \to 2^\omega$ is a Borel function such that g-images of rectangles (or squares)* *with I-positive analytic sides have nonempty interior. Then I has a rectangular (or square) coding function.*

Proof Find a Borel set $D \subset 2^\omega \times 2^\omega$ such that its vertical sections are pairwise disjoint dense subsets of 2^ω and $\mathrm{rng}(D)$ is a Borel set. Define $f : X \times X \to 2^\omega$ by $f(x, y) = z$ if $g(x, y) \in D_z$; $f(x, y) = 0$ if $g(x, y) \notin \mathrm{rng}(D)$. This is the desired rectangular (or square) coding function. If $B, C \subset X$ are I-positive analytic sets, then $g''(B \times C)$ has nonempty interior. Thus, for every $z \in 2^\omega$ there is a point $v \in D_z$ in the interior, and there are points $x \in B$ and $y \in C$ such that $g(x, y) = v$ and so $f(x, y) = z$. □

A number of results in this book and others is connected together via the notion of *Katětov ordering* of σ-ideals.

Definition 2.68 Let I, J be σ-ideals on respective Polish spaces X, Y. We say that I is *Katětov above* J, $I \geq_K J$, if there is a Borel map $g : X \to Y$ such that preimages of sets in J are in I.

Proposition 2.69 *Let I, J be σ-ideals on respective Polish spaces X, Y. If $I \geq_K J$ and J has a (square or rectangular) coding function defined on the whole space, then I has a (square or rectangular) coding function defined on the whole space.*

Proof Let $g : X \to Y$ be the Borel Katětov function witnessing $I \geq_K J$. Let $f : Y^2 \to 2^\omega$ be a Borel (square or rectangular) coding function for J. We claim that the Borel function $f \circ g$ is then a (square or rectangular) coding function for I. To verify the coding property, consider the rectangular case. If $B, C \subset X$ are analytic I-positive sets, then their g-images $g''B, g''C$ are analytic and J-positive by the definition of Katětov order. Let $z \in 2^\omega$ be an arbitrary point. Since f is a rectangular coding function for J, there must be points $y_0 \in g''B$ and $y_1 \in g''C$ such that $z = f(y_0, y_1)$. Let $x_0 \in B$ and $x_1 \in C$ be points such that $g(x_0) = y_0$ and $g(x_1) = y_1$, and observe that $f(g(x_0), g(x_1)) = z$. This completes the proof in the rectangular case; the square case is similar. □

Proposition 2.70 *Let I, J be σ-ideals on respective Polish spaces X, Y, and assume that every analytic J-positive set has a Borel J-positive subset, and the quotient P_J is proper. If $I \geq_K J$ and E is in the spectrum of the σ-ideal J as witnessed by an equivalence relation on the whole space, then E is also in the spectrum of the σ-ideal I as witnessed by an equivalence relation defined on the whole space.*

Proof Let F be an equivalence relation on the space Y such that for every Borel J-positive set $C \subset Y$, $F \restriction C$ is bireducible with E. Let $g : X \to Y$ be a

Borel Katětov function witnessing $I \geq_K J$. Consider the pullback equivalence relation $G = g^{-1}F$ on the space X: $x_0 \; G \; x_1$ if $g(x_0) \; F \; g(x_1)$. The equivalence relation G is certainly reducible to E as the Borel function g reduces it to F. We will show that E is Borel reducible to $G \restriction B$ for every I-positive Borel set $B \subset X$; this will complete the proof of the proposition.

Let $B \subset X$ be a Borel I-positive set. The set $g''B \subset Y$ is analytic and J-positive by the definition of the Katětov order. By the assumptions, it has a Borel J-positive subset $C \subset Y$. The properness of the quotient poset P_J together with Fact 2.51(*ii*) shows that thinning out C if necessary, we may find a Borel function $h : C \to B$ such that $g \circ h = \text{id}$. Clearly, the function h reduces $F \restriction C$ to $G \restriction B$. Since $F \restriction C$ is bireducible with the equivalence relation E, it must be the case that E is Borel reducible to $G \restriction B$ as desired.

□

3

Analytic equivalence relations and models of set theory

Let I be a σ-ideal on a Polish space X. The key idea underlying much of the current of thought in this book is the correspondence between Borel equivalence relations on X and intermediate forcing extensions of the generic extension given by the quotient poset P_I of Borel I-positive sets ordered by inclusion. The correspondence is easiest to illustrate on smooth equivalence relations. If E is a smooth equivalence on X and $f : X \to 2^\omega$ is a Borel function reducing it to the identity, then the model $V[f(\dot{x}_{\text{gen}})]$ depends only on E and not on the choice of the reduction. However, we need to find a sufficiently general definition that covers the nonsmooth case, where most of the interest and difficulty lies.

The main application of the concepts developed in this chapter is the following immediate consequence of the more technical trichotomy theorem (Theorem 3.5):

Theorem 3.1 *Suppose that I is a σ-ideal on a Polish space X such that the poset P_I is proper and the P_I-extension is a minimal forcing extension. For every I-positive Borel set $B \subset X$ and every analytic equivalence relation E on B, there is an I-positive Borel subset $C \subset B$ such that $E \restriction C = \text{id}$ or $E \restriction C$ is ergodic.*

3.1 The model $V[x]_E$

If $x \in X$ is a V-generic point, we consider a ZFC model $V[x]_E$ that depends on the E-equivalence class of x, not on x itself. The correct formalization:

Definition 3.2 Let E be an analytic equivalence relation on a Polish space X. Let $x \in X$ be a V-generic point. Let κ be a cardinal greater than $(2^{|P|})^V$ where $P \in V$ is a poset from which the generic extension is derived, and let $H \subset \text{Coll}(\omega, \kappa)$ be a generic filter over $V[x]$. The model $V[x]_E$ is the collection of all sets hereditarily definable from parameters in V and from the equivalence class $[x]_E$ in the model $V[x][H]$.

It is clear that $V[x]_E$ is a model of ZFC by basic facts about HOD-type models (Jech 2002, theorem 13.26). The basic observation:

Theorem 3.3 *Let X be a Polish space, E an analytic equivalence relation on it, and let $x \in X$ be a V-generic point. Then:*

(i) $V[x]_E \subset V[x]$;
(ii) *the definition of $V[x]_E$ does not depend on the choice of the cardinal κ or the filter H;*
(iii) *if $y \in X$ is a V-generic point such that $x\ E\ y$, then $V[x]_E = V[y]_E$.*

The correct way to interpret (*iii*) is that V is an inner model of some larger universe containing the points x, y. We only need V to be a transitive model of some large fragment of ZFC; in particular, the case where V is perhaps countable in the larger universe is allowed. A typical context may arise when V is the transitive collapse of some countable elementary submodel of a large structure.

Proof For the first two items, let P be a partial order, τ a P-name for its generic point in X, and $\mu = |\mathcal{P}(P)|$. Let $x \in X$ be a P-generic point over V, and work in the model $V[x]$. For any poset Q, let Φ be a definition for the class of sets hereditarily ordinally definable from $[x]_E$ and parameters in V in the Q-extension of $V[x]$. We will show that in $V[x]$ the following is true:

(a) for every cardinal $\kappa > \mu$, we have $\mathrm{Coll}(\omega, \kappa) \Vdash \forall a\ \Phi(a) \to a \in V[x]$;
(b) for every set $a \in V[x]$ and every two cardinals $\kappa, \lambda > \mu$, $\mathrm{Coll}(\omega, \kappa) \Vdash \Phi(a)$ iff $\mathrm{Coll}(\omega, \lambda) \Vdash \Phi(a)$.

This will prove both (*i*) and (*ii*), since by the usual homogeneity arguments as in Fact 2.25, $V[x]_E = \{a \in V[x] : V[x] \models \mathrm{Coll}(\omega, \mu^+) \Vdash \Phi(a)\}$ must hold.

To prove (a), it is enough to consider sets of ordinals. Suppose ϕ is a formula and $\vec{v} \in V$ is a finite sequence; we must argue that $\mathrm{Coll}(\omega, \kappa)$ in $V[x]$ forces $\{\alpha \in \beta : \phi(\vec{v}, [x]_E, \alpha)\} \in V[x]$. By homogeneity, for every ordinal $\alpha \in \beta$ the statement $\phi(\vec{v}, [x]_E, \alpha)$ is decided by the largest condition in $\mathrm{Coll}(\omega, \kappa)$. Thus, writing $b = \{\alpha \in \beta : \mathrm{Coll}(\omega, \kappa) \Vdash \phi(\vec{v}, [x]_E, \alpha)\}$, we have $\mathrm{Coll}(\omega, \kappa) \Vdash \check{b} = \{\alpha \in \beta : \phi(\vec{v}, [x]_E, \alpha)\} \in V[x]$ as desired.

The proof of (b) is more slippery, and uses the following claim:

Claim 3.4 *For every finite sequence $\vec{v} \in V$, every ordinal β and every formula ϕ, there is another formula ψ and another finite sequence $\vec{w} \in V$ such that for* *cardinals* $\kappa, \lambda > \mu$ *and* $\alpha \in \beta$, $\text{Coll}(\omega, \kappa) \Vdash \phi(\vec{v}, [x]_E, \alpha)$ *iff* $\text{Coll}(\omega, \lambda) \Vdash \psi(\vec{w}, \kappa, [x]_E, \alpha)$.

Proof Let $D = \langle D_\alpha : \alpha \in \beta \rangle$ be the sequence of subsets of P in V defined by $p \in D_\alpha$ if $V \models p \Vdash \text{Coll}(\omega, \kappa) \Vdash \phi(\vec{v}, [\tau]_E, \alpha)$. Consider the formula $\psi(D, \kappa, [x]_E, \alpha)$ saying "there exists a V-generic filter $g \subset P$ such that $\tau/g \in [x]_E$ and $g \cap D_\alpha \neq 0$." We claim that ψ works as desired.

First of all, observe that $\text{Coll}(\omega, \kappa) \Vdash \forall \alpha \in \beta\ \phi(\alpha) \leftrightarrow \psi(\alpha)$. To verify this formula, let $G \subset \text{Coll}(\omega, \kappa)$ be a filter generic over $V[x]$ and work in $V[x][G]$. Clearly, if $\phi(\alpha)$ holds, then $\psi(\alpha)$ holds, as witnessed by the V-generic filter $g \subset P$ given by the point $x \in X$. On the other hand, suppose that $\psi(\alpha)$ holds as witnessed by some V-generic filter $h \subset P$ and its associated point $y \in X$. Use Fact 2.25 to find a filter $H \subset \text{Coll}(\omega, \kappa)$ generic over $V[y]$ such that $V[x][G] = V[y][H]$, use the forcing theorem to argue that $V[y][H] \models \phi(\vec{V}, [y]_E, \alpha)$, and since $x \; E \; y$, it must be the case that $V[x][G] \models \phi(\vec{V}, [x]_E, \alpha)$ as well.

Second, we will argue that the formula ψ is absolute between all transitive models of ZFC containing V and x in which $\mathcal{P}(P) \cap V$ is countable, in particular between the $\text{Coll}(\omega, \kappa)$ and $\text{Coll}(\omega, \lambda)$ extensions of $V[x]$. This will complete the proof of the claim.

To prove the absoluteness, observe that in any transitive model extending V, containing x, and containing an enumeration of $\mathcal{P}(P) \cap V$ in ordertype ω, for any fixed ordinal $\alpha \in \beta$, the formula $\psi(\vec{D}, [x]_E, \alpha)$ is a statement about illfoundedness of a certain tree. Namely, let $\{C_n : n \in \omega\}$ enumerate all open dense subsets of P in V, let $A \subset X \times X \times \omega^\omega$ be a closed set coded in V that projects to the equivalence E, and let T be the tree of all finite sequences $\langle p_i, O_i : i \in j \rangle$ such that the conditions $p_i \in P$ form a decreasing sequence such that $p_0 \in D_\alpha$, $p_i \in C_i$, and $O_i \subset X \times X \times \omega^\omega$ form a decreasing sequence of basic open neighborhoods coded in V of radius $< 2^{-i}$ with nonempty intersection with A, such that x belongs to the projection of O_i to the first coordinate, and $p_i \Vdash \tau$ belongs to the projection of O_i in the second coordinate. It is immediate that $\psi(\vec{D}, [x]_E, \alpha)$ is equivalent to the statement that T contains an infinite branch.

Now, observe that the validity of the illfoundedness statement does not depend on the particular enumeration of open dense subsets of P chosen–if one gets an infinite branch with one enumeration, then a subsequence of the conditions in P used in this branch will yield a branch for a different enumeration. Therefore, the formula $\psi(D, \kappa, [x]_E, \alpha)$ is absolute between all transitive models of set theory containing V, x in which the set $\mathcal{P}(P) \cap V$ is countable. This completes the proof of the claim. □

The proof of (b) immediately follows. It is again enough to check it only for sets of ordinals. Suppose that κ, λ are cardinals larger than μ and a is a set of ordinals such that $\text{Coll}(\omega, \kappa) \Vdash \Phi(\check{a})$. There must be a defining formula for the set a, say $\text{Coll}(\omega, \kappa) \Vdash \check{a} = \{\alpha \in \beta : \phi(\vec{v}, [x]_E, \alpha)\}$. Use Claim 3.4 to find a formula ψ that will define a in $\text{Coll}(\omega, \lambda)$ extension from $[x]_E$. Thus, $\text{Coll}(\omega, \lambda) \Vdash \Phi(\check{a})$ as well, and (b) has been verified.

For *(iii)*, suppose that $x \mathrel{E} y \in X$ are V-generic E-related points. Let $P, Q \in V$ be posets generating them, via a P-name τ and a Q-name σ respectively. To see that $V[x]_E = V[y]_E$, it is just enough to argue that every set of ordinals in $V[x]_E$ is in $V[y]_E$ and finish by symmetry. Let $a \in V[x]_E$ be a set of ordinals, let $\kappa \in V$ be a cardinal larger than both $\beta = \sup(a)$ and the sizes of the posets P and Q. Let $G \subset \text{Coll}(\omega, \kappa)$ be a filter generic over both $V[x]$, and let ϕ be a defining formula for a in $V[x][G]$. Let $D = \langle D_\alpha : \alpha \in \beta \rangle$ be the collection of subsets of Q defined by $q \in D_\alpha$ if $V \models q \Vdash \text{Coll}(\omega, \kappa) \Vdash \phi(\vec{v}, [\sigma]_E, \alpha)$.

Note that $V[x][G]$ satisfies the sentence ψ: "for every filter $h \subset Q$ generic over V such that $\sigma/h \mathrel{E} x$, for every $\alpha \in \beta$ $h \cap D_\alpha \neq 0$ if and only if $\alpha \in a$." This again follows from the homogeneity in Fact 2.25: in $V[x][G]$, for every filter $h \subset Q$ generic over V there is a filter $H \subset \text{Coll}(\omega, \kappa)$ generic over $V[h]$ such that $V[x][G] = V[h][H]$. If $x \mathrel{E} \sigma/h$, then for every ordinal $\alpha \in \beta$ the equal models $V[x][G]$ and $V[h][H]$ satisfy the equal formulas $\phi(\vec{v}, [x]_E, \alpha)$ and $\phi(\vec{v}, [\tau/h]_E, \alpha)$ in the same way, and so $\alpha \in a$ if and only if $h \cap D_\alpha \neq 0$.

Now, just as in Claim 3.4, the sentence ψ is equivalent to an illfoundedness statement about certain tree which is absolute between all transitive models of ZFC containing V and x in which the set $\mathcal{P}(Q) \cap V$ is countable. In particular, ψ is absolute between $V[x]$ and some outer model containing V, x, y in which the set $\mathcal{P}(Q) \cap V$ is countable. Since y is Q-generic over V, this means that the sentence ψ can be applied to the filter $h \subset Q$ generic over V such that $\sigma/h = y$ to show that the set a is defined by the formula $\phi(\vec{v}, [y]_E)$ in every $\text{Coll}(\omega, \kappa)$-extension of $V[y]$. By *(ii)* applied to $V[y]$, $a \in V[y]_E$ as desired. □

The definition and properties of the model $V[x]_E$ may be somewhat mysterious. Several more or less trivial examples will serve as a good illustration of what can be expected. Let $E = E_0$ and consider the cases of the Sacks real, the Cohen real, and the Silver real. In the case of Sacks real $x \in 2^\omega$, a simple density argument will show that there will be a perfect tree T in the Sacks generic filter whose branches are pairwise E_0-inequivalent. Then, x can be defined from $[x]_{E_0}$ as the only point of $[T] \cap [x]_{E_0}$ and therefore $V[x] = V[x]_{E_0}$. In the case of Cohen generic real $x \in 2^\omega$, the rational translations of 2^ω that generate E_0 can transport any Cohen condition to any other one, and the equivalence class $[x]_{E_0}$ is invariant under all of them, leading to the conclusion that $V[x]_{E_0} = V$. In the case of Silver generic real x, the model $V[x]_{E_0}$ is a nontrivial σ-closed generic extension of V generated by the quotient of Silver forcing in which E_0-equivalent conditions are identified, roughly speaking (Theorem 8.7). On general grounds (Jech 2002, lemma 15.43) $V[x]_E$ is a generic extension of V, but it is not so easy to compute the forcing that induces it. In a typical case, it is the poset of Borel E-invariant sets positive with respect to some σ-ideal as in Theorem 3.25.

The importance of the model $V[x]_E$ is clarified in the main result of this section, a trichotomy theorem:

Theorem 3.5 *Suppose that X is a Polish space, I is a σ-ideal on it such that the quotient forcing P_I is proper, and E is an analytic equivalence relation on X. Let $x_{\text{gen}} \in X$ be a P_I-generic point over V. One of the following holds:*

(i) (ergodicity) $V = V[x_{\text{gen}}]_E$, and then there is a Borel I-positive set $B \subset X$ such that $E \restriction B$ is ergodic;
(ii) (intermediate extension) or the model $V[x_{\text{gen}}]_E$ is strictly between V and $V[x_{\text{gen}}]$;
(iii) (canonization) or $V[x_{\text{gen}}]_E = V[x_{\text{gen}}]$ and then there is a Borel I-positive set $B \subset X$ such that $E \restriction B = \text{id}$.

A trivial way to satisfy the first item is to find an I-positive equivalence class of E. A typical special case in which the ergodicity case holds nontrivially is that of Laver forcing and the E_{K_σ} equivalence relation (see Section 6.3). The Laver forcing generates a minimal forcing extension, prohibiting the second item from ever occurring, but still there is an equivalence relation that cannot be simplified to id or ev on any Borel I-positive set.

In the second item we get an intermediate forcing extension strictly between V and the P_I-extension $V[x_{\text{gen}}]$. For many σ-ideals, the existence of such an intermediate model is excluded purely on forcing grounds, such as the ideal of countable sets (the Sacks forcing; Jech 2002, theorem 15.34) or σ-compact subsets of ω^ω (the Miller forcing; Miller 1984), or the more general Theorem 5.25. In such cases, we get a neat dichotomy. It is not true that every intermediate extension is necessarily obtained from a Borel equivalence relation in this way, as Section 8.1 shows. In special cases though (such as the countable length $\alpha \in \omega_1$ iteration of Sacks forcing), the structure of intermediate models is well known (each of them is equal to $V[x_\gamma : \gamma \in \beta]$ for some ordinal $\beta \leq \alpha$) and they exactly correspond to certain critical equivalence relations (such as the relations E_β on $(2^\omega)^\alpha$ connecting two sequences if they are equal on their first β entries). We cannot compute the forcing responsible for the extension $V[x_{\text{gen}}]_E$ in a fully general situation, but in all specific cases we *can* compute, it is the poset of E-invariant I-positive Borel sets ordered by inclusion, as described in Theorem 3.25 and Section 3.3. In the common situation that the model $V[x_{\text{gen}}]_E$ does not contain any new reals, we can compute the forcing leading from $V[x_{\text{gen}}]_E$ to $V[G]$–it is a quotient poset of the form P_{I^*} for a suitable σ-ideal $I^* \supset I$ which has the ergodicity property of the first item of the theorem.

Proof of Theorem 3.5 Consider the first case, $V[x_{\text{gen}}]_E = V$. In this case, consider the collection K of those Borel sets $C \subset X$ coded in V such that $C \cap [x_{\text{gen}}]_E \neq 0$. This collection is certainly in the model $V[x_{\text{gen}}]_E$ as we have just defined it from the class $[x_{\text{gen}}]_E$, and therefore it is in V. There must be a condition $B \in P_I$ and an element $J \in V$ such that $B \Vdash \dot{K} = \check{J}$. If there were two Borel I-positive sets $C_0, C_1 \subset B$ with $[C_0]_E \cap [C_1]_E = 0$, then at most one of these sets can belong to the set J. Suppose that $C_1 \notin J$; then $C_1 \Vdash C_1 \in \dot{K} \setminus \check{J}$! This contradiction verifies the conclusion of the first case of the theorem.

The second possibility is that $V \subset V[x_{\text{gen}}]_E \subset V[x_{\text{gen}}]$ and these inclusions are proper. In this case, we are content to fall into the second item of the theorem.

Lastly, assume that $V[x_{\text{gen}}]_E = V[x_{\text{gen}}]$; in particular, $x_{\text{gen}} \in V[x_{\text{gen}}]_E$. Then, return to the ground model V and pick large enough cardinals $\lambda \in \kappa$. There must be a condition $B \in P_I$ forcing $\dot{x}_{\text{gen}} \in \dot{V}[x_{\text{gen}}]_E$; so, in the product $P_I \times \text{Coll}(\omega, \kappa)$, the condition $\langle B, 1\rangle$ forces $\dot{x}_{\text{gen}}$ to be the only element $x \in X$ satisfying $\phi(x, [\dot{x}_{\text{gen}}]_E, \vec{v})$ for some sequence of parameters $\vec{v} \in V$ and a formula ϕ. We must find an I-positive Borel set $C \subset B$ consisting of pairwise E-inequivalent points.

Let H be a $\text{Coll}(\omega, \kappa)$-generic filter over V. Use Theorem 2.49 to find in $V[H]$ an elementary embedding $j : V \to N$ to a possibly illfounded generic ultrapower N such that $M = j''H_\lambda$ is a countable set in N. Recall that the wellfounded part of the model N is identified with its transitive isomorph. Use Theorem 2.49 to choose an interpretation of the space X in $V[H]$ so that $j(X) \subset X$ and for every analytic set A coded in the ground model, $j(A) \subseteq A^{V[H]}$.

Working in N, let $C' = \{x \in j(B) : x \text{ is } j(P_I) = P_{j(I)}\text{-generic over } M\}$. This is a $j(I)$-positive Borel set by Fact 2.50 applied in the model N: the set $M = j''H_\lambda$ is a countable elementary submodel of $j(H_\lambda)$ by the Tarski–Vaught criterion, and the poset $j(P_I) = P_{j(I)}$ is proper in N by elementarity of j. We shall show that $N \models C'$ consists of pairwise $j(E)$-inequivalent points. The desired set $C \in V$ is then obtained through elementarity of the embedding j.

Claim 3.6 *Every point of the set $C' \cap N$ is P_I-generic over V.*

Proof Let $x \in C' \cap N$ be arbitrary. For every maximal antichain $A \subset P_I$ in V, $j(A)$ is a maximal antichain of $j(P_I)$, and since the point x is $P_{j(I)}$-generic over M, there must be a set $D \in A$ such that $N \models x \in j(D)$. However, the model N evaluates the membership in ground model Borel sets correctly by Theorem 2.49(*ii*), so $x \in D$. As $A \subset P_I$ was an arbitrary maximal antichain in V, this proves the claim. □

Suppose $N \models x \neq y \in C'$. Both points x, y are P_I-generic over V and also belong to $B^{V[H]}$ and $V[H]$. Find filters $H_x, H_y \subset \mathrm{Coll}(\omega, \kappa)$ generic over the models $V[x]$ and $V[y]$ respectively such that $V[x][H_x] = V[y][H_y] = V[H]$. If the points $x, y \in X$ were E-equivalent in $V[H]$, then the formula $\phi(\cdot, [x]_E, \vec{v}) = \phi(\cdot, [y]_E, \vec{v})$ would have to define both x and y in the model $V[H]$ by the forcing theorem by the choice of ϕ. This is impossible as $x \neq y$. Thus, $V[H] \models \neg x\ E\ y$. Since $j(E) \subset E \cap V[H]$, we conclude that $N \models \neg x\ j(E)\ y$ as desired. □

Theorem 3.7 *Suppose that X is a Polish space, I a σ-ideal on X such that the quotient forcing P_I is proper. Let x be a P_I-generic point. Then, $2^\omega \cap V[x]_E = \{f(x) : f : X \to 2^\omega$ is a partial Borel function coded in V which is E-invariant on its domain and $x \in \mathrm{dom}(f)\}$.*

Proof For the first part, if $f : B \to 2^\omega$ is a Borel function which is constant on each E-equivalence class, then certainly $B \Vdash \dot{f}(\dot{x}_{\mathrm{gen}}) \in V[x]_E$. Namely, $\dot{f}(\dot{x}_{\mathrm{gen}})$ will be definable from $[\dot{x}_{\mathrm{gen}}]_E$ as the unique value of f on $\dot{B} \cap [\dot{x}_{\mathrm{gen}}]_E$. Note that the constant property of f survives to every forcing extension by coanalytic absoluteness. This proves the right-to-left inclusion.

For the opposite inclusion, proceed as in the third case of the trichotomy theorem. Let $\lambda \in \kappa$ be large regular cardinals. Suppose that $\langle B, p\rangle \in P_I \times \mathrm{Coll}(\omega, \kappa)$ is a condition that forces $\dot{y} \in V[\dot{x}_{\mathrm{gen}}]_E \cap 2^\omega$. Strengthening the condition $\langle B, p\rangle$ we may assume that there is a formula ϕ and a sequence $\vec{v} \in V$ of parameters such that it forces that $\dot{y}$ is the unique element of 2^ω satisfying $\phi(\dot{y}, [\dot{x}_{\mathrm{gen}}]_E, \vec{v})$. Since $V[x_{\mathrm{gen}}]_E \subset V[x_{\mathrm{gen}}]$, we may also strengthen the condition if necessary to find a Borel function $f : X \to 2^\omega$ such that it is forced that $\dot{y} = \dot{f}(\dot{x}_{\mathrm{gen}})$. We will now find a Borel I-positive set $C \subset B$ on which the function f is E-invariant. This will certainly complete the proof of the proposition, since then $\langle C, p\rangle$ forces that $\dot{y}$ is the image of the generic real by an E-invariant function.

Let $H \subset \text{Coll}(\omega, \kappa)$ be a generic filter. Use Theorem 2.49 to find in $V[H]$ an elementary embedding $j : V \to N$ such that $M = j''H_\lambda$ is a countable set in the model N and an interpretation of the space X so that $j(X)$ is a subset of it and for every analytic set $A \subset X$ in the ground model, $j(A) \subset A^{V[H]} \cap j(X)$. Working in the generic ultrapower N, let $C' = \{x \in j(B) : x \text{ is } P_{j(I)}\text{-generic over } M\}$; this is a Borel $j(I)$-positive set. We will show that in N, the function $j(f)$ is $j(E)$-invariant on the set C'; the desired set C is then obtained by the elementarity of the embedding j. Thus, let $x_0, x_1 \in C'$ be $j(E)$-equivalent points in the model N. The two points are E-equivalent in $V[H]$ as $j(E) \subset E$ by Theorem 2.49(*ii*), and they are also P_I-generic over V by Claim 3.6. Find filters $H_0, H_1 \subset \text{Coll}(\omega, \kappa)$ generic over $V[x_0]$ and $V[x_1]$ respectively such that $V[H] = V[x_0][H_0] = V[x_1][H_1]$. By the forcing theorem, it must be the case that the formula $\phi(\cdot, [x_0]_E, \vec{v}) = \phi(\cdot, [x_1]_E, \vec{v})$ must define a point $y \in 2^\omega$ which is the joint value of $f(x_0)$ and $f(x_1)$ in the model $V[H]$. Now, by the definitions in the model N, the points x_0, x_1 are in $j(B)$ and therefore in the domain of the function $j(f)$. Since $j(f) \subset f^{V[H]}$, it follows that y must be the unique value of $j(f)(x_0)$ and $j(f)(x_1)$ in the model N. We conclude that in the model N, the function $j(f)$ is $j(E)$-invariant on the set C', and the theorem follows. □

Let I be a σ-ideal on a Polish space X such that the quotient poset P_I is proper. Let $x \in X$ be a P_I-generic point over V. On general forcing grounds (Jech 2002, lemma 15.43), it must be the case that $V[x]_E$ is a generic extension of V, and $V[x]$ is a generic extension of $V[x]_E$. We do not know which poset is responsible for the former extension, even though there is a canonical candidate isolated in the next section which works in all cases computed in this book. In the common case where $V[x]_E$ has the same reals as V, we can compute the poset responsible for the latter extension.

Definition 3.8 Let I be a σ-ideal on a Polish space X, let E be an analytic equivalence relation, and let $x \in X$ be a P_I-generic point. In the model $V[x]$, let I^x be the collection of ground model coded Borel sets B such that for every cardinal κ, $\text{Coll}(\omega, \kappa)$ forces that $\dot{B}$ contains no point that is simultaneously P_I-generic over V and E-equivalent to x.

It should be clear that $I \subset I^x$, $I^x \in V[x]_E$, and if $V[x]_E$ contains no new reals then I^x is a σ-ideal of Borel sets there. Moreover, if $2^\omega \cap V[x]_E = 2^\omega \cap V$, the ideal I^x has the ergodicity property in $V[x]_E$: any pair of I^x-positive Borel sets contains a pair of E-connected points. This follows from the fact that in some large collapse, each of the sets in the pair contains some points equivalent to x; in particular, such points are equivalent to each other, and analytic absoluteness transports this feature back to $V[x]_E$.

Theorem 3.9 *Suppose that X is a Polish space, I a σ-ideal on X such that the quotient forcing P_I is proper. Let x be a P_I-generic point. If $2^\omega \cap V = 2^\omega \cap V[x]_E$, then $V[x]_E = V[I^x]$ and $V[x]$ is a P_{I^x}-extension of $V[x]_E$.*

Proof We will first show that x is a P_{I^x} generic point over $V[x]_E$. Move back to the ground model, and assume for contradiction that a condition $\langle B, p\rangle \in P_I \times \mathrm{Coll}(\omega, \kappa)$ forces that $\dot{D} \subset P_{I^x}$ is an open dense set in the model $V[x_{\mathrm{gen}}]_E$ such that the point $\dot{x}_{\mathrm{gen}}$ does not belong to any of its elements. Strengthening the condition if necessary, we may find a formula $\phi(u, [\dot{x}_{\mathrm{gen}}]_E, \vec{v})$ with parameters in $V \cup \{[\dot{x}_{\mathrm{gen}}]_E\}$ defining the set $\dot{D}$: $\dot{D} = \{C \in P_{I^{\dot{x}_{\mathrm{gen}}}} : \phi(C, [\dot{x}_{\mathrm{gen}}]_E, \vec{v})\}$. Of course, the condition $\langle B, p\rangle$ forces $B \in P_{I^{\dot{x}_{\mathrm{gen}}}}$, so there must be a strengthening $B' \subset B$, $p' \leq p$ and a Borel set $C \subset B$ such that $\langle B', p'\rangle \Vdash C \in \dot{D}$. Note that $C \in P_I$ must be an I-positive Borel set in the ground model, since $V[x]_E \cap 2^\omega = V \cap 2^\omega$.

Find mutually generic filters $G \subset P_I$, $H \subset \mathrm{Coll}(\omega, \kappa)$ with $B' \in G$, $p' \in H$, let $x \in X$ be the point associated with the filter G, use the definition of the ideal I^x to find a point $y \in C \cap V[G, H]$ which is P_I-generic over V and equivalent to x. Use Fact 2.25 to find mutually generic filters $\bar{G} \subset P_I$, $\bar{H} \subset \mathrm{Coll}(\omega, \kappa)$ such that y is the generic point associated with $\bar{G}$ (so $C \in \bar{G}$), $p \in \bar{H}$, and $V[G, H] = V[\bar{G}, \bar{H}]$.

Since the points x, y are E-equivalent, the definition of the set $\dot{D}$ is evaluated in the same way with either x or y plugged into the definition. Applying the forcing theorem to the filters $\bar{G}, \bar{H}$, there must be a condition in the filter $\bar{G} \times \bar{H}$ stronger than $\langle B, p\rangle$ which forces $C \in \dot{D}$ and $\dot{x}_{\mathrm{gen}} \in C$. However, this directly contradicts the statement forced by the weaker condition $\langle B, p\rangle$!

To show that $V[x]_E = V[I^x]$, note that $I^x \in V[x]_E$, so the right to left inclusion is immediate. For the opposite inclusion, assume that $a \in V[x]_E$ is a set of ordinals, perhaps defined as $a = \{\alpha : \phi(\alpha, \vec{v}, [x]_E)\}$ in the model $V[x, H]$, where $\vec{v}$ are parameters in the ground model and $H \subset \mathrm{Coll}(\omega, \kappa)$. Let $b = \{\alpha : \exists B \in P_I^V, B \notin I^x\ B \Vdash \mathrm{Coll}(\omega, \kappa) \Vdash \phi(\check{\alpha}, \vec{v}, [\dot{x}_{\mathrm{gen}}]_E)\}$. Note that $b \in V[I^x]$, and work to show that $a = b$. It is immediate that $a \subset b$ since the filter defined by the P_I-generic point $x \in X$ has empty intersection with I^x. On the other hand, if $\alpha \in b$ as witnessed by a set B then there is a P_I-generic point $y \in B \cap [x]_E$ over V. By the usual homogeneity argument there is a $V[y]$-generic filter $H_y \subset \mathrm{Coll}(\omega, \kappa)$ such that $V[y][H_y] = V[x][H]$, by the forcing theorem $V[y][H_y] \models \phi(\alpha, \vec{v}, [x]_E)$, and therefore $\alpha \in a$. The theorem follows. □

3.2 The model $V[[x]]_E$

The model $V[x]_E$ has a larger, less well understood, possibly choiceless companion $V[[x]]_E$. In spirit, this larger model is the intersection of all models containing V and a point E-equivalent to x. In formal language:

Definition 3.10 Let E be an analytic equivalence relation on a Polish space X. Let $x \in X$ be a V-generic point. The model $V[[x]]_E$ consists of all sets $a \in V[x]$ such that the model $V[x]$ satisfies: for every ordinal κ, $\mathrm{Coll}(\omega, \kappa)$ forces that for every $y \in X$ with $x \mathrel{E} y$, $a \in V[y]$.

Theorem 3.11 *Let E be an analytic equivalence relation on a Polish space X, and let $x \in X$ be a V-generic point. Then:*

(i) $V[[x]]_E$ is a model of ZF with $V[x]_E \subseteq V[[x]]_E \subseteq V[x]$;
(ii) if $x, y \in X$ are V-generic E-equivalent points, then $V[[x]]_E = V[[y]]_E$.

Proof Start with the second item. For every pair of V-generic points $x, y \in X$ such that $x \mathrel{E} y$ and every transitive set $a \in V[[x]]_E$, we must show that $a \in V[y]$ holds – by the arbitrary choice of y this will show that $a \in V[[y]]_E$ as well. Let Q be the partial order generating y, let σ be the Q-name that generates y, and let κ be a cardinal larger than $\mathcal{P}(Q)$ and the rank of the set a. The definition of the class $V[[x]]_E$ shows that $V[x] \models \mathrm{Coll}(\omega, \kappa) \Vdash \psi$, where ψ is the sentence "for every filter $g \subset Q$ generic over V, if $\sigma/g \mathrel{E} x$ then there is $\tau \in V$, a Q-name of rank α, such that $a = \tau/g$." Just as in Claim 3.4, ψ is a statement about wellfoundedness of a certain tree, and therefore absolute between the models $V[x][G]$ and W, where $G \subset \mathrm{Coll}(\omega, \kappa)$ is a filter generic over $V[x]$ and W is any outer model containing x, G, y. However, in the model W, the sentence ψ applied to the filter $g \subset V$ generic over V that $y = \sigma/g$ shows that $a \in V[y]$ as desired.

To show that $V[[x]]_E$ is a model of ZF, argue that $L(V[[x]]_E) = V[[x]]_E$. Indeed, for every ordinal κ it is the case that $\mathrm{Coll}(\omega, \kappa)$ forces that $L(V[[x]]_E)$ is contained in every model $V[y]$ where $y \mathrel{E} x$, since as argued in the previous paragraph, the models $V[y]$ all contain $V[[x]]_E$ as a class. But then, every element of $L(V[[x]]_E)$ belongs to $V[[x]]_E$ by the definition of $V[[x]]_E$.

To show that $V[x]_E$ is a subset of $V[[x]]_E$, let $a \in V[x]_E$ be a set of ordinals and let $V[G]$ be a generic extension of V inside a generic extension of $V[x]$ containing a point $y \in X$ equivalent to x; we must argue that $a \in V[G]$. Let P be a poset generating $V[G]$, and let $\kappa = |\mathcal{P}(P)|$. There is a common $\mathrm{Coll}(\omega, \kappa)$ extension $V[H]$ of both $V[G]$ and $V[x]$. Since $a \in V[x]_E$, the set a is definable in $V[H]$ from parameters in V and perhaps the additional parameter $[x]_E = [y]_E$. Since $V[H]$ is an extension of the model $V[G]$ by a weakly homogeneous forcing and all parameters of the definition are themselves definable from parameters in $V[G]$, it must be the case that $a \in V[G]$ as desired by Fact 2.24. The proof is complete. □

Theorem 3.12 *Let I be a σ-ideal on a Polish space X such that the quotient forcing P_I is proper. Let E be an analytic equivalence relation on X, and let $x \in X$ be a P_I-generic point. Then $2^\omega \cap V[[x]]_E = \{f(x) : f$ is a Borel function coded in V, which on its domain attains only countably many values on each equivalence class and $x \in \mathrm{dom}(f)\}$.*

An important observation complementary to the wording of this theorem: the statement that a given Borel function f attains only countably many values on each E-equivalence class is Π^1_1 in the code for the function and the equivalence relation, and therefore it is invariant under forcing by Shoenfield absoluteness. To verify the complexity estimate, let $z \in 2^\omega$ be a point coding f, E. If $x \in X$ and $[x]_E$ is countable, then as $[x]_E$, $f''[x]_E$ are both countable $\Sigma^1_1(x, z)$ sets, they contain only $\Delta^1_1(x, z)$ points by Fact 2.8(*ii*). Thus, f attains only countably many values on each E-equivalence class if and only if for every pair of points $x, y \in \mathrm{dom}(f)$, either x, y are not equivalent or $f(x) \in \Delta^1_1(y, z)$. This is a coanalytic formula by Fact 2.8(*i*).

Proof First, observe that if $f : B \to 2^\omega$ is a Borel function attaining only countably many values on each equivalence class, and $y \in X$ is any point in a forcing extension of the ground model V, then the model $V[y]$ contains some enumeration of the countable set $f''([y]_E \cap B)$ by Shoenfield absoluteness, and therefore contains all elements of this countable set. This proves the right-to-left inclusion.

For the opposite inclusion, suppose that τ is a P_I-name and $B \subset X$ is an I-positive Borel set forcing $\tau \in V[[\dot{x}_{\mathrm{gen}}]]_E \cap 2^\omega$. Let M be a countable elementary submodel of a large structure containing B and the name τ, and let $C \subset B$ be the Borel I-positive set of all P_I-generic points over M. Let $f : C \to 2^\omega$ be the function defined by $f(x) = \tau/x$; thus f is a Borel function and $C \Vdash \dot{f}(\dot{x}_{\mathrm{gen}}) = \tau$. We will prove that f attains only countably many values at every E-equivalence class in C. By Theorem 3.11(*ii*) applied to the transitive isomorph of M, whenever $x_0, x_1 \in C$ are E-related points, then $V[[x_0]]_E^{M[x_0]} = V[[x_1]]_E^{M[x_1]}$. The model $V[[x_0]]_E^{M[x_0]}$ is countable, and it must contain the real τ/x_1. Thus the countable model $V[[x_0]]_E^{M[x_0]}$ contains all reals τ/x where $x \in C$ is a point E-related to x_0, and the proof is complete. □

3.3 The poset P_I^E

In all special cases investigated in this book, the intermediate extension $V[x_{\text{gen}}]_E$ of Definition 3.2 is generated by a rather natural regular subordering of P_I, denoted by P_I^E.

Let X be a Polish space and I a σ-ideal on it. We would like to define P_I^E to be the partial ordering of I-positive E-saturated Borel sets ordered by inclusion. The problem here is that E-saturations of Borel sets are in general analytic, and in principle there could be very few Borel E-saturated sets. In order to steer clear of this problem, we will work only with the very broad class of σ-ideals I such that every analytic set is either in the ideal I or it contains an I-positive Borel subset. Then the quotient P_I is dense in the partial order of analytic I-positive sets, and we can define P_I^E as the ordering of analytic E-saturated I-positive sets.

Definition 3.13 Let X be a Polish space and I a σ-ideal on it such that every analytic I-positive set contains a Borel I-positive subset. P_I^E is the poset of I-positive E-saturated analytic sets ordered by inclusion. If $C \subset X$ is a Borel I-positive set then $P_I^{E,C}$ is the poset of relatively E-saturated analytic subsets of C ordered by inclusion.

By a slight abuse of notation, if every I-positive analytic set contains a Borel I-positive set, we will write P_I for the poset of all analytic I-positive sets ordered by inclusion, so that $P_I^E \subset P_I$. The difficulty with finding densely many Borel E-saturated sets inside P_I^E is conveniently resolved in a broad class of cases:

Proposition 3.14 *Suppose that I is a σ-ideal on a Polish space X such that every analytic I-positive set contains a Borel I-positive subset, and the poset P_I is proper. Let E be an equivalence relation on X Borel reducible to an orbit equivalence. Then every analytic I-positive set has a Borel I-positive subset whose E-saturation is Borel. In particular, every analytic I-positive E-saturated set has a Borel I-positive E-saturated subset.*

Proof Let $B \subset X$ be an analytic I-positive set. Use the assumptions to thin it down to a Borel I-positive set. Let M be a countable elementary submodel of a large structure, and let $C \subset B$ be the Borel set of all points in B which are P_I-generic over M; this is I-positive by properness and Fact 2.50. Theorem 2.29 applied to $P_I \restriction B$ and E then says that the E-saturation of the set C is Borel as required. □

The previous discussion seems to have little to offer for the solution of the main problem associated with the topic of this book:

Question 3.15 Suppose that I is a σ-ideal on a Polish space X such that the poset P_I is proper. Is $P_I^{E,C}$ a regular subposet of P_I for some Borel I-positive set $C \subset X$?

We will first duly produce the counterexamples showing that restricting the domain of the ideal I and the properness assumption on P_I are necessary for a general positive answer.

Proposition 3.16 *There is a Polish space X, a σ-ideal I on it such that P_I is proper, and a countable Borel equivalence relation E on X such that P_I^E is not a regular subposet of P_I.*

Proof Consider $X = \omega^\omega \times 2$ and the σ-ideal I generated by sets $B \subset X$ such that $\{y \in \omega^\omega : \langle y, 0\rangle \in B\}$ is compact and $\{y \in \omega^\omega : \langle y, 1\rangle \in B\}$ is countable. The quotient forcing P_I is clearly just a disjoint union of Sacks and Miller forcing. Consider the equivalence relation E on X defined as the equality on the first coordinate. Then E has countable classes, indeed each of its classes has size 2, and P_I^E is not a regular subposet of P_I. To see that, consider a maximal antichain A in Sacks forcing consisting of uncountable compact sets, and let $\bar{A} = \{C \times 2 : C \in A\}$. It is not difficult to see that $\bar{A}$ is a maximal antichain in P_I^E, but the set $\omega^\omega \times 1 \subset X$ is Borel, I-positive, and has an I-small intersection with every set in $\bar{A}$. □

Proposition 3.17 *There is a σ-ideal I on a Polish space X and a smooth equivalence relation E on X such that $P_I^{E,C}$ is not regular subposet of $P_I \restriction C$ for any I-positive Borel $B \subset X$.*

Proof We will need a family $\{I_r : r \in (0, 1)\}$ of σ-ideals on some Polish space Y and a σ-ideal J on $(0, 1)$ containing all singletons such that:

- $r < s$ implies that $I_r \subset I_s$, and every Borel I_r-positive set has a Borel I_r-positive subset in I_s;
- each I_r has the rectangular Ramsey property with the σ-ideal J.

There are very many options; we suggest setting I_r to be the σ-ideal of subsets of $Y = [0, 1]$ of σ-finite r-dimensional Hausdorff measure zero, and J be the σ-ideal of countable sets. The verification of the first item is a routine exercise in geometric measure theory; the second item follows, for example, from Zapletal (2008, theorem 4.4.8).

Now let $X = (0, 1) \times Y$ and let I be the σ-ideal of Borel sets $A \subset X$ containing no *positive rectangle*, which is a set $C_0 \times C_1$ where $C_0 \subset (0, 1)$ is Borel I-positive and $C_1 \subset Y$ is a Borel I_r-positive set for some $r \geq \sup(C_0)$ – the number r will be called a *witness* for the rectangle. This is a σ-ideal: if

$C_0 \times C_1$ is a positive rectangle with witness r and $C_0 \times C_1 = \bigcup_n A_n$ is a countable union of Borel sets, then the rectangular Ramsey property between J and I_r shows that one of the sets A_n contains a positive rectangle with the same witness r.

Now let E be the smooth equivalence relation on Y relating two points of X if they have the same Y-coordinate. This is the desired equivalence relation. Let C be an I-positive Borel set; thinning out C if necessary, we may assume that C is a positive rectangle $C_0 \times C_1$. To show that $P_I^{E,C}$ is not regular in $P_I \restriction C$, we must produce a maximal antichain in $P_I^{E,C}$ which is not maximal in $P_I \restriction C$. Let $s_0 = \max\{s \in (0,1) : C_0 \cap (0,s) \in J\}$, let $t \in (0,1)$ be some number such that both $C_0 \cap (0,t)$ and $C_0 \cap (t,1)$ are J-positive, and let $K = \bigcap_{u>s} I_u$. Use the first item above to find a maximal antichain A in $P_K \restriction C_1$ consisting of elements of I_t. Write $A^E = \{C_0 \times D' : D' \in A\}$. It is not difficult to verify that A^E is a maximal antichain in $P_I^{E,C}$, while the condition $(C_0 \cap (t,1)) \times C_1 \in P_I \restriction C$ is incompatible with every element of A^E. The proposition follows. □

In all specific cases investigated in this book, the answer to Question 3.15 turns out to be positive, the set C is invariably equal to the whole space, and the saturation is the projection on a dense set of conditions in P_I. We will include one rather simple case criterion for regularity that will be used throughout the book:

Definition 3.18 Let I be a σ-ideal on a Polish space X, E an analytic equivalence relation on X, and let $B \subset X$ be an analytic set. We say that B is an *E-kernel* if for every analytic set $C \subset B$, $[C]_E \cap B \in I$ implies $[C]_E \in I$.

Theorem 3.19 *Let I be a σ-ideal on a Polish space X such that every I-positive analytic set has a Borel I-positive subset, and let E be an analytic equivalence relation on X. The poset P_I^E is regular in P_I if and only if every Borel I-positive set has an analytic I-positive subset which is also an E-kernel.*

The main point here is that in the cases encountered in this book, the kernels are typically very natural closed sets. There are two immediate corollaries:

Corollary 3.20 *Suppose that I is a σ-ideal on a Polish space X such that every I-positive analytic set has a Borel I-positive subset, and E is an analytic equivalence relation on X. If E-saturations of sets in I are still in I, then the poset P_I^E is regular in P_I.*

To prove this, observe that every analytic set is a kernel.

Corollary 3.21 *Suppose that I is a σ-ideal on a Polish space X. Suppose that G is a countable group acting on X in a Borel, I-preserving way. Let E be the orbit equivalence relation. Then the poset P_I^E is regular in P_I.*

Note that in this case E-saturations of Borel sets are Borel, and E-saturations of I-small sets are still I-small as the acting group is countable.

Proof of Theorem 3.19 Suppose first that every I-positive Borel set contains a Borel I-positive kernel; we must show that P_I^E is regular in P_I. Let $A \subset P_I^E$ be a maximal antichain; we must show that A is also maximal in P_I. Suppose not, and let $B \in P_I$ be an analytic set incompatible with all elements of the antichain A, and thin it out if necessary to a kernel. Then, for every set $C \in A$, $B \cap C \in I$, and so $[B \cap C]_E \in I$ by the assumptions. However, $[B \cap C]_E = [B]_E \cap C$ since the set C is E-invariant, and so $[B]_E$ is a condition in P_I^E incompatible with all elements of the antichain A, contradicting its maximality in P_I^E.

On the other hand, suppose that P_I^E is regular in P_I, and $B \subset X$ is a Borel I-positive set. To produce an I-positive E-kernel in it, consider the open dense set $D \subset P_I^E$ consisting of those analytic E-invariant sets with I-small intersections with B and those sets which do not have an E-invariant I-positive analytic subset with I-small intersection with B. Since the poset P_I^E is regular in P_I, there must be a set $B' \in D$ compatible with B, so $B' \cap B \notin I$. Note that this means that no E-invariant I-positive subset of B' has I-small intersection with B. We claim that $B \cap B'$ is the requested E-kernel. And indeed, if $C \subset B \cap B'$ is an analytic set such that $[C]_E \cap B \cap B' \in I$, then $[C]_E \subset B'$ has I-small intersection with the set B and therefore has to be I-small! □

The poset P_I^E is frequently $\aleph_0$-distributive, and yields the model $V[\dot{x}_{\mathrm{gen}}]_E$. The following provide criteria for this to happen:

Definition 3.22 The equivalence E on a Polish space X is *I-dense* if for every I-positive Borel set $B \subset X$ there is a point $x \in B$ such that $[x]_E \cap B$ is dense in B.

Proposition 3.23 *If the poset P_I has the continuous reading of names, P_I^E is a regular subposet of P_I, and E is I-dense, then P_I^E is $\aleph_0$-distributive.*

Proof Since P_I is proper, so is P_I^E, and it is enough to show that P_I^E does not add reals. Suppose that $B \in P_I$ forces that τ is a name for a real in the P_I^E-extension. Let M be a countable elementary submodel of a large enough structure, and let $C \subset B$ be the I-positive set of all M-generic points for P_I. If $x \in C$ is a point, write $H_x \subset P_I^E$ for the M-generic filter of all sets $D \in P_I^E$ such that $[x]_E \subset D$. It is clear that the map $x \mapsto H_x$ is constant on equivalence classes, and so is the Borel map $x \mapsto \tau/H_x$. Thin out the set C if necessary to make sure that the map $x \mapsto \tau/H_x$ is continuous on C. By the density property, there is a point $x \in C$ such that $[x]_E \cap C$ is dense in C. Thus, the map $x \mapsto \tau/H_x$ is continuous on C and constant on a dense subset of C, therefore constant on C. The condition C clearly forces τ to be equal to the single point in the range of this function. □

Note that typically the poset P_I^E is fairly simply definable. If suitable large cardinals exist, then the descending chain game on the poset P_I^E is determined. Thus, when the poset is $\aleph_0$-distributive, Player I has no winning strategy in this game, so it must be Player II who has a winning strategy and the poset is strategically σ-closed. It is not entirely clear whether there may be cases in which P_I^E is $\aleph_0$-distributive but fails to have a σ-closed dense subset. In many situations, one can find forcing extensions in which P_I^E is even $\aleph_1$-distributive or more.

In all situations described in the later chapters of this book, the forcing P_I^E generates the $V[\dot{x}_{\text{gen}}]_E$ extension of Theorem 3.5. To prove the appropriate general theorem, we need a definition:

Definition 3.24 Let I be a σ-ideal on a Polish space X and let E be an equivalence relation on X. We say that E is *I-homogeneous* if, for every pair of I-positive analytic sets $B, C \subset X$:

(i) either there is a Borel I-positive subset $B' \subset B$ such that $[B']_E \cap C \in I$;
(ii) or there is a Borel I-positive subset $B' \subset B$ and an injective Borel map $f : B' \to C$ such that $f \subset E$ and f preserves the ideal I: images of I-small sets are I-small, and images of I-positive sets are I-positive.

This property may not always be completely straightforward to verify, but it is satisfied in all natural cases.

Theorem 3.25 *Suppose that I is a σ-ideal on a Polish space X such that the quotient forcing P_I is proper, every analytic I-positive set contains a Borel I-positive subset, and P_I^E is regular in P_I. If E is I-homogeneous, then the model $V[\dot{x}_{\text{gen}}]_E$ is forced to be equal to the P_I^E-extension.*

Proof For the right-to-left inclusion, the homogeneity assumption is not needed. Let $G \subset P_I$ be a generic filter and $H = G \cap P_I^E$. It is clear that $V[H] \subset V[\dot{x}_{\text{gen}}]_E$, since H is definable from $[\dot{x}_{\text{gen}}]_E$ in every further forcing extension as the collection of those Borel I-positive E-saturated sets coded in the ground model containing the class $[\dot{x}_{\text{gen}}]_E$ as a subset.

The left-to-right inclusion is harder. Suppose that $\phi(\alpha, \vec{v}, [\dot{x}_{\text{gen}}]_E)$ is a formula with an ordinal parameter, other parameters in V, and another parameter

$[\dot{x}_{\text{gen}}]_E$, and let κ be a cardinal. A density argument shows that there must be a condition $B \in P_I$ such that the condition $\langle B, 1\rangle \in P_I \times \text{Coll}(\omega, \kappa)$ decides the truth value of ϕ, and H contains a condition below the pseudoprojection of B to P_I^E. We will show that for every two such conditions, the decision of the truth value must be the same. Thus, the set of ordinals defined by ϕ in the $P_I \times \text{Coll}(\omega, \kappa)$ must be already in the model $V[H]$, defined as the set of those ordinals α such that for some condition $B \in P_I$ whose pseudoprojection to P_I^E belongs to the filter H, we have $\langle B, 1\rangle \Vdash \phi(\alpha, \vec{v}, [\dot{x}_{\text{gen}}]_E)$.

So suppose for contradiction that there are conditions $C_0, C_1 \in P_I$ and $B \in P_I^E$ such that B is below the pseudoprojection of both C_0, C_1 into P_I^E, and $\langle C_0, 1\rangle \Vdash \phi(\alpha, \vec{v}, [\dot{x}_{\text{gen}}]_E)$ and $\langle C_1, 1\rangle \Vdash \neg\phi$. Since B is below the pseudoprojections, the first item of Definition 3.24 is excluded for B, C in both cases $C = C_0$ or $C = C_1$; indeed, the set $[B']_E$ would be an extension of B in P_I^E incompatible with the condition C_0 or C_1. So the second case must happen, and strengthening twice, we get a Borel I-positive set $B' \in P_I$ below B and Borel injections $f_0 : B' \to C_0$ and $f_1 : B' \to C_1$ as postulated in the assumptions. Since these functions preserve the ideal I and are subsets of the equivalence E, the condition B' forces in the forcing P_I that the points $\dot{x}_{\text{gen}} \in \dot{B}'$, $\dot{f}_0(\dot{x}_{\text{gen}}) \in \dot{C}_0$ and $\dot{f}_1(\dot{x}_{\text{gen}}) \in \dot{C}_1$ are E-equivalent P_I-generic points over V, giving the same generic extension. Thus, if $G \subset P_I$ is a filter generic over V containing the condition B' and $K \subset \text{Coll}(\omega, \kappa)$ is a filter generic over $V[G]$, the forcing theorem applied to $f_0(x_{\text{gen}})$ and $f_1(x_{\text{gen}})$ implies that $V[G, K] \models \phi(\alpha, \vec{v}, [f_0(x_{\text{gen}})]_E) \wedge \neg\phi(\alpha, \vec{v}, [f_1(x_{\text{gen}})]_E)$, which is impossible in view of the fact that $f_0(x_{\text{gen}})\ E\ f_1(x_{\text{gen}})$. $\square$

4

Classes of equivalence relations

4.1 Smooth equivalence relations

In the case that the equivalence relation E is smooth, the model $V[x_{\text{gen}}]_E$ as described in Theorem 3.5 takes on a particularly simple form.

Proposition 4.1 *Let I be a σ-ideal on a Polish space X such that the quotient forcing P_I is proper, and suppose that E is a smooth equivalence relation on the space X. Then $V[x_{\text{gen}}]_E = V[f(\dot{x}_{\text{gen}})]$ for every ground model Borel function f reducing E to the identity.*

Proof Choose a Borel function $f : X \to 2^\omega$ that reduces the equivalence E to the identity. Let $G \subset P_I$ and $H \subset \text{Coll}(\omega, \kappa)$ be mutually generic filters, and let $x \in X$ be a point associated with the filter G. First of all, $f(x) \in 2^\omega$ is definable from the equivalence class $[x]_E$ in the model $V[x][H]$: it is the unique value of $f(x)$ for all $x \in [x_{\text{gen}}]_E$. Thus, $V[f(x)] \subset V[x]_E$. On the other hand, the equivalence class $[x]_E$ is also definable from $f(x_{\text{gen}})$, the model $V[G, H]$ is an extension of $V[f(x)]$ via a weakly homogeneous notion of forcing $\text{Coll}(\omega, \kappa)$, and therefore by Fact 2.24, $V[x_{\text{gen}}]_E \subset V[f(x)]$. □

The total canonization of smooth equivalence relations has an equivalent restatement with the quotient forcing adding a minimal real degree. It is necessary though to discern between the forcing adding a minimal real degree, and the forcing producing a minimal extension. In the former case, we only ascertain that $V[y] = V$ or $V[y] = V[x_{\text{gen}}]$ for every set $y \subset \omega$ in the generic extension. In the latter case, this dichotomy in fact holds for every set y of ordinals in the extension. For example, the Silver forcing adds a minimal real degree, while it produces many intermediate σ-closed extensions, as shown in Section 8.2.

Proposition 4.2 *Let X be a Polish space and I be a σ-ideal on X such that the quotient forcing P_I is proper. The following are equivalent:*

(i) P_I *adds a minimal real degree;*
(ii) I *has total canonization for smooth equivalence relations.*

Proof First, assume that the canonization property holds. Suppose that $B \in P_I$ forces that τ is a new real; we must show that some stronger condition forces $V[\tau] = V[\dot{x}_{gen}]$. Thinning out the set B if necessary we may find a Borel function $f : B \to 2^\omega$ such that $B \Vdash \tau = \dot{f}(\dot{x}_{gen})$. Note that singletons must have I-small f-preimages, because a large preimage of a singleton would be a condition forcing τ to be that ground model singleton. Consider the smooth equivalence relation E on B, $E = f^{-1}\text{id}$. Since it is impossible to find a Borel I-positive set $C \subset B$ such that $E \restriction B = \text{ev}$, there must be a Borel I-positive set $C \subset B$ such that $E \restriction C = \text{id}$, in other words $f \restriction C$ is an injection. The function $f \restriction C$ is an injection even in the model $V[\dot{x}_{gen}]$ by analytic absoluteness between V and $V[\dot{x}_{gen}]$. Therefore, $C \Vdash \dot{x}_{gen} \in V[\tau]$ since $\dot{x}_{gen}$ can be recovered in $V[\tau]$ as the unique point $x \in C$ such that $f(x) = \tau$.

Now suppose that P_I adds a minimal real degree, and let $B \in P_I$ be a Borel I-positive set and E a smooth equivalence relation on B with I-small classes. We need to produce a Borel I-positive set $C \subset B$ consisting of pairwise E-inequivalent elements. Let $f : B \to 2^\omega$ be a Borel function reducing E to the identity. Note that since f-preimages of singletons are in the ideal I, $\dot{f}(\dot{x}_{gen})$ is a name for a new real and therefore it is forced that $V[\dot{x}_{gen}] = V[\dot{f}(\dot{x}_{gen})]$. Let M be a countable elementary submodel of a large structure, and let $C \subset B$ be the Borel I-positive set of all M-generic points in B for the poset P_I as guaranteed by Fact 2.50. Fact 2.51(*iii*) shows that we may thin out C' if necessary so that the set $f''C'$ becomes Borel. Consider the Borel set $D \subset f''C' \times C'$ defined by $\langle y, x\rangle \in D$ if and only if $f(x) = y$. We claim that the set D has countable vertical sections. If $f(x) = y$ then by the forcing theorem applied in the model M it must be the case that $M[x] = M[y]$ and thus the vertical section D_y is a subset of the countable model $M[y]$. Use Luzin–Novikov uniformization (Fact 2.3) to find countably many graphs of Borel functions $\{g_n : n \in \omega\}$ whose union is D. The sets $\{\text{rng}(g_n) : n \in \omega\}$ are Borel since the functions g_n are Borel injections, and they cover the whole set C'. Thus there is a number $n \in \omega$ such that the set $C = \text{rng}(g_n)$ is I-positive. The function $f \restriction C$ is an injection as desired. □

4.2 Countable equivalence relations

Borel countable equivalence relations are a frequent guest in mathematical practice. Their treatment in this book is greatly facilitated by essentially complete analysis of the associated forcing extensions. The purpose of this section is to prove the following canonization theorem:

Theorem 4.3 *Let I be a σ-ideal on a Polish space X such that the forcing P_I is proper, nowhere c.c.c., and adds a minimal forcing extension. Then I has total canonization for essentially countable equivalence relations.*

This result will be strengthened in Section 4.3 to include all equivalence relations classifiable by countable structures; there, several illustrative examples are discussed. However, the critical case of essentially countable equivalence relations has to be treated separately using the methods of this section. The theorem will be a consequence of two results of independent interest analyzing the model $V[x]_E$ for countable equivalence relations E.

Theorem 4.4 *Suppose that I is a σ-ideal on a Polish space X such that the quotient forcing is proper. Suppose that E is a countable Borel equivalence relation on X. Then for every Borel I-positive set $B \subset X$ there is a Borel I-positive set $C \subset B$ such that:*

(i) $P_I^{E,C}$ is a regular subposet of P_I below C;
(ii) C forces the model $V[\dot{x}_{\text{gen}}]_E$ to be equal to the $P_I^{E,C}$ extension;
(iii) the remainder forcing $(P_I \restriction C)/P_I^{E,C}$ is weakly homogeneous and c.c.c.

If E-saturations of I-small sets are I-small and $B = X$, then in fact $C = X$ works. In particular, P_I^E is regular in P_I and the model $V[\dot{x}_{\text{gen}}]_E$ is forced to be equal to the P_I^E extension.

Here, $P_I^{E,C}$ is the poset of I-positive Borel subsets of C which are relatively E-saturated inside C. The set $C \subset X$ may not be the whole space, as Example 3.16 shows.

Proof The set $C \subset B$ is easy to identify:

Claim 4.5 *There is a Borel I-positive set $C \subset B$ such that the E-saturation of any I-small Borel subset of C has I-small intersection with C.*

Proof Use the Feldman–Moore theorem (Fact 2.13) to find a Borel action of a countable group G on B whose orbit equivalence relation is exactly E. Let M be a countable elementary submodel of a large structure containing B and the action, and let $C = \{x \in B : x \text{ is } P_I\text{-generic over } M\}$. The set C is Borel and I-positive by Fact 2.50. We claim that this set works.

Suppose for contradiction that there is an I-small Borel subset D of C whose E-saturation has an I-positive intersection with C. Note that $[D]_E$ is a Borel set again as E is a countable Borel equivalence relation. Thinning out the set

$C \cap [D]_E$ we may find a Borel I-positive subset C' and a group element $g \in G$ such that $gC' \subset D$.

Now consider the set $O = \{A \in P_I :$ if there is a Borel I-positive subset of A whose shift by g is I-small, then A is such a set$\}$. Clearly, this is an open dense subset of P_I below B in the model M, and therefore there is $A \in M \cap O$ such that $A \cap C' \notin I$. Now $g(A \cap C') \subset D$ and therefore this set is in the ideal I, and by definition of the set O it must be the case that $gA \in I$. But as $gA \in I \cap M$, it must be the case that $C \cap gA = 0$, which contradicts the fact that $g(A \cap C')$ is a nonempty subset of C. □

Fix a set $C \subset B$ as in Claim 4.5; we will show that it has the properties required in the theorem. Note that if E-saturations of I-small sets were I-small, we could put $C = B$, thus arguing for the last two sentences of the theorem. Item *(i)* holds by Theorem 3.20. For *(iii)*, let $K \subset P_I$ be a generic filter containing the condition C with its associated generic point $x \in X$, and let $H = K \cap P_I^{E,C}$. To study the remainder forcing $Q = (P_I \restriction C)/H \in V[H]$, note first that it consists of exactly those Borel sets $D \subset C$ in V such that $C \cap [D]_E \in H$, or equivalently $[x]_E \cap D \neq 0$.

For the c.c.c. of Q, suppose that $\{D_\alpha : \alpha \in \omega_1\}$ is a putative antichain in Q in the model $V[H]$. Since $[x]_E$ is countable and $\aleph_1$ is preserved in $V[K]$, there must be $\alpha \neq \beta \in \omega_1$ such that D_α, D_β contain the same element of the equivalence class $[x]_E$. It follows that $D_\alpha \cap D_\beta$ is a lower bound of D_α, D_β in Q.

For the weak homogeneity of Q, back in V use the Feldman–Moore theorem again to find a Borel action of a countable group G on the set C such that $E \restriction C$ is the corresponding orbit equivalence relation. By Claim 4.5, we now see that the action is I-preserving, as saturations of I-small subsets of C have I-small intersection with C. Suppose that $D_0, D_1 \in Q$ are conditions. Then there are elements $y, z \in [x]_E$ such that $y \in D_0$ and $z \in D_1$. Let $g \in G$ be the group element such that $g(y) = z$, and consider the map $D \mapsto gD$. It is easily checked that this is an automorphism of the poset Q and gD_0 is compatible with D_1, since $gD_0 \cap D_1$ contains the point $z \in [x]_E$.

Finally, item *(ii)* follows from Theorem 3.25. To verify the I-homogeneity of E back in the ground model, note that if $D_0, D_1 \in P_I \restriction C$ are sets such that $[D_0]_E \cap D_1 \notin I$, then by σ-additivity of I there must be $g \in G$ such that the Borel set $\{y \in D_0 : gy \in D_1\}$ is I-positive. The injection $y \mapsto gy$ verifies the I-homogeneity of E as in Definition 3.24(ii). □

In the event that the poset P_I adds a minimal real degree, there is an alternative description of the model $V[\dot{x}_{\mathrm{gen}}]_E$ for countable Borel equivalence relations E which is essentially independent of the choice of E:

Theorem 4.6 *Suppose that I is a σ-ideal on a Polish space X such that P_I is proper and adds a minimal real degree. Suppose that E is a countable Borel equivalence relation on an I-positive Borel set $B \subset X$. Then B forces $V[\dot{x}_{\text{gen}}]_E$ to be either the whole $V[\dot{x}_{\text{gen}}]$ or else the largest intermediate extension between V and $V[\dot{x}_{\text{gen}}]$.*

It is a part of the statement of the theorem that if $V[\dot{x}_{\text{gen}}]_E \neq V[\dot{x}_{\text{gen}}]$, then the largest intermediate extension exists.

Proof Let $G \subset P_I$ be a generic filter containing a condition B, with an associated generic point $x \in X$. Assume that $V[x]_E \neq V[x]$. By the minimal real degree assumption, this implies that in fact $V[x]_E$ contains no new reals (any new real would construct x) and the properness assumption then implies that $V[x]_E$ contains no new countable sets of ordinals (any such new set would be covered by a ground model countable set and so give a new real (Jech 2002, lemma 31.4)). We need to show that if $V[H]$ is a model of ZFC such that $V \subseteq V[H] \subseteq V[G]$, then either $x \in V[H]$ or $H \in V[x]_E$. We may assume that $H \subset \kappa$ for some ordinal κ. Fix a P_I-name $\dot{H}$ for the set H. If there is a countable set $a \subset \kappa$ in V such that $H \cap a \notin V$ then $x \in V[H]$ by the minimal real degree assumption. Suppose then that for every countable set $a \in V$, it is forced that $\dot{H} \cap \check{a} \in V$. Consider the names for countably many sets $\{\dot{H}/y : y \in [\dot{x}_{\text{gen}}]_E$ and y is P_I-generic over $V\}$. There must be a countable set $b \subset \kappa$ such that if any two of these sets differ at some ordinal, then they differ at an ordinal in b. By properness, this set b has to be covered by a ground model countable set $a \subset \kappa$. Suppose that $C \subset B$ in P_I is a condition which forces this property of the set a, and moreover decides the value of $\dot{H} \cap \check{a}$ to be some definite set $c \in V$. Then C forces that H is definable from $[x]_E$, C, c, and $\dot{H}$ in the model $V[G]$ by the formula ϕ saying "H is the unique value of $\dot{H}/y$ for all V-generic points $y \in [x]_E$ such that $\dot{H}/y \cap a = c$". We claim that ϕ defines H in every further generic extension of $V[G]$. The equivalence class $[x]_E$ is countable in $V[G]$ and therefore evaluated the same in $V[G]$ as in the further extension by coanalytic absoluteness. Therefore, the quantifiers in ϕ range over the same set whether applied in $V[G]$ or any further extension. Thus, $H \in V[x]_E$ holds and the proof is complete. □

There is an important heuristic consequence of the last sentence of this theorem: if the poset P_I adds a minimal real degree, we cannot use the model $V[\dot{x}_{\text{gen}}]_E$ to discern between the various nonsmooth countable equivalence relations on the underlying Polish space X, because they all happen to generate the same model. This is apparently not to say that every two such countable equivalence relations must be similar or of the same complexity at some I-positive Borel set.

Proof of Theorem 4.3 Let $B \subset X$ be a Borel I-positive set, and E an essentially countable Borel equivalence relation on B. Let $f : B \to Y$ be a Borel function reducing E to some countable Borel equivalence relation F on a Polish space Y. Use the minimal degree assumption to thin out B so that f is either constant or one-to-one. In the former case, B has only one E-class and we are done. In the latter case, $E \restriction B$ has countable classes. There are two cases:

Case 1 If $V[\dot{x}_{\text{gen}}]_E = V[\dot{x}_{\text{gen}}]$ then by the trichotomy theorem (Theorem 3.5) there is a Borel I-positive set $C \subset B$ such that $E \restriction C = \text{id}$ and we are done. There is also an alternative argument avoiding the use of the trichotomy theorem. By Theorem 4.4, thinning out the set B if necessary, we know that the model $V[\dot{x}_{\text{gen}}]_E$ is forced to be a $P_I^{E,B}$ extension. Fix a $P_I^{E,B}$-name τ for the point $\dot{x}_{\text{gen}}$, a countable elementary submodel M of a large structure containing B, E, τ, and let $C = \{x \in B : x \text{ is } P_I\text{-generic over } M\}$. This is an I-positive Borel set, the evaluation $x = \tau/x$ must be E-invariant on it since τ is a $P_I^{E,B}$-name, and therefore $E \restriction C = \text{id}$.

Case 2 If $V[\dot{x}_{\text{gen}}]_E \subsetneq V[\dot{x}_{\text{gen}}]$, then the minimal forcing assumption extension implies $V[\dot{x}_{\text{gen}}]_E = V$. However, Theorem 4.4 shows that $V[\dot{x}_{\text{gen}}]$ is a c.c.c. extension of $V[\dot{x}_{\text{gen}}]_E = V$, contradicting the assumption that the poset P_I (which generates the extension of $V[\dot{x}_{\text{gen}}]$ from V) is nowhere c.c.c. by Corollary 2.27. Thus, this case is impossible, completing the proof. □

The final observation in this section concerns the simplification within the class of countable equivalence relations. The class of Borel countable equivalence relations is very complex and extensively studied. It is possible that many forcings have highly nontrivial intersection of their spectrum with this class. In the fairly common case that the forcing preserves Baire category, however, all that complexity boils down to E_0.

Theorem 4.7 *Whenever I is a σ-ideal on a Polish space X such that the forcing P_I is proper and preserves Baire category, then*

$$\text{essentially countable} \to_I \leq_B E_0.$$

Proof Suppose $B \in P_I$ is an I-positive Borel set and $f : B \to Y$ is a Borel map into a Polish space with a countable equivalence relation E on it. Kechris and Miller (2004, theorem 12.1) show that there is a Borel set $D \subset \omega^\omega \times Y \times Y$

such that for every $z \in \omega^\omega$ the section D_z is a Borel hyperfinite equivalence relation included in E, and for every $y \in Y$ the set $A_y = \{z \in \omega^\omega : [y]_E = [y]_{D_z}\}$ is comeager in ω^ω. Now the set $D' \subset B \times \omega^\omega$ of all pairs x, z such that $z \in A_{f(x)}$, has comeager vertical sections. Since the forcing P_I preserves the Baire property, it must be the case that there is a point $z \in \omega^\omega$ such that the corresponding horizontal section $C \subset B$ is I-positive. But then, $D_z = E$ on $f''C$, and therefore the pullback $f^{-1}E$ is on C reducible to a hyperfinite equivalence relation D_z and therefore to E_0. □

4.3 Equivalence relations classifiable by countable structures

This is a class of equivalence relations very common in mathematical practice.

Definition 4.8 An equivalence relation E on a Polish space X is *classifiable by countable structures* if there is a Borel map $f : X \to \Pi_n \mathcal{P}(\omega^n)$ such that elements $x, y \in X$ are E-related if and only if $f(x), f(y)$ are isomorphic as relational structures on ω.

An isomorphism of structures on ω is naturally induced by a permutation of ω, and therefore every equivalence relation classifiable by countable structures is Borel reducible to an orbit equivalence generated by a continuous action of a closed subgroup of the infinite permutation group S_∞. The converse implication also holds by a result of Becker and Kechris (Kanovei 2008, theorem 12.3.3): every orbit equivalence relation generated by a Polish action of the infinite permutation group or its closed subgroups is classifiable by countable structures. Every countable Borel equivalence relation is classifiable by countable structures as well (Kanovei 2008, lemma 6.1.3).

The main result of the present section is the following elaboration of Theorem 4.3. We say that a forcing *adds finitely many real degrees* if in its extension $V[G]$ there is a finite set $\{z_i : i \in n\}$ of elements of 2^ω such that for every other element $z \in 2^\omega \cap V[G]$ there is $i \in n$ such that $V[z_i] = V[z]$.

Theorem 4.9 *If I is a σ-ideal on a Polish space X such that the quotient P_I is proper and adds only finitely many real degrees, then*

$$\text{classifiable by countable structures} \to_I \text{essentially countable.}$$

Corollary 4.10 *Let I be a σ-ideal on a Polish space X such that the poset P_I is proper, nowhere c.c.c. and adds a minimal forcing extension. Then I has total canonization for equivalences classifiable by countable structures.*

Corollary 4.11 *Let I be a σ-ideal on a Polish space X such that the poset P_I is proper. If I has total canonization for essentially countable equivalence relations, then it has total canonizations for equivalence relations classifiable by countable structures.*

In fact, the method of proof yields total canonization for a wider class of equivalence relations including E_1, $=_J$ for the indecomposable ideals or Weiss ideals of Kanovei and Reeken (2000) and others. Within the class of orbit equivalence relations though, the conclusion of the theorem seems to be optimal. In Section 6.4, we will produce a proper poset P_I producing a minimal forcing extension such that E_2 is in the spectrum. The whole treatment is parallel to the approach to turbulence in Kanovei (2008, chapter 13).

The strength of the above statements is best illustrated using a couple of examples. The following σ-ideals have total canonization for equivalence relations classifiable by countable structures:

Example 4.12 The Laver ideal I on ω^ω. The quotient poset P_I contains a dense subset naturally isomorphic to the Laver forcing, therefore it is proper and nowhere c.c.c. It generates a minimal forcing extension by a theorem of Groszek (1987), and therefore satisfies the assumptions of Corollary 4.10.

Example 4.13 The σ-ideal generated the sets of continuity of a fixed Borel function f between Polish spaces X and Y. There is a standard example of a function f for which this σ-ideal does not contain the whole space X – the *Pawlikowski function* (Cichon *et al.*, 1991). The quotient poset P_I is proper and preserves Baire category and outer Lebesgue measure (Zapletal 2008, section 4.2.3). The σ-ideal I is a porosity ideal by Example 6.40. By Theorem 6.43(*ii*), every nontrivial intermediate extension must be given by a single Cohen real, but this is impossible as Cohen forcing does not preserve outer Lebesgue measure. Thus, the assumptions of Corollary 4.10 are satisfied. Steprāns showed earlier and in a different language that the poset P_I has the Laver property and therefore does not add Cohen reals (Steprans 1993).

Example 4.14 Let X be a compact metric space, and let I be the σ-ideal generated by sets of finite Hausdorff measure of a fixed dimension. The poset P_I is $< \omega_1$-proper, bounding, and adds no independent real by Zapletal (2008, corollary 4.4.10). The σ-ideal I is a game ideal, the game was isolated in Zapletal (2008, theorem 4.4.5), and it is a $\mathbf{\Pi}^1_1$ on $\mathbf{\Sigma}^1_1$ σ-ideal. Thus, all intermediate extensions of the P_I extension must be c.c.c. by Theorem 5.21, but no such extension can be nontrivial by Corollary 2.63. Thus, the assumptions of Corollary 4.11 are satisfied.

Example 4.15 Let X be a compact metric space, let ϕ an outer regular subadditive capacity, and let I be the σ-ideal of sets of ϕ-mass zero. Zapletal (2008, section 4.3) studied the forcing properties of the quotient forcing P_I and defined the notion of stable and Ramsey capacity. If the capacity is stable and Ramsey, then the quotient forcing P_I is $< \omega_1$-proper, bounding, and adds no independent real (Zapletal 2008, theorems 4.3.23 and 4.3.25). The σ-ideal I is $\mathbf{\Pi}^1_1$ on $\mathbf{\Sigma}^1_1$ and it is a game ideal (Zapletal 2008, theorem 4.3.21). The discussion from the previous example then can be repeated to show that the assumptions of Corollary 4.11 are satisfied.

The proof of Theorem 4.9 requires a couple of definitions of independent interest.

Definition 4.16 (Kanovei 2008, definition 13.5.2) The class $\mathbf{K}$ of equivalence relations is the smallest class containing the identity on 2^ω and closed under the following operations:

(i) Borel reduction;
(ii) countable union of equivalence relations: if $\{E_n : n \in \omega\}$ are equivalence relations on respective pairwise disjoint domains X_n, then their union is the equivalence relation on $\bigcup_n X_n$ connecting two points x, y if they belong to the same domain X_n and $x \mathrel{E_n} y$;
(iii) product modulo the Fréchet ideal: if $\{E_n : n \in \omega\}$ are equivalence relations on respective Polish spaces X_n, then their *product* is the equivalence relation E on $\prod_n X_n$ defined by $x \mathrel{E} y$ if the set $\{n \in \omega : x(n) \mathrel{E_n} y(n)\}$ is finite;
(iv) the Friedman–Stanley jump operation: if E is an equivalence relation on X then its *jump* is an equivalence relation E^+ on X^ω defined by $x \mathrel{E^+} y$ if for every $n \in \omega$ there is $m \in \omega$ such that $x(n) \mathrel{E} y(m)$, and vice versa, for every $m \in \omega$ there is $n \in \omega$ such that $x(n) \mathrel{E} y(m)$.

The equivalence relations in the class $\mathbf{K}$ are all Borel. They include all Borel equivalence relations classifiable by countable structures as well as E_1, and the equivalence relation modulo J where J is one of the Weiss ideals or indecomposable ideals of Kanovei and Reeken (2000). The key feature of the class $\mathbf{K}$ is that all equivalence relations in it satisfy a technical variation of the pinned condition of Kanovei (2008, section 17.1). Foreman and Magidor (1995) call a generic extension $V[G]$ of V *reasonable* if every countable set of ordinals in $V[G]$ is covered by a countable set of ordinals in V; in particular, every proper forcing extension is reasonable.

Definition 4.17 An analytic equivalence relation E is *trim* if in every forcing extension, for every pair x, y of points of X such that $x \; E \; y$ and $V[x]$ is a reasonable extension of V, if $V[x] \cap V[y] = V$, then there is $z \in X \cap V$ such that $x \; E \; y \; E \; z$.

Proposition 4.18 *Every equivalence relation in* **K** *is trim.*

Proof The identity equivalence relation is trim. Thus, we must only prove that the trim property persists through the operations indicated.

For the Borel reduction, suppose that E is an equivalence relation on X, F is an equivalence relation on Y, $f : X \to Y$ is a Borel reduction of E to F, and F is trim. Let $x, z \in X$ be V-generic points such that $x \; E \; z$, $V[x] \cap V[z] = V$, and $V[x]$ is a reasonable extension of V. Then $V[f(x)] \subset V[x]$, $V[f(z) \subset V[z]$, and so $V[f(x)] \cap V[f(z)] = V$ and $V[f(x)]$ is a reasonable extension of V. By the trim property of F, there is $y \in Y \cap V$ which is F-equivalent to both $f(x)$ and $f(z)$. By the Shoenfield absoluteness, there is $v \in X \cap V$ such that $f(v) \; F \; y$. It follows that v is E-equivalent to both x, z and E is trim.

For the product, suppose that E_n are equivalence relations on respective Polish spaces X_n, each trim. Let E be the product of the equivalence relations E_n for $n \in \omega$; we must show that E is trim. Let $x, y \in \prod_n X_n$ be V-generic points such that $x \; E \; y$ and x is in a reasonable extension of V and $V[x] \cap V[y] = V$; in particular, for every $n \in \omega$ we have $V[x(n)] \cap V[y(n)] = V$. There is a number $n_0 \in \omega$ such that for every $n > n_0$ it is the case that $x(n) \; E_n \; y(n)$. For every such $n \in \omega$, there is $v \in X_n \cap V$ which is E_n-related to both $x(n)$ and $y(n)$. Let $R = \{\langle n, v\rangle : n > n_0, v \in X_n \cap V \text{ and } v \; E_n \; x(n)\}$. This is a relation with nonempty vertical sections which belongs to both $V[x]$ and $V[y]$, therefore it belongs to V. In V, an application of countable choice shows that there is $z \in \prod_n X_n$ such that for every $n > n_0$, $\langle n, z(n)\rangle \in R$. The point $z \in V$ is then E-related to both x and y as required for the trim property.

For the union, suppose that $\{E_n : n \in \omega\}$ are trim equivalence relations on pairwise disjoint Polish spaces $\{X_n : n \in \omega\}$, and let E be their union. To see that E is trim, let $x \; E \; y$ be E-related V-generic points such that $V[x] \cap V[y] = V$. Then there is n such that $x, y \in X_n$, and by the trim property of E_n, it is the case that $x \; E \; y \; E \; z$ for some $z \in V \cap X_n$. Then $z \; E \; x$ and the trim property of E follows.

For the jump operation, suppose that E is a trim Borel equivalence relation on a Polish space X. Suppose that $x, y \in X^\omega$ are E^+-related points in some forcing extension such that $V[x]$ is a reasonable extension of V. Suppose that $V[x] \cap V[y] = V$ and work towards proving that there is $z \in X^\omega \cap V$ such that $x \; E^+ \; y \; E^+ \; z$.

First, note that for every $n \in \omega$ there is $z_n \in X \cap V$ such that $x(n)\ E\ z_n$. To find the point z_n, choose $m \in \omega$ such that $y(m)\ E\ x(n)$ and apply the trim property the equivalence relation E to the pair $x(n), y(m)$. $V[x(n)] \subset V[x]$ and therefore $V[x(n)]$ is still a reasonable extension of V, $V[y(m)] \subset V[y]$ and therefore $V[x(n)] \cap V[y(m)] = V$ still holds, and as E is trim, there must be $z_n \in X \cap V$ such that $x(n)\ E\ y(m)\ E\ z_n$.

Second, working in $V[x]$, choose the points $z_n \in X \cap V$ as indicated in the previous paragraph for each $n \in \omega$. Since $V[x]$ is a reasonable extension of V, there must be a countable set $a \subset X$ in V such that $\{z_n : n \in \omega\} \subset a$. Consider the set $b = \{u \in a : \exists n\ x(n)\ E\ u\}$. Since $x\ E^+\ y$ holds, $b = \{u \in a : \exists n\ y(n)\ E\ u\}$, and so this set is both in $V[x]$ and $V[y]$. By the assumptions, $b \in V$, and as a subset of the countable set a it is countable in V. Let $z \in V \cap X^\omega$ enumerate the elements of b and conclude that $x\ E^+\ y\ E^+\ z$ as desired. □

Proof of Theorem 4.9 We will first argue that the canonization holds for all equivalence relations in the class **K**. Let B be an I-positive Borel set and E an equivalence relation on B with $E \in \mathbf{K}$. Since P_I adds only finitely many real degrees, there is a name τ for an element of B such that $B \Vdash \tau\ E\ \dot{x}_{\text{gen}}$ and τ is of minimal possible degree. Let M be a countable elementary submodel of a large enough structure containing B, and let $C \subset B$ be the Borel I-positive set of points in B which are P_I-generic over M. Let $f : C \to B$ be the Borel map defined by $f(x) = \tau/x$. We will show that this is a Borel reduction of $E \restriction C$ to $E \restriction \text{rng}(f)$, and $E \restriction \text{rng}(f)$ has countable classes; this will prove the theorem.

First of all, the function f as well as its range are Borel by Fact 2.51. Second, $f \subset E$: if $x \in C$ then $M[x] \models f(x) = \tau/x\ E\ x$ by the forcing theorem, and $f(x)\ E\ x$ by the Mostowski absoluteness between $M[x]$ and V. The key point will be to show that $E \restriction \text{rng}(f)$ has countable classes.

Claim 4.19 *Whenever $x, y \in C$ are E-related points, then $f(x) \in M[f(y)]$.*

Restated, $M[f(x)] = M[f(y)]$ holds. In fact, a more careful computation shows that $M[f(x)] = M[[x]_E]$, and then the claim follows from Theorem 3.11(*ii*).

Proof Suppose for contradiction that $f(x) \notin M[f(y)]$. Since the model $M[f(x)]$ contains only finitely many real degrees, there is a point $z \in M[f(x)] \cap M[f(y)] \cap X$ of the largest possible real degree. Clearly, every point of $M[f(x)] \cap M[f(y)] \cap X$ is in the model $M[z]$. Now, let $g \subset \text{Coll}(\omega, \lambda)$ be a generic filter over $M[x]$ for some large cardinal λ. Note that $\mathcal{P}(P_I)^M$ becomes countable in the model $M[x][g]$, and the quantification over filters on P_I generic over M turns into a quantification over elements of a Polish space in $M[x][g]$. By analytic absoluteness between $M[x][g]$ and V, there is a point $y' \in X$ P_I-generic over M such that $x \; E \; y'$, $z \in M[f(y')]$ and $f(x) \notin M[f(y')]$, since the point $y \in X$ is such.

Now, apply Proposition 4.18 in the model $M[z]$. Note that $M[z] \subset M[x][g]$, $M[x][g]$ is a generic extension of $M[z]$ by Jech (2002, lemma 15.43), and $M[z] \models E \in \mathbf{K}$. So, for the pair $f(x) \; E \; f(y')$ in $M[x][g]$, it must be the case that there is some point $v \in X \cap M[z]$ such that $v \; E \; f(x)$. However, such a point v must be of real degree strictly smaller than $f(x)$, and this contradicts the minimal degree choice of $f(x)$. □

It follows that $E \restriction \text{rng}(f)$ has countable classes: all E-equivalents of any point $f(y)$ in $\text{rng}(f)$ must belong to the countable set $X \cap M[f(y)]$.

We can now proceed to treat the case of equivalence relations classifiable by countable structures. Let $B \subset X$ be a Borel I-positive set and E an equivalence relation on B classifiable by countable structures. Use Theorem 4.26 to thin the set B such that $E \restriction B$ is Borel. The analysis of Borel equivalence relations classifiable by countable structures using the Friedman–Stanley jump shows that they all belong to the class $\mathbf{K}$. Thus, the previous argument yields a Borel I-positive set $C \subset B$ on which $E \restriction C$ is essentially countable. □

In view of the real degree structure occurring in countable support iteration extensions (Chapter 10), it would be highly desirable to know the answer to the following:

Question 4.20 Can the assumptions of Theorem 4.9 be weakened to "P_I is proper and adds a countable set of degrees" or to "P_I is proper and adds a wellordered set of degrees"?

Proof of Corollary 4.10 Let I be a σ-ideal on a Polish space X such that the quotient P_I is proper, nowhere c.c.c. and adds a minimal forcing extension. Let $B \subset X$ be a Borel I-positive set and E an equivalence relation on it, classifiable by countable structures. Since P_I adds a minimal real degree, by Theorem 4.9 we can thin out the set B if necessary so that $E \restriction B$ is essentially countable. Finally, Theorem 4.3 can be applied so as to conclude that there is an I-positive Borel set C on which either $E \models \text{id}$ or $E \restriction C = \text{ev}$. □

Proof of Corollary 4.11 Let I be a σ-ideal on a Polish space X such that the quotient P_I is proper, and I has total canonization for essentially countable equivalence relations. Let $B \subset X$ be a Borel I-positive set and E an equivalence relation on it, classifiable by countable structures. Now, I has total canonization for smooth equivalences, so P_I adds a minimal real degree; by

Theorem 4.9 we can thin out the set B if necessary so that $E \restriction B$ is essentially countable. The total canonization for essentially countable equivalence relations can be applied again to canonize E further on an I-positive Borel set $C \subset B$ to id or ev. □

We will want to prove canonization theorems for equivalence relations classifiable by countable structures even in situations where the poset P_I adds a complicated degree structure, and equivalence relations like E_1 are in the spectrum – for example, canonizing on infinite products of perfect sets in Section 9.3. The approach here uses the following forcing property of the poset P_I:

Definition 4.21 A poset P has the *separation property* if it forces that for every countable set $a \subset 2^\omega$ and every point $y \in 2^\omega \setminus a$ there is a ground model coded Borel set $B \subset 2^\omega$ separating y from a: $y \in B$ and $a \cap B = 0$.

Theorem 4.22 *Suppose that I is a σ-ideal on a Polish space X such that P_I is proper.*

(i) If I has total canonization for essentially countable Borel equivalence relations, then P_I has the separation property.
(ii) If P_I has the separation property, then for every analytic equivalence relation E, the models $V[\dot{x}_{\text{gen}}]_E$ and $V[[\dot{x}_{\text{gen}}]]_E$ contain the same hereditarily countable sets.
(iii) If P_I has the separation property then

$$\text{classifiable by countable structures} \rightarrow_I \text{smooth.}$$

Note that the conjunction of (*i*) and (*iii*) gives a different argument for Corollary 4.11.

Proof For (*i*), suppose that the assumptions hold and $B \Vdash \dot{a} \subset 2^\omega$ is a countable set. We will find a stronger condition $D \subset B$ and a countable ground model list of Borel sets $\{B_n : n \in \omega\}$ such that D forces $\{B_n : n \in \omega\}$ to *separate* the set $\dot{a}$ in the sense that every element of $\dot{a}$ belongs to at least one Borel set on the countable list while no Borel set on the list contains more than one element of $\dot{a}$. This certainly implies the separation property. Without loss of generality we may assume that $V \cap \dot{a} = 0$ is forced as well. (If not, use properness to enclose the countable set $\dot{a} \cap V$ into a ground model countable set, and separate it from the rest by a countable collection of singletons.) Thinning out the set B we may assume that there are Borel functions $\{f_n : n \in \omega\}$ on the set B such that $B \Vdash \dot{a} = \{\dot{f}_n(\dot{x}_{\text{gen}}) : n \in \omega\}$. Let M be a countable elementary submodel of a large enough structure, and let $C \subset B$ be the I-positive Borel set of all M-generic points for P_I in the set B.

Observe that each of the functions f_n is countable-to-one on the set C: since the poset P_I adds a minimal real degree by Proposition 4.2, the set $D_n \subset P_I$ consisting of those conditions on which the function f_n is one-to-one is dense in P_I, and $C \subset \bigcup(M \cap D_n)$. This means that the equivalence relation E generated by the relation $x_0 E x_1 \leftrightarrow \exists n_0 \exists n_1\ f_{n_0}(x_0) = f_{n_1}(x_1)$ has countable equivalence classes; it is easily verified that it is Borel as well. By the canonization assumption, there must be then a Borel I-positive set $D \subset C$ such that $E \restriction D = \mathrm{id}$. This means that for distinct points $x_0, x_1 \in D$ the countable sets $\{f_n(x_0) : n \in \omega\}$ and $\{f_n(x_1) : n \in \omega\}$ are disjoint; in particular, the functions f_n are all injections on D. Let $\bar{f}_n = (f_n \restriction D) \setminus \bigcup_{m \in n} f_m$ and let $B_n = \mathrm{rng}(\bar{f}_n)$. The sets B_n are ranges of Borel injections and as such they are Borel, and for every point $x \in D$ their list separates the set $\{f_n(x) : n \in \omega\}$. It follows that $D \Vdash$ the sets in $\{B_n : n \in \omega\}$ separate the set a.

For (*ii*), first argue that $V[x]_E$ and $V[[x]]_E$ contain the same reals. As $V[x]_E \subset V[[x]]_E$ by Theorem 3.11(*i*), it is enough to show that every real in $V[[x]]_E$ is in $V[x]_E$. If $B \in P_I$ is a condition forcing $\tau \in V[[x]]_E \cap 2^\omega$, then by Theorem 3.12 we can thin out the set B if necessary to find a Borel function $f : B \to 2^\omega$ which attains only countably many values at each equivalence class, such that $B \Vdash \tau = \dot{f}(\dot{x}_{\mathrm{gen}})$. Use the separation property of P_I to thin out the condition B if necessary and find a Borel set $C \subset 2^\omega$ such that $B \Vdash \dot{C}$ separates $\dot{f}(\dot{x}_{\mathrm{gen}})$ from the rest of the countable set $f''([\dot{x}_{\mathrm{gen}}]_E \cap B)$. Then $B \Vdash \tau \in V[\dot{x}_{\mathrm{gen}}]_E$, since τ is defined in any further generic extension as the only element of the countable set $f''([\dot{x}_{\mathrm{gen}}]_E \cap B)$ that belongs to the set C.

Now, the properness of the P_I extension can be used to show that every countable subset of $V[x]_E$ in $V[[x]]_E$ is already in $V[x]_E$. If $a \subset V[x]_E$ is countable and in $V[[x]]_E$, it is covered by a countable set $b \in V[x]_E$. The characteristic function of a inside b is (coded as) a real in $V[[x]]_E$, so it has to be in $V[x]_E$ by the previous paragraph, and so $a \in V[x]_E$. (*ii*) now follows – if there was a hereditary countable set in $V[[x]]_E \setminus V[x]_E$, then $\in$-minimal such a set a would have to be a countable subset of $V[x]_E$, and therefore an element of $V[[x]]_E$, as we have just proved.

For (*iii*), let $B \subset X$ be an I-positive Borel set and E be an equivalence relation on B classifiable by countable structures. We will show that B forces $\dot{x}_{\mathrm{gen}}$ to have an E-equivalent in the model $V[\dot{x}_{\mathrm{gen}}]_E$. Then, use Theorem 3.7 to find a Borel I-positive set $C \subset B$ and a Borel function $f : C \to B$ such that f is constant on E-classes of C, and for every $x \in C$, $f(x)\ E\ x$. It is immediate that f reduces $E \restriction C$ to identity.

By Theorem 4.26, the set B can be thinned down so that the equivalence relation $E \restriction B$ is Borel. By the standard analysis of Borel equivalence relations classifiable by countable structures using the Friedman–Stanley jump (Kanovei 2008, theorem 12.5.2), there is a countable ordinal α such that $E \restriction B$ is Borel reducible to isomorphism of countable structures with one binary relation satisfying the axiom of extensionality, which is wellfounded of rank $\leq \alpha$. Let g be the Borel reduction and let $x \in B$ be a P_I-generic point. Let a be the transitive isomorph of the structure $g(x)$ and observe that $a \in V[y]$ for every y such that $x \; E \; y$. Thus, $a \in V[[x]]_E$, by (*ii*) $a \in V[x]_E$, and by analytic absoluteness between the models $V[x]_E$ and $V[x]$, there is a point $y \in B$ in $V[x]_E$ such that the structure $g(y)$ is isomorphic to a. Clearly, $x \; E \; y$, and the proof is complete. □

4.4 Hypersmooth equivalence relations

Hypersmooth equivalence relations are connected to decreasing sequences of real degrees in the generic extension. A *decreasing sequence of real degrees* is a sequence $\langle x_n : n \in \omega \rangle$ of infinite binary sequences such that whenever $m \in n$ then $V[x_n] \subsetneq V[x_m]$. A *strict* decreasing sequence of degrees is a decreasing sequence such that moreover for every number m, the sequence $\langle x_n : m \in n \rangle$ belongs to $V[x_m]$.

Theorem 4.23 *Suppose that X is a Polish space and I is a σ-ideal such that the forcing P_I is proper. The following are equivalent:*

(*i*) *E_1 is in the spectrum of I;*
(*ii*) *some condition in P_I forces that in the generic extension there is an infinite decreasing sequence of real degrees;*
(*iii*) *some condition in P_I forces that in the generic extension there is a strict infinite decreasing sequence of real degrees.*

Proof For (*i*)→(*ii*) implication, suppose that E_1 is in the spectrum of I, as witnessed by a Borel I-positive set $B \subset X$ an equivalence relation E on B, and a Borel reduction $f : B \to (2^\omega)^\omega$ of E to E_1. We claim that B forces that in the generic extension, for every number $n \in \omega$ there is $m \in \omega$ such that $f(x_{\text{gen}}) \restriction [n, \omega) \notin V[f(x_{\text{gen}}) \restriction [m, \omega)]$; this will prove (*ii*).

Suppose for contradiction that some condition $C \subset B$ forces the opposite for some definite number $n \in \omega$. Let M be a countable elementary submodel of a large enough structure containing C and f, and let $D \subset C$ be the Borel I-positive set of all M-generic points in the set C. Consider the Borel set $Z \subset$

$(2^\omega)^{\omega\setminus n}$ defined by $z \in Z$ if and only if for every number $m > n$ there is a Borel function $g \in M$ such that $g(z \restriction [m, \omega)) = z$. It is not difficult to see that E_1 restricted to Z has countable classes, since there are only countably many functions in the model M. Now, whenever $x \in D$ then $f(x) \restriction [n, \omega) \in Z$ by the forcing theorem, and for $x, y \in D$, $f(x)\ E_1\ f(y)$ if and only if $f(x) \restriction [n, \omega)\ E_1\ f(y) \restriction [n, \omega)$. This reduces $E \restriction D$ to a Borel equivalence relation with countable classes, which contradicts the assumption that E_1 is embeddable into it.

Now, in the generic extension $V[x]$ define a function $h \in \omega^\omega$ by induction: $h(0) = 0$, and $h(n+1)$ is the least number $m \in \omega$ such that $f(x) \restriction [h(n), \omega) \notin V[f(x) \restriction [m, \omega)]$. The function h clearly belongs to all models $V[f(x) \restriction [m, \omega)]$, and therefore the sequence $\langle f(x) \restriction [h(n), \omega) : n \in \omega\rangle$ forms the required strictly decreasing sequence of real degrees.

To show *(ii)* implies *(iii)*, suppose that some condition forces $\langle \dot{x}_n : n \in \omega\rangle$ to be the decreasing sequence in 2^ω. The strict decreasing sequence is obtained by a simple adjustment. Just replace $\langle x_n : n \in \omega\rangle$ by a decreasing sequence $\langle \bar{x}_n : n \in \omega\rangle$ such that $V[\bar{x}_n] \models \bar{x}_{n+1}$ is the $\leq_n$-least point below which there is an infinite decreasing sequence of degrees and $V[\bar{x}_{n+1}] \neq V[\bar{x}_n]$. Here, $\leq_n$ is any well-ordering of $2^\omega \cap V[\bar{x}_n]$ simply definable from a wellordering of V and $\bar{x}_n$; for example, let Q be the least poset of size $\leq \mathfrak{c}$ in V for which $\bar{x}_n$ is generic, and then turn any well-order of Q-names for reals into a well-order of $V[\bar{x}_n]$.

To prove *(iii)* implies *(i)*, Fact 2.51 yields a Borel I-positive set B and Borel functions $f_n : B \to 2^\omega$, for each $n \in \omega$, representing the points on the strict decreasing sequence. We claim that the product of these functions, $g : B \to (2^\omega)^\omega$ defined by $g(x)(n) = f_n(x)$, has the property that E_1 is Borel reducible to the pullback $E = g^{-1}E_1$ restricted to any Borel I-positive set. This will prove the theorem. Suppose for contradiction that there is a Borel I-positive set $C \subset B$ such that $E_1 \not\leq E \restriction B$. By the hypersmooth dichotomy (Fact 2.18) it must be the case that $E \restriction C \leq E_0$ via some Borel map $g : C \to 2^\omega$. Move to the forcing extension $V[x_{\text{gen}}]$ for some P_I-generic point $x_{\text{gen}} \in C$.

Claim 4.24 *For every number $n \in \omega$, $g(x_{\text{gen}}) \in V[f_n(x_{\text{gen}})]$.*

Proof Working in $V[f_n(x_{\text{gen}})]$, we see that there is a point $x \in C$ such that $\langle f_m(x) : m > n\rangle E_1 \langle f_m(x_{\text{gen}}) : m > n\rangle$, since such an x exists in the model $V[x_{\text{gen}}]$ and analytic absoluteness holds between the two models. Note that the sequence $\langle f_m(x_{\text{gen}}) : m > n\rangle$ is in the model $V[f_n(\dot{x}_{\text{gen}})]$. Now all such numbers x must be mapped by g into the same E_0-equivalence class. This class is countable and contans $g(x_{\text{gen}})$. The claim follows. □

Now in the model $V[g(x_{gen})]$, there is a point $x \in C$ such that $g(x) = g(x_{gen})$, since there is such a point x in the model $V[x_{gen}]$ and analytic absoluteness holds between the two models. The sequence $\langle f_n(x) : n \in \omega \rangle$ must be E_1-equivalent to the sequence $\langle f_n(x_{gen}) : n \in \omega \rangle$; let us say that they are equal from some m on. But then, $f_m(x) = f_m(x_{gen}) \in V[g(x_{gen})] \subset V[f_{m+1}(x_{gen})]$, a contradiction! □

4.5 Analytic vs. Borel equivalence relations

One of the most basic concerns regarding the complexity of equivalence relations is whether they are analytic or Borel. If the relation on the whole space X is analytic, can we find a large set $B \subset X$ on which it is Borel? In this generality, the answer is negative, as exhibited by the following example:

Example 4.25 Let K be an analytic non-Borel ideal on ω and the equivalence relation E_K on $X = (2^\omega)^\omega$ defined by $\vec{x} E_K \vec{y}$ if $\{n \in \omega : \vec{x}(n) \neq \vec{y}(n)\} \in K$. Consider the ideal I on X associated with the countable support product of Sacks forcing as in Section 9.3. Then the equivalence E_K is non-Borel, and since $E_K \leq E_K \restriction B$ for every Borel I-positive set $B \subset X$, it is also the case that $E_K \restriction B$ is non-Borel.

In more specific situations, the answer may be positive for large classes of analytic equivalence relations. We include two results:

Theorem 4.26 *Suppose that X is a Polish space, I a σ-ideal on it such that the quotient poset P_I is proper, $B \in P_I$ an I-positive Borel set, and E an equivalence relation on the set B reducible to an orbit equivalence relation of a Polish action. Then there is a Borel I-positive set $C \subset B$ such that $E \restriction C$ is Borel. If the ideal I is c.c.c. then in fact the set C can be found so that $B \setminus C \in I$.*

Proof Suppose that a Polish group G acts on a Polish space Y, and $f : B \to Y$ is a Borel reduction of E to the orbit equivalence relation E_G of the action. By a result of Becker and Kechris (1996, theorem 7.3.1), the space Y can be decomposed into a union of a collection $\{A_\alpha : \alpha \in \omega_1\}$ of pairwise disjoint invariant Borel sets such that $E_G \restriction A_\alpha$ is Borel for every ordinal α; moreover, this decomposition is absolute between various transitive models of ZFC with the same $\aleph_1$.

Let M be a countable elementary submodel of a large enough structure and let $C \subset B$ be the Borel I-positive subset of all P_I-generic points over M in the set B. This is a Borel I-positive set by the properness of the quotient poset

P_I, and if I is c.c.c. then in fact $B \setminus C \in I$. For every point $x \in C$, we have $M[x] \models f(x) \in \bigcup\{A_\alpha : \alpha \in M \cap \omega_1\}$. Absoluteness considerations imply that $f''C \subset \bigcup_{\alpha \in \omega_1 \cap M} A_\alpha$. The equivalence relation E_G restricted on the latter set is Borel, and so $E \restriction C$ is Borel. □

Theorem 4.27 *Suppose that X is a Polish space, I a σ-ideal on it such that the quotient poset P_I is proper, $B \in P_I$ an I-positive Borel set, and E an analytic equivalence relation on B with all classes countable. Then there is a Borel I-positive set $C \subset B$ such that $E \restriction C$ is Borel.*

Proof Note that the statement "all classes of E are countable" is $\mathbf{\Pi}^1_2$, and therefore holds in the P_I extension as well by the Shoenfield absoluteness. Thus, $B \Vdash [\dot{x}_{\text{gen}}]_E$ is countable, and by Fact 2.51(*i*) we may thin out B if necessary to find a Borel function $f : B \to B^\omega$ such that $B \Vdash \dot{f}(\dot{x}_{\text{gen}})$ enumerates $[\dot{x}_{\text{gen}}]_E$. Let M be a countable elementary submodel of a large structure containing B, and let $C = \{x \in B : x \text{ is } P_I\text{-generic over } M\}$; the set C is Borel and I-positive by the properness assumption. We claim that for all $x \in C$, $\text{rng}(f(x)) = [x]_E$, showing that $E \restriction C$ is Borel: for every $x, y \in C$, $x \; E \; y$ holds if and only if $y \in \text{rng}(f(x))$. Indeed, if $x \in C$ then $f(x)$ enumerates the set $[x]_E$ in the model $M[x]$ by the forcing theorem, and $[x]_E^{M[x]} = [x]_E$ by the absoluteness of the coanalytic statement "$\forall y \in X \; (\neg x \; E \; y) \vee y \in \text{rng}(f(x))$" between $M[x]$ and V. □

The difference between Theorems 4.27 and 4.26 and Example 4.25 is that in the theorems the equivalence classes are Borel. This motivates the following question:

Question 4.28 Suppose that X is a Polish space, I a σ-ideal on it such that the quotient poset P_I is proper, $B \in P_I$ an I-positive Borel set, and E an analytic equivalence relation with Borel classes. Is there a Borel I-positive set $C \subset B$ such that $E \restriction C$ is Borel?

5

Games and the Silver property

5.1 Integer games connected with σ-ideals

Many of the σ-ideals considered in this book have integer games associated with them. As a result, they satisfy several interconnected properties, among them the selection property that will be instrumental in upgrading the canonization results to Silver-style dichotomies for these σ-ideals.

The subject of integer games and σ-ideals was treated in Zapletal (2008) rather extensively, but on a case-by-case basis. In this section, we provide a general framework, show that it is closely connected with uniformization theorems, and prove a couple of dichotomies under the assumption of the Axiom of Determinacy.

To help motivate the following definitions, we will consider a simple task. Let I be a collection of subsets of a Polish space X, closed under subsets. Suppose that I has a *basis*, a Borel set $B \subset \omega^\omega \times X$ such that a subset of X is in I iff it is covered by a vertical section of B. Let $A \subset X$ be a set, and consider an infinite game in which Player I produces a point $y \in \omega^\omega$ and Player II a point $x \in X$. Player II wins if $x \in A \setminus B_y$. Certainly, if $A \in I$ then Player I has a winning strategy that completely disregards moves of Player II – just producing the vertical section of the basis which covers the set A. We want to set up the game so that if $A \notin I$ then Player I has no winning strategy. In order to do this, we must mask Player II's point so that essentially nothing can be inferred about it from Player II's moves. To do this, we use a *scrambling tool*, a Borel function $f : \omega^\omega \to X$ (possibly of high Baire class) and let Player II play an integer sequence $z \in \omega^\omega$ and recover his point x as $x = f(z)$. We will also need to consider the unraveled version of this game as in the following definition.

Definition 5.1 Let $B \subset \omega^\omega \times X$ be a set, $f : \omega^\omega \to X \times \omega^\omega$ be a function, and $C \subset X \times \omega^\omega$ be a set. The game $G(C)$ (the parameters B, f are typically clear from the context) between Players I and II is played as an integer game in which Player I produces a sequence $y \in \omega^\omega$, Player II produces a function $z \in \omega^\omega$, and Player I wins if either $f(z) \notin C$ or the first coordinate of the point $f(z)$ belongs to B_y.

Definition 5.2 Let I be a collection of subsets of a Polish space X closed under subset. We say that I is a *game ideal* if there is:

(i) a *Borel basis* $B \subset \omega^\omega \times X$ for I – a Borel set all of whose vertical sections are in I and such that every set in I is a subset of one of the vertical sections in B;
(ii) a *scrambling tool* – a Borel function $f : \omega^\omega \to X \times \omega^\omega$,

such that for every closed set $C \subset X \times \omega^\omega$, Player I has a winning strategy in the game $G(C)$ if and only if the projection of the set C to the space X is in I.

Note that the games $G(C)$ as in the definition above are all Borel and therefore determined by a result of Martin (1985). The notion is going to be used only for σ-ideals in this book, but it appears to be useful even in the more general context of hereditary collections of subsets of Polish spaces.

The meaning of the definition must be illustrated using examples. The existence of a Borel basis for a σ-ideal is typically easy to verify. Thus, the σ-ideal of countable subsets of $X = 2^\omega$ has a Borel basis B naturally indexed by the space $(2^\omega)^\omega$, and $\langle y, x\rangle \in B \leftrightarrow \exists n\ x = y(n)$. To produce a Borel basis B for the σ-ideal of meager subsets of $X = 2^\omega$, note that the collection Z of closed nowhere dense subsets of 2^ω is G_δ in $K(2^\omega)$ and therefore Polish, index the basis by the space Z^ω, and let $\langle y, x\rangle \in B \leftrightarrow \exists n\ x \in y(n)$. There are many σ-ideals of Borel sets without a Borel basis. One way how that can happen is when the σ-ideal in question is not σ-generated by Borel sets of fixed Borel rank. Such is the case of the E_0-ideal of Section 7.1. A more subtle failure may occur in the case when I is σ-generated by Borel sets of a fixed Borel rank (e. g., closed sets), but the generators cannot be organized into a Borel collection. Such is the case of the σ-ideal σ-generated by closed sets of uniqueness as proved in Debs and Saint-Raymond (1987).

For many σ-ideals – such as those investigated in Zapletal (2008, chapter 4) – an integer game has been manually produced independently of the concerns of the present book, and these games can be trivially restated in the language of a basis and a scrambling tool. The restatement will be made explicit in Example 5.10. We will first prove two general theorems that produce the scrambling tools for various ideals just from certain abstract descriptive properties of the ideals. The construction of the tools is closely connected with

Borel uniformization theorems. The first example we give is associated with the Arsenin–Kunugui theorem (Kechris 1994, theorem 18.18).

Theorem 5.3 *Suppose that I be a σ-ideal on a Polish space X which has a Borel basis consisting of G_δ-sets and, moreover, if $C \subset X \times \omega^\omega$ is a closed set with I-positive projection then C has a compact subset with an I-positive projection. Then I is a game ideal.*

For example, the σ-ideal of ϕ-null sets for any subadditive outer regular capacity ϕ on a Polish space is a game ideal by the Choquet theorem (Kechris 1994, theorem 30.13) and the above theorem.

Proof Let $B \subset \omega^\omega \times X$ be a Borel set with I-small G_δ vertical sections such that every I-small sets is a subset of a vertical section of B. Let W be the space of all compact subsets of $X \times \omega^\omega$. Use the Arsenin–Kunugui uniformization theorem (Kechris 1994, theorem 18.18) to find a Borel function $g : \omega^\omega \times W \to X \times \omega^\omega$ such that whenever $\langle y, w\rangle \in \omega^\omega \times W$ is a pair such that B_y does not cover the projection of w then $g(y) \in w$ is a point whose first coordinate is not in B_y. Let S be the space of all strategies for Player I in integer games, and let $h : \omega^\omega \to S \times W$ be a Borel bijection. Finally, let $f(z) = g(\sigma * z, w)$ for every $z \in \omega^\omega$, where w is such that $h(z) = \langle \sigma, w\rangle$. We claim that f works as required.

Indeed, if $C \subset X \times \omega^\omega$ is a closed set, then either its projection is in the σ-ideal I, in which case Player I can win $G(C)$ just by producing $y \in \omega^\omega$ such that $p(C) \subset B_y$. Or, the projection of C is I-positive, and by the assumptions C has a compact subset w with I-positive projection. In such a case, no strategy σ can be winning for Player I in the game $G(C)$: if $z \in \omega^\omega$ is a point such that $h(z) = \langle \sigma, w\rangle$, then the definition of the function f shows that z is a winning counterplay for Player II against the strategy σ. □

The uniformization theorems for sets with nonmeager or non-null sections (Kechris 1994, theorem 18.6) yields the following:

Theorem 5.4 *Let I be a σ-ideal on a Polish space X with a Borel basis. If, on the collection of all analytic sets, I is the intersection of meager ideals for some collection of Polish topologies yielding the same Borel structure, then I is a game ideal. Similarly, if I is the intersection of null ideals for some collection of Borel probability measures, then I is a game ideal.*

Ideals which on the collection of analytic sets are intersections of null ideals are frequently called *polar* in the literature (Debs 1995).

Proof Both the meager and null case are similar. We start with the meager case. Let $B \subset \omega^\omega \times X$ be the Borel basis for the σ-ideal I. Let W be the space of all partial continuous maps from ω^ω to $X \times \omega^\omega$ with G_δ domain with a standard Borel structure on it as in Kechris (1994, definition 2.5 and proposition 2.6). Use the nonmeager uniformization theorem (Kechris 1994, theorem 18.6) to find a Borel function $g : \omega^\omega \times W \to \omega^\omega$ such that whenever $\langle y, w\rangle \in \omega^\omega \times W$ is a pair such that the set $\{u \in \omega^\omega$: the first coordinate of $w(u)$ is not in $B_y\}$ is nonmeager in ω^ω, then the functional value $g(y, w)$ picks an element from this nonmeager set. Let S be the space of all strategies for Player I in integer games, and let $h : \omega^\omega \to S \times W$ be a Borel bijection. Let $f(z) = g(\sigma * z, w)$ for every $z \in \omega^\omega$, where $h(z) = \langle \sigma, w\rangle$. The function f is the scrambling tool we need.

To verify this, suppose that $C \subset X \times \omega$ is a closed set whose projection is not in I, and suppose that $\sigma \in S$ is a strategy for Player I; we must produce a winning counterplay for Player II in the game $G(C)$. The assumption on the σ-ideal I yields a topology t on the space X with the same Borel structure such that the σ-ideal I_t of t-meager sets is a superset of I, and $\text{proj}(C) \notin I_t$. The set $\text{proj}(C)$ is analytic, it has the Baire property in the topology t, and so it is comeager in some nonempty t-open set O. There is a Borel isomorphism $k : \omega^\omega \to O$ which carries the meager ideal on ω^ω to I_t. The Jankov–von Neumann uniformization theorem yields a partial Baire-measurable function $l : \omega^\omega \to \omega^\omega$ such that whenever $u \in \omega^\omega$ is a point with $k(u) \in \text{proj}(C)$ then $l(u)$ is defined and $\langle k(u), l(u)\rangle \in C$. There is a dense G_δ set P on which both k, l are continuous maps (the continuity of k is considered with respect to the original topology on X). Let $w \in W$ be the function $u \mapsto \langle k(u), l(u)\rangle$ restricted to P. Find $z \in \omega^\omega$ such that $h(z) = \langle \sigma, w\rangle$ and show that z is the desired winning counterplay for Player II. Indeed, the set $B_{\sigma * z}$ is in the σ-ideal I, so in I_t, the set $\{u \in \omega^\omega : g(u) \notin B_{\sigma * z}\}$ is comeager in P, and therefore the functional value $f(z)$ is defined and picks a winning point in C.

The null case is similar, using the isomorphism theorem for measures (Kechris 1994, theorem 17.41) and the Borel uniformization for Borel sets with vertical sections of nonzero mass (Kechris 1994, corollary 18.7). □

It is now time to expose a long and not at all exhaustive list of game σ-ideals.

Example 5.5 Every $\mathbf{\Pi}^1_1$ on $\mathbf{\Sigma}^1_1$ σ-ideal I such that I has a Borel basis, every I-positive analytic set has a Borel I-positive subset, and the quotient poset P_I is proper and preserves Baire category, is a game ideal.

Proof Zapletal (2008, corollary 3.5.5) shows that on the collection of analytic sets, the σ-ideal I is an intersection of meager ideals. Theorem 5.4 applies to show that I is a game ideal. □

Example 5.6 Whenever E is a Borel equivalence relation on a Polish space X and I is the σ-ideal generated by the equivalence classes of E, then I is a game σ-ideal.

Proof The σ-ideal I certainly has a Borel basis. I is also the intersection of meager ideals as a very special case of Example 5.5. This can be also proved directly from the Silver dichotomy. We must show that for every analytic I-positive set $A \subset X$ there is a Polish topology τ yielding the same Borel structure such that every I-small set is τ-meager and A is not τ-meager. In order to do this, invoke Corollary 2.15 to find a nonempty perfect set $C \subset A$ consisting of pairwise E-inequivalent elements. Let $f : X \to 2^\omega$ be any Borel bijection such that $f''C$ is a dense G_δ subset of 2^ω, and let τ be the pullback of the topology on 2^ω to X. Thus, C is comeager in this topology, singletons are τ-meager, and every E-equivalence class has intersection of size at most 1 with the comeager set τ and therefore it is τ-meager. Theorem 5.4 shows that I is a game ideal. □

Example 5.7 The σ-ideal I σ-generated by sets of finite h-dimensional Hausdorff measure of any fixed dimension h, on any fixed compact metric space, is a game ideal.

Proof It is not difficult to check from the definitions that the σ-ideal I has Borel basis. The integer game has been isolated in Zapletal (2008, claim 4.4.6). In a great number of cases, one can provide a different abstract argument. For example, if the underlying compact metric space is Euclidean, then every analytic set of positive mass has a compact subset of finite positive mass, and this implies that the σ-ideal I is an intersection of null ideals by Zapletal (2008, example 3.6.4). Theorem 5.4 then implies that I is a σ-ideal. □

Example 5.8 The σ-ideal of sets of zero mass for an outer regular capacity c on a Polish space X is a game ideal.

Proof The σ-ideal I has a Borel basis consisting of G_δ sets. The Choquet capacitability theorem (Kechris 1994, theorem 30.13) shows that the assumptions of Theorem 5.3 are satisfied and so I is a game ideal. In certain special cases, another argument is possible. Consider the special case of a strongly subadditive capacity c. Another theorem of Choquet (Debs 1995) shows that c is an envelope of measures, which implies that I is an intersection of null ideals. Theorem 5.4 then implies that I is a game ideal. Many related integer games have been manually produced in Zapletal (2008, section 4.3). □

Example 5.9 The σ-ideal I of σ-porous sets on any Polish metric space is a game ideal.

Proof The σ-ideal has a Borel basis essentially by definition. The requisite integer games have been produced manually in Zapletal (2008, theorem 4.2.3). A different abstract argument will first show that the quotient poset P_I is proper and preserves Baire category, so the σ-ideal I is an intersection of meager ideals by Example 5.5, and so it is a game ideal by Theorem 5.4. Humke and Preiss (1985) show that the σ-porous ideal on the real line is not an intersection of null ideals. □

Example 5.10 The Laver ideal I on ω^ω is a game ideal.

Proof This is an example of a σ-ideal which does not satisfy the assumptions of Theorem 5.3 or 5.4 and so it is necessary to resort to manual construction of the integer games. This has been done, for example, in Zapletal (2008, section 4.5.2). We will take the opportunity to show how this construction can be translated into the language of Definition 5.2.

The σ-ideal I is generated by sets $A_g = \{x \in \omega^\omega : \exists^\infty n\ x(n) \notin g(x \restriction n)\}$ as g varies over all functions from $\omega^{<\omega}$ to ω. This obviously yields a Borel basis for I: just find a bijection between ω and $\omega^{<\omega}$, extend it to a continuous bijection h between ω^ω and $\omega^{\omega^{<\omega}}$, and let $B \subset \omega^\omega \times \omega^\omega$ be described by the formula $B_y = A_{h(y)}$.

Now revisit the Laver game exhibited for example in Zapletal (2008, section 4.5.2). Suppose that $C \subset \omega^\omega \times \omega^\omega$ is a closed set, and consider the game $G^*(C)$ in which Player II first indicates a finite sequence $t \in \omega^\omega$, and then in each round i Player I plays $n_i \in \omega$, Player II responds with $m_i > n_i$, and at some rounds of his choice Player II also produces another natural number k_i. Player II wins if $\langle (t^\frown m_0, m_1, \ldots), (k_0, k_1, \ldots)\rangle \in C$.

Claim 5.11 (Zapletal 2003) *Player I has a winning strategy in the game $G^*(C)$ if and only if the projection of C is I-positive.*

Proof If $p(C) \in I$, $p(C) \subset A_g$ for some function $g : \omega^{<\omega} \to \omega$, then Player I wins by simply indicating $m_i = g(t_i)$ where $t_i = t_0^\frown \langle m_j : j \in i\rangle$. On the other hand, if σ is a winning strategy for Player I then let $g : \omega^{<\omega} \to \omega$ be any function such that $g(t)$ is a number larger than all values of $\sigma(p)$ as p ranges over the finitely many plays in which Player II produces the sequence t as the concatenation of his moves and all the numbers k_i played are smaller than $|t|$. We claim that $p(C) \subset A_g$, and that will complete the proof of the claim.

Suppose for contradiction that there is some point $\langle x, y\rangle \in C$ such that $x \notin A_g$, and let $j \in \omega$ be such that $\forall i \geq j\ x(i) > g(x \restriction i)$. Consider the counterplay against the strategy σ in which Player II starts with $t = x \restriction j$ and then proceeds to create points x, y, playing the numbers $k_i = y(i)$ only at rounds of index greater than k_i. The definition of the function g shows that all moves of Player II in this counterplay are legal, and in the end Player II wins as $\langle x, y\rangle \in C$. This contradicts the assumption that σ is a winning strategy. □

To translate the game G^* into the present context, let $\pi : \omega^{<\omega} \to \omega$ be an arbitrary bijection, and define the basis $B \subset \omega^\omega \times \omega^\omega$ for the σ-ideal I by $\langle y, x\rangle \in B$ if $x \in A_{y\circ\pi}$. The scrambling tool f is defined as $f(z) = \langle x, y\rangle$ where $x(i) = z'(2i)-1$, $y(i) = z'(2i+1)-1$ and z' is the sequence of nonzero entries of $z \in \omega^\omega$. It is a mechanical matter to translate the proof of the previous claim to see that this basis and scrambling tool work as desired. □

Faced with such a powerful machinery to produce integer games, it appears hard to identify a σ-ideal with a Borel basis that is *not* a game ideal, and we have no such example. An obvious complexity computation shows that σ-ideals with Borel basis are in general $\mathbf{\Pi}^1_2$ on $\mathbf{\Sigma}^1_1$, while the game ideals are simpler:

Definition 5.12 A σ-ideal I on a Polish space X is $\mathbf{\Delta}^1_2$ *on* $\mathbf{\Sigma}^1_1$ if for every analytic set $A \subset 2^\omega \times X$ the collection $\{y \in 2^\omega : A_y \in I\}$ is $\mathbf{\Delta}^1_2$.

Theorem 5.13 *All game ideals are* $\mathbf{\Delta}^1_2$ *on* $\mathbf{\Sigma}^1_1$.

Proof Let $A \subset 2^\omega \times X$ be an analytic set, with $C \subset 2^\omega \times X \times \omega^\omega$ a closed set projecting to it. Then, for every $z \in 2^\omega$, the vertical section A_z is in the ideal I if and only if Player I has a winning strategy in the game $G(C_z)$ if and only if Player II has no winning strategy in the game $G(C_z)$. These are a $\mathbf{\Sigma}^1_2$ and a $\mathbf{\Pi}^1_2$ formula respectively. □

Question 5.14 Is every σ-ideal with a Borel basis a game ideal? Is every $\mathbf{\Delta}^1_2$ on $\mathbf{\Sigma}^1_1$ σ-ideal with Borel basis a game ideal? Is there a simple uniformization-type property of a σ-ideal that is equivalent to an existence of a game for I?

5.2 Determinacy conclusions

In this section, we will use the games to derive a couple of interesting corollaries in the context of mild large cardinal assumptions or the Axiom of Determinacy.

Theorem 5.15 (ZF + Axiom of Determinacy) *Let I be a game ideal on a Polish space X. If every coanalytic set is either in I or contains an analytic subset not in I, then* every *set is either in I or contains an analytic subset not in I.*

Proof Let X be the Polish domain of the σ-ideal I, let $B \subset \omega^\omega \times X$ be a Borel basis of I and f a scrambling tool as in Definition 5.2. Let $A \subset X$ be an arbitrary set, consider the set $C = A \times \omega^\omega$, and the associated game $G(C)$. By the determinacy assumptions, either Player I has a winning strategy or Player II does.

If Player II has a winning strategy σ, let $A' = \{x \in X : \exists y, z, v \in \omega^\omega\ z = \sigma * y \text{ and } f(z) = (x, v)\}$. This is clearly an analytic set, and since the strategy σ is winning, it must be a subset of the set A. It also must be I-positive, since whenever B_y is a generator of the σ-ideal I for some $y \in \omega^\omega$, the first coordinate of the point $f(\sigma * y)$ is in the set A' and not in the set B_y. Thus, the set A contains an I-positive analytic subset.

If Player I has a winning strategy σ, let $A'' = \{x \in X : \forall y, z, v \in \omega^\omega\ y = \sigma * z \text{ and } f(z) = (x, v) \text{ implies } x \in B_y\}$. This is a coanalytic set, and certainly $A \subset A''$ since σ was a winning strategy for Player I. We claim that that $A'' \in I$, and so $A \in I$. Otherwise, by the assumption of the theorem, A'' would contain an analytic I-positive subset, which would be a projection of a closed set $D \subset X \times \omega^\omega$. Player II would have a winning strategy τ in the game $G(D)$. The struggle between the strategies σ and τ in the game $G(D)$ would result in a point $x \in A''$ contradicting the definition of the set A''.

In both cases, we verified the conclusion of the theorem. □

Theorem 5.15 has a substantial weakness in that for all game ideals we know, the integer games can be readjusted to prove the assumption of the theorem under AD. In the special but frequent case of σ-ideals such that the quotient forcing is proper and preserves Baire category, the assumption on coanalytic sets can be proved to hold from AD as shown in Clemens and Zapletal (2012). However, we do not know for example whether for every outer regular capacity ϕ, every coanalytic subset of ϕ-positive capacity contains an analytic subset of positive capacity.

We will now discuss one interesting case where the verification of the assumption of Theorem 5.15 requires a nontrivial amount of work: the σ-ideal I_E generated by the equivalence classes of E, where E is a given Borel equivalence relation on a Polish space X.

Theorem 5.16 *Suppose that ω_1 is inaccessible to the reals. Let E be a Borel equivalence relation on a Polish space X. Then every $\mathbf{\Sigma}^1_2$ set either intersects only countably many equivalence classes of E, or else contains a nonempty perfect subset of pairwise inequivalent elements.*

Note that the large cardinal assumption is necessary to obtain the conclusion already for coanalytic sets. If ω_1 is not inaccessible to the reals, then there is an uncountable coanalytic set containing no perfect subset, and so the statement will fail already for $E = \mathrm{id}$. Unlike the Silver dichotomy, the conclusion fails for coanalytic equivalence relations. Consider the space X of all trees on ω, and let E be the coanalytic relation defined by $x\ E\ y$ if and only if $x = y$ or there is no level preserving homomorphic embedding of x into any of the trees $y \restriction t$ where $t \in y$ is a node distinct from the root, and vice versa, there is no such embedding of y into any of the trees $x \restriction t$. It is not difficult to see that this is an equivalence relation that connects wellfounded trees with the same rank, and illfounded trees form singleton equivalence classes. The set A of wellfounded trees is coanalytic and contains uncountably many equivalence classes. Moreover, every perfect subset of it meets only countably many of the classes by the boundedness theorem, and so it cannot consist of pairwise inequivalent elements.

The proof of Theorem 5.16 uses the following result of Stern. Whenever M is a transitive model of set theory, $P \in M$ a forcing notion and $\dot{B} \in M$ a P-name for a Borel subset of X, write $\mathrm{Re}(M, P, \dot{B})$ for the set $\{x \in X : M[x] \models P \Vdash \check{x} \in \dot{B}\}$.

Fact 5.17 (Stern (Stern 1984)) *Let M be a transitive model of set theory, $P \in M$ a forcing notion, $\beta \in M$ an ordinal, an $\dot{B} \in M$ a P-name for a Borel set of rank $\leq \check{\beta}$. Then:*

(i) the set $\mathrm{Re}(M, P, \dot{B}) \subset X$ *is Borel;*
(ii) if $h \subset \mathrm{Coll}(\omega, \beth^M_{\beta+1})$ is a filter generic over M, then the set $\mathrm{Re}(M, P, \dot{B})$ *has a Borel code in the model $M[h]$.*

Proof of Theorem 5.16 We will start with the case of a coanalytic set $A \subset X$. Fix a real parameter $z \in 2^\omega$ such that $E \in \Delta^1_1(z)$, $A \in \Pi^1_1(z)$ and let $\leq$ be a $\Pi^1_1(z)$ rank on A.

Consider the model $L[z]$. The large cardinal assumption implies that ω_1^V is an inaccessible cardinal in $L[z]$: it is certainly regular, and if it was a successor of some cardinal γ of $L[z]$, then $\omega_1^V = \omega_1^{L[z][y]}$ for any real y coding a wellordering of ω in ordertype γ, contradicting the large cardinal assumption. This means that for every ordinal $\alpha \in \omega_1$, the set $\mathcal{P}(\alpha) \cap L[z]$ is countable, and so the set $\{g \in \alpha^\omega : g \text{ is } \mathrm{Coll}(\omega, \alpha)\text{-generic over } L[z]\}$ is a dense G_δ subset of the space α^ω. Here α is considered as a discrete space and α^ω as a space with the product topology.

Given a countable ordinal $\alpha \in \omega_1$ and a function $g \in \alpha^\omega$ which is $\mathrm{Coll}(\omega, \alpha)$-generic over $L[z]$, the model $L[z][g]$ contains a Borel code for the Borel set A_α of all points of A of $\leq$-rank $< \alpha$. The Silver dichotomy applied in the model $L[x][g]$ to the equivalence relation E on A_α yields a split into cases:

Case 1 There is a countable ordinal $\alpha \in \omega_1$ and a function $g \in \alpha^\omega$ $\mathrm{Coll}(\omega, \alpha)$-generic over $L[z]$ such that $L[z][g] \models$ "A_α contains a perfect set of pairwise E-unrelated elements." In this case, note that the statement to the right of the $\models$ sign is $\mathbf{\Sigma}^1_2$, and by the Shoenfield absoluteness it transfers to V, showing that the set A contains a perfect set of pairwise E-unrelated elements.

Case 2 If Case 1 fails, then for every countable ordinal $\alpha \in \omega_1$ it is the case that $L[z] \models \mathrm{Coll}(\omega, \alpha) \Vdash$ "$\dot{A}_\alpha$ intersects only countably many $\dot{E}$-classes." For every $\alpha \in \omega_1$ fix $\mathrm{Coll}(\omega, \alpha)$-names $\langle \tau_{n,\alpha} : n \in \omega \rangle$ in the model $L[z]$ such that $L[z] \models \mathrm{Coll}(\omega, \alpha) \Vdash \dot{A}_\alpha \subset \dot{B}_\alpha = \bigcup_n [\tau_{n,\alpha}]_E$. Note that if E is a Borel equivalence relation of some fixed rank β, then $\dot{B}_\alpha \in L[z]$ is a $\mathrm{Coll}(\omega, \alpha)$-name for a Borel set of rank at most $\beta + 1$ (independently of α).

Let $h \subset \mathrm{Coll}(\omega, \beth^{L[z]}_{\beta+2})$ be a filter generic over $L[z]$. By Fact 5.17, the set $\mathrm{Re}(L[z], \mathrm{Coll}(\omega, \alpha), \dot{B}_\alpha) \subset X$ is Borel and has a Borel code in the model $L[z][h]$. Since ω_1^V is still inaccessible in $L[z][h]$, there are only countably many reals in $L[z][h]$ and only countably many Borel codes as well. By a cardinality argument, there is an unbounded set $R \subset \omega_1$ and a fixed Borel set C such that for all ordinals $\alpha \in R$, it is the case that $C = \mathrm{Re}(M, \mathrm{Coll}(\omega, \alpha), \dot{B}_\alpha)$. We will show that $A \subset C$ and that C intersects only countably many E-classes. This will conclude the proof for the case of a coanalytic set A.

For the inclusion $A \subset C$, let $x \in A$ be arbitrary. There is an ordinal $\alpha \in R$ such that the $\leq$-rank of x is smaller than α. Whenever g is a $\mathrm{Coll}(\omega, \alpha)$-generic filter over $L[z][x]$, then g is also $\mathrm{Coll}(\omega, \alpha)$-generic over the smaller model $L[z]$, and $L[z][g] \models$ "every element of A_α is equivalent to one of $\tau_{n,\alpha}/y$." By coanalytic absoluteness this also holds in $L[z][x][g]$, and therefore x is equivalent to one of $\tau_{n,\alpha}/y$. As g was an arbitrary generic, this means that $x \in \mathrm{Re}(M, \mathrm{Coll}(\omega, \alpha), \dot{B}_\alpha)$ and so $x \in C$. Thus, $A \subset C$ as desired.

To prove that the set $C \subset X$ intersects only countably many E-equivalence classes, assume for contradiction that it does not, and use the Silver dichotomy to produce a perfect subset $P \subset C$ of its pairwise E-inequivalent elements. Fix a countable ordinal $\alpha \in \omega_1$ such that $C = \mathrm{Re}(L[z], P, \dot{B}_\alpha)$. Consider the Polish space $Y = \alpha^\omega$ where α has the discrete topology and Y has the product topology, and the maps $f_n : Y \to X$ given by $f_n(g) = \tau_{n,\alpha}/g$. Thus, the maps f_n are defined and continuous on the dense G_δ set of filters generic over $L[z]$. Consider also the set $D \subset Y \times X$ of those points $\langle g, x \rangle$ such that $\exists n\ f_n(g)\ E\ x$. The definition of the set $\mathrm{Re}(L[z], \mathrm{Coll}(\omega, \alpha), \dot{B}_\alpha)$ shows that it is equal to the set $\{x \in X : D^x \text{ is comeager in } Y\}$, while for every $g \in Y$, the vertical section

$D_g \cap P$ can be at most countable. Therefore the set $D \cap (Y \times P)$ contradicts the Kuratowski–Ulam theorem for the product $Y \times P$.

This completes the proof for a coanalytic set A. If A is instead $\mathbf{\Sigma}^1_2$, then it is the image of some coanalytic set A' under a continuous function $f : \omega^\omega \to X$. The conclusion then follows from the application of the coanalytic case to the pullback equivalence relation $E' = f^{-1}E$ and the coanalytic set A'. □

Corollary 5.18 (ZF + DC + Axiom of Determinacy) *Let E be a Borel equivalence relation on a Polish space X. Whenever $A \subset X$ is a set, then either A is covered by countably many E-equivalence classes or A contains a perfect subset of pairwise E-inequivalent elements.*

Proof Write I_E for the σ-ideal generated by the equivalence classes of the relation E. By the Silver dichotomy or more precisely Corollary 2.15, it is enough to show that every I_E-positive set contains an I_E-positive analytic subset. Since the σ-ideal I_E is a game ideal by Example 5.6, Theorem 5.15 shows that it is enough to verify this for coanalytic sets $A \subset X$. However, this is exactly the contents of Theorem 5.16. □

5.3 The selection property

The most important application of game ideals and integer games in this book is the following concept. It will lead towards many Silver-type dichotomies for various σ-ideals.

Definition 5.19 A σ-ideal I on a Polish space X has the *selection property* if for every analytic set $A \subset 2^\omega \times X$ with I-positive sections there is a partial Borel map $g : 2^\omega \to X$ such that $g \subset A$ and $\text{rng}(g) \notin I$.

Theorem 5.20 *All game ideals have the selection property.*

Proof Let I be a game σ-ideal on a Polish space X. Let $B \subset \omega^\omega \times X$ be the Borel basis for I, and $f : \omega^\omega \to X$ be the scrambling tool as in Definition 5.2. For a closed set $C \subset X \times \omega^\omega$ write $G(C)$ for the resulting integer game.

Let $A \subset 2^\omega \times X$ be an analytic set with all vertical sections I-positive, A equal to the projection of some closed set $C \subset 2^\omega \times X \times \omega^\omega$. We will first find a perfect set $Z \subset 2^\omega$ and a continuous function τ with domain Z, assigning each element of Z a winning strategy $\tau(z)$ for Player II in the game $G(C_z)$, where $C_z \subset X \times \omega^\omega$ is the vertical section of C associated with z.

To do this, consider the Sacks forcing P adding the generic point $\dot{z}_{\text{gen}}$ to 2^ω. Since the statement "Player I has no winning strategy in neither game

$G(C_z)$ for $z \in 2^\omega$" is $\mathbf{\Pi}^1_2$ and true in the ground model by the assumptions, it is also true in the Sacks extension by the Shoenfield absoluteness 2.34. In particular, $P \Vdash$ "Player I has no winning strategy in the game $G(C_{\dot{z}_{\mathrm{gen}}})$." Since the game is determined, there is a name $\dot{\tau}$ for a winning strategy for Player II in this game. By the continuous reading of names for Sacks forcing, this name below some condition reduces to a continuous function. Let M be a countable elementary submodel of a large enough structure, and let $Z \subset 2^\omega$ be a perfect set consisting of M-generic Sacks reals below this condition only; thus, the name $\dot{\tau}$ can be viewed as a continuous function e on the set Z. To see that e, Z work as required, let $z \in Z$ and look in the model $M[z]$. There, by the forcing theorem, $e(z)$ is a winning strategy for Player II in the game $G(C_z)$. This is a coanalytic statement about $z, e(z)$ and therefore is absolute between the wellfounded model $M[z]$ and V. Thus, $e(z)$ indeed is a winning strategy for Player II in the game $G(C_z)$.

Now, find any Borel bijection $h : Z \to \omega^\omega$ and let $g : Z \to X$ be the function defined in the following way: $g(z)$ is the first coordinate of the point $f(v)$, where $v \in \omega^\omega$ is the sequence resulting from the application of the strategy $e(z)$ for Player II to the counterplay $h(z)$ of Player I, $v = e(z) * h(z)$. We claim that the function g is as required. It is certainly Borel. For every $z \in Z$, the strategy $e(z)$ is winning for Player II and therefore it produces a point in A_z, so $g \subset A$. Finally, $\mathrm{rng}(g) \notin I$, since whenever B_y is some generator of the σ-ideal I and $y = h(z)$ for some $z \in Z$, the point $g(z) \in X$ belongs to $\mathrm{rng}(g)$ and not to B_y. □

The main application of the selection property in this book – cranking up the total canonization to the Silver property for various σ-ideals – is postponed to Section 5.4. Here, we will show that the selection property can be used to prove total canonization itself in a certain context. The argument is very abstract and uses a number of forcing tricks. First, in a rather surprising turn of events, we show that the selection property rules out most nontrivial intermediate extensions of the P_I extension. Then, in another unexpected twist, the trichotomy theorem (Theorem 3.5) shows that equivalence relations are directly connected to such intermediate extensions, and we conclude that the lack of such extensions implies the total canonization.

In the early versions of this book, this was the path through which total canonization was proved for a great number of σ-ideals. Later, in many cases we found direct combinatorial arguments that avoid this path and shed more light on the whole context.

Theorem 5.21 *Suppose that I is a σ-ideal on a Polish space X such that the quotient forcing P_I is $< \omega_1$-proper in all σ-closed extensions. If I has the* *selection property, then every intermediate forcing extension of the P_I extension is either a c.c.c. extension of the ground model, or it is equal to the whole generic extension.*

The wording of this theorem must be parsed carefully. First of all, σ-closed forcing extensions add no new reals, and therefore the space X, σ-ideal I and the poset P_I remain the same in such extension. Thus, we do not need any definability of the σ-ideal I to interpret it in σ-closed forcing extensions. Second, a forcing P is said to be $< \omega_1$*-proper* if for every countable ordinal α and every $\in$-tower $\langle M_\beta : \beta \in \alpha \rangle$ of countable elementary submodels of a large structure such as $\langle V_\kappa, \in \rangle$ for a sufficiently large ordinal κ, with $P \in M_0$, for every condition $p \in P \cap M_0$ there is a strengthening $q \leq p$ which is simultaneously M_β-master for each ordinal $\beta \in \alpha$. This is a technical strengthening of properness due to Shelah; Ishiu (2005) proved that $< \omega_1$-properness of a poset is equivalent to the Axiom A property of a partial order in forcing sense equivalent to the original poset. It is satisfied for all σ-ideals in this book as proved by a standard transfinite induction on α in each case, with the exception of certain ideals of the form I_S of Section 7.2, where the quotient P_{I_S} is not even proper. Moreover, in each case the proofs just use the definition of the σ-ideals which remains the same in any σ-closed forcing extension, and therefore yield $< \omega_1$-properness in all σ-closed extensions. For the purposes of the theorem, this property does not seem to be replaceable by anything descriptive set theoretic in nature. The conclusion cannot be strengthened to rule out intermediate c.c.c. extensions as such examples as the Cohen forcing (associated with the meager ideal, which is a game ideal and therefore does have the selection property) will show.

Proof Every intermediate forcing extension of the P_I-extension is a forcing extension (Jech 2002, lemma 15.43): there is a complete subalgebra A of the completion $\hat{P}_I$ of P_I such that the intermediate forcing extension the result of adjoining $H \cap A$ to the ground model, where $H \subset \hat{P}_I$ is the generic filter. To prove the theorem, let $p \in P_I$ be a condition and A a complete subalgebra of $\hat{P}_I$; we must find either a condition $q \leq p$ below which A is c.c.c. or otherwise a way to recover the generic filter H from $H \cap A$. To simplify the notation, assume that p is the largest condition and the algebra A is already nowhere c.c.c..

Let $\lambda = \mathfrak{c}$. We claim that A contains a term for a function $f : \lambda \to \lambda$ such that for every condition $p \in P_I$ there is $\beta \in \lambda$ such that the set $\{\gamma \in \lambda : \exists q \leq p\ \dot{f}(\beta) = \gamma\}$ is uncountable. To find the term, let $\langle p_\beta : \beta \in \lambda \rangle$ be an enumeration of P_I, let A_β be an uncountable antichain of conditions in A still compatible with p_β with some fixed enumeration, and let $\dot{f}(\beta) = \gamma$ if the generic filter contains the γ-th element of A_δ, and $\dot{f}(\beta) = \text{trash}$ if the generic filter contains no elements of A_β. We will show that P_I forces that $\dot{x}_{\text{gen}}$ can be recovered from $\dot{f}$; this will be enough. The rest of the argument does not use the complete algebra A anymore.

The following claim describes the way in which $\dot{x}_{\text{gen}}$ is recovered from f:

Claim 5.22 *If $B \subset X$ is a Borel I-positive set, $a \subset \lambda$ a countable set and $g : B \to a^a$ a Borel injection such that $B \Vdash \dot{f} \restriction a = \dot{g}(\dot{x}_{\text{gen}})$, then $B \Vdash V[\dot{f}] = V[\dot{x}_{\text{gen}}]$.*

Proof It is only necessary to argue that $V[f]$ contains the generic point x_{gen}. Note that in the model $V[x_{\text{gen}}]$, the function g is still an injection by analytic absoluteness between V and $V[x_{\text{gen}}]$. By analytic absoluteness between $V[f]$ and $V[x_{\text{gen}}]$, the model $V[f]$ contains a point y such that $g(y) = f \restriction a$—such y exists in the model $V[x_{\text{gen}}]$, namely $y = x_{\text{gen}}$. Now, it must be the case that the point $y \in V[f]$ is equal to x_{gen} since g is injective, and so $x_{\text{gen}} \in V[f]$. □

Choose a σ-closed forcing extension $V[G]$ satisfying the continuum hypothesis; thus $|\lambda| = \aleph_1$ there. In $V[G]$, we will find a Borel set $B \in P_I$, a countable set $a \subset \lambda$, and a Borel injection $g : B \to a^a$ such that $B \Vdash \dot{f} \restriction a = \dot{g}(\dot{x}_{\text{gen}})$. Once this is done, we will finish the proof as follows. First of all, as the extension $V[G]$ contains no new ω-sequences, the objects B, a, g belong to V. Second, since the poset P_I is the same in V and $V[G]$ and the statement to the right of the $\Vdash$ sign is a Δ_0 statement of the P_I-forcing language using names in V, it must be the case that in V it is also the case that $B \Vdash \dot{f} \restriction a = \dot{g}(\dot{x}_{\text{gen}})$. Claim 5.22 applied in V then completes the proof.

The rest of the argument takes place in $V[G]$. The Borel set B is obtained as a set of generics for a countable model. Suppose that M is a countable elementary submodel of a large structure containing $I, \dot{f}$, write $a = M \cap \lambda$, write B_M for the set of all points in X P_I-generic over M, and write $g_M : B_M \to a^a$ for the function defined by $g_M(x) = \{\langle \delta, \gamma \rangle \in a \times a : \text{for some condition } B' \in P_I \cap M \text{ we have } x \in B' \text{ and } B' \Vdash \dot{f}(\delta) = \gamma\}$.

Claim 5.23
(i) B_M is a Borel set and g_M is a Borel function.
(ii) $B_M \Vdash \dot{f} \restriction a = \dot{g}_M(\dot{x}_{\text{gen}})$.

To make sense of the complexity estimate for the function g_M, topologize the set a with discrete topology and the set a^a with the product topology.

Proof (*i*) follows immediately from the definitions. For (*ii*) note that $B_M \Vdash \dot{x}_{\text{gen}} \in \dot{B}_M$ and that the definition of the function g_M is identical to the evaluation function of the name $\dot{f}$. □

We will find a countable elementary submodel M and a Borel I-positive set $B \subset B_M$ such that the function $g_M \restriction B$ is injective. The set B is obtained by an application of the selection property of I. Note that the selection property is absolute between V and $V[G]$ as it speaks only about Borel sets, and these are the same in both models. We will find a Borel set $C \subset 2^\omega \times X$ such that the vertical sections of C are I-positive and are subsets of B_M, such that for distinct vertical sections of C have disjoint g_M-images; in particular, these vertical sections are pairwise disjoint. Once this is done, the selection property yields a partial Borel function $h \subset C$ whose range is I-positive. Since the vertical sections of the set C are pairwise disjoint, the function h is injective, and therefore its range B is a Borel I-positive set. Obviously, $g_M \restriction B$ is injective, and that will complete the proof of the theorem.

The search for C uses a preliminary definition. Say that two countable continuous $\in$-towers $\vec{M}_0$ and $\vec{M}_1$ of countable elementary submodels of a large structure with respective largest models $\max(\vec{M}_0)$ and $\max(\vec{M}_1)$ are *perpendicular* if $\lambda \cap \max(\vec{M}_0) = \lambda \cap \max(\vec{M}_1)$, and for every $x \in X$ which is P_I-generic over each model on $\vec{M}_0$ and every $y \in X$ which is P_I-generic over each model on $\vec{M}_1$, it is the case that $g_{\max(\vec{M}_0)}(x) \neq g_{\max(\vec{M}_1)}(y)$. Note that the functions $g_{\max(\vec{M}_0)}(x)$, $g_{\max(\vec{M}_1)}(y)$ have the same domain, so they must have disagreement in the values.

Claim 5.24 *For every $b \in H_{\mathfrak{c}^+}$, there are perpendicular countable towers $\vec{M}_0$ and $\vec{M}_1$ of countable elementary submodels of $H_{\mathfrak{c}^+}$ such that $b \in \vec{M}_0(0), \vec{M}_1(0)$.*

Proof Let $\vec{M}'_0$ be any continuous ω_1 tower of countable elementary submodels of some large structure such that the first model contains the set b, and let $\vec{M}_1$ be a tower of length 1, containing just one model $N = \max(\vec{M}_1)$, which in turn contains the tower $\vec{M}'_0$. Let $\delta \in \omega_1$ be the intersection of this single model with ω_1 and let $\vec{M}_0 = \vec{M}'_0 \restriction \delta + 1$. We claim that the two towers $\vec{M}_0, \vec{M}_1$ work as required in the claim.

First of all, observe that $\lambda \subset \bigcup\{\vec{M}_0(\alpha) : \alpha \in \omega_1\}$. As $|\lambda| = \aleph_1$ in $V[G]$, the first model in the tower must contain some bijection ρ between ω_1 and λ. Since the union contains ω_1 as a subset, it must also contain $\lambda = \rho''\omega_1$ as a subset by elementarity. This observation is the only reason for passing to the extension $V[G]$.

It follows that $\lambda \cap \max(\vec{M}_0) = \lambda \cap N$ holds. Since $\bigcup\{\vec{M}_0(\alpha) : \alpha \in \omega_1\}$ contains λ as a subset, the elementarity of the model N implies that it contains exactly the same elements of λ as $\bigcup_{\alpha\in\delta} \vec{M}_0(\alpha) = \max(\vec{M}_0)$.

Second, consider the set $D = \{\vec{M}'_0(\alpha) \cap \lambda : \alpha \in \omega_1\}$. This is a closed unbounded set of countable subsets of λ. We claim that

$P_I \Vdash \exists a \in D\ \dot{f}''(a) \not\subset a$. In V, by the choice of the function $\dot{f}$, for every condition $p \in P$ there is an ordinal $\beta \in \lambda$ such that $\dot{f}(\beta)$ has still uncountably many possible values below P, and the set of possible values is still uncountable in $V[G]$. Thus, we can choose any $a \in D$ containing β and a condition $q \leq p$ such that $q \Vdash \dot{f}(\check{\beta}) \notin \check{a}$, confirming the forcing statement.

It follows that for every P_I-generic point $x \in X$ over the model N, there is $\alpha \in \delta$ such that the set $\vec{M}_0'(\alpha) \cap \lambda$ is not closed under the function $\dot{f}/x$, since N contains the set D as an element. On the other hand, if y is P_I-generic point for all the models on the sequence $\vec{M}_1$, all the sets $\{\vec{M}_0(\alpha) \cap \lambda : \alpha \in \delta\}$ are closed under the function $\dot{f}/y$ by the genericity. This concludes the proof of the claim. □

Now, by induction on $n \in \omega$ build perpendicular countable towers $\vec{M}_{n,0}$ and $\vec{M}_{n,1}$ and the towers $\vec{M}_{n-1,0}, \vec{M}_{n-1,1}$ belong to the first models on the n-indexed towers. Let M be the union of all models on the towers; this is an increasing union of a sequence of countable elementary submodels of the large structure, and so is elementary as well. Define $C \subset 2^\omega \times X$ by $\langle y, x\rangle \in C$ if x is P_I-generic over every model on all the towers $\vec{M}_{n,y(n)}$ for $n \in \omega$. This is a Borel set. The poset $P_I^V = P_I^{V[G]}$ is $< \omega_1$-proper by the assumptions, and so the vertical sections of the set C are I-positive. We claim that vertical sections of C have disjoint g_M-images as desired. To verify the disjointness, let $y_0 \neq y_1$ and $x_0, x_1 \in X$ be such that $\langle y_0, x_0\rangle, \langle y_1, x_1\rangle \in C$ and $n \in \omega$ is such that $y_0(n) = 0 \neq 1 = y_1(n)$, then $g_{\max(\vec{M}_n^0)}(x_0) \subset g_M(x_0)$, $g_{\max(\vec{M}_n^1)}(x_1) \subset g_M(x_1)$. By perpendicularity, $g_{\max(\vec{M}_n^0)}(x_0) \neq g_{\max(\vec{M}_n^1)}(x_1)$, and so $g_M(x_0) \neq g_M(x_1)$. This ends the proof. □

Note that (regardless of the selection property) the proof yields a perfect collection of pairwise disjoint conditions in P_I as soon as P_I is nowhere c.c.c. This should be compared with the work of Judah *et al.* (1994, theorem 6.1), where a non-c.c.c. suitably definable poset is produced which has no perfect antichain, and the construction can produce a δ-proper poset for any ordinal $\delta \in \omega_1$ fixed beforehand.

Theorem 5.25 *Suppose that I is a $\mathbf{\Pi}^1_1$ on $\mathbf{\Sigma}^1_1$ σ-ideal on a Polish space X such that the quotient forcing P_I is $< \omega_1$-proper in all σ-closed extensions. If I has the rectangular Ramsey and selection properties, then I has total canonization for analytic equivalence relations.*

The $\mathbf{\Pi}^1_1$ on $\mathbf{\Sigma}^1_1$ part of the assumptions is used only to weed out possible c.c.c. intermediate extensions of the P_I-extension in the application of the trichotomy theorem. The selection property assumption allows for cranking up the conclusion to the much stronger Silver property of the σ-ideal I, as spelled out in Section 5.4.

Proof Let $B \subset X$ be a Borel I-positive set and E an analytic equivalence relation on it. Let $x \in B$ be a V-generic point for the poset P_I. The trichotomy Theorem 3.5 yields three possibilities:

Case 1 If $V[x] = V[x]_E$ then there is an I-positive Borel set $C \subset B$ consisting of pairwise inequivalent points.

Case 2 $V \neq V[x]_E \neq V[x]$ is impossible, since the assumptions imply that there is no model of ZFC strictly between the ground model and its P_I-extension. Namely, Theorem 5.21 (using the selection property) shows that such an intermediate model would have to be c.c.c. and Corollary 2.63 (using the rectangular Ramsey property) shows that a nontrivial c.c.c. extension cannot occur.

Case 3 If $V[x]_E = V$, then there is a Borel I-positive set $C \subset B$ on which E is ergodic. We claim that C must contain a Borel I-positive subset consisting of pairwise E-equivalent elements. For if every class of $E \restriction C$ was in the σ-ideal I, the rectangular property would have to produce I-positive Borel sets $C_0, C_1 \subset C$ such that $(C_0 \times C_1) \cap E = 0$; this would contradict the ergodicity.

In all three cases, the proof of the theorem is complete. □

Reviewing the assumptions of the previous theorem, it becomes clear that very many interesting ideals are $\mathbf{\Pi}^1_1$ on $\mathbf{\Sigma}^1_1$, $< \omega_1$-proper, and have the selection property by virtue of being game ideals. Such ideals appear in all the main sections of Zapletal (2008, chapter 4), and include, among others, the σ-ideals generated by Borel collection of closed sets, σ-ideals generated by the sets of finite mass for a fixed Hausdorff measure, or the σ-ideals of sets of zero mass for a fixed pavement submeasure. The main problem on the path to total canonization for these σ-ideals appears to be the verification of the rectangular Ramsey property. The list of game ideals for which we cannot decide status of the rectangular Ramsey property and the total canonization includes among many others, the σ-ideal σ-generated by sets of finite mass for a fixed Hausdorff measure, the σ-ideal σ-generated by sets of continuity of the Pawlikowski function, and the σ-ideal of sets of Newtonian capacity zero.

However, the connection between the rectangular Ramsey property and canonization is not as tight as the previous theorem may suggest. Total canonization may hold even where the rectangular property fails – such is the case of the σ-ideal σ-generated by closed sets of zero Lebesgue mass, see

Theorem 6.11. Rectangular property and total canonization may fail and still there may be useful canonization features – such is the case for the Laver ideal of Section 6.3.

5.4 A Silver-type dichotomy for a σ-ideal

The main purpose of the selection property and the integer games in this book is that they can be used to upgrade total canonization results to Silver-style dichotomies, which are significantly more striking. Recall from the introduction:

Definition 5.26 A σ-ideal I on a Polish space X has the *Silver property* for a class $\mathbf{E}$ of equivalence relations if for every I-positive analytic set $B \subset X$ and every equivalence relation $E \in \mathbf{E}$ on the set B, either B can be decomposed into countably many equivalence classes and an I-small set or there is a Borel I-positive set $C \subset B$ such that $E \restriction C = \text{id}$. If $\mathbf{E}$ is equal to the class of all Borel equivalence relations, it is dropped from the terminology and we speak of the Silver property of I instead.

This should be compared with the classical Silver dichotomy (Fact 2.14), which establishes the Silver property for the ideal of countable sets. Observe that unlike total canonization, the Silver property introduces a true dichotomy: the two options cannot coexist. It also has consequences for undefinable sets: if an equivalence relation $E \in \mathbf{E}$ has an I-positive set $A \subset X$ consisting of pairwise E-inequivalent points, then it has an I-positive *Borel* set $B \subset X$ consisting of pairwise E-inequivalent points, simply because the first clause of the dichotomy cannot hold in such circumstances.

Let us offer a perhaps artificial reading of the dichotomy, which nevertheless fits well with the techniques developed in this book and Zapletal (2008). Given a σ-ideal I on a Polish space X and a Borel equivalence E on X, consider the ideal $I^* \supset I$ σ-generated by E-equivalence classes and sets in I. Then either I^* is trivial, containing the whole space, or else the quotient forcing P_{I^*} is equal to P_I below some condition.

Theorem 5.27 *Suppose that I is a $\mathbf{\Pi}^1_1$ on $\mathbf{\Sigma}^1_1$ σ-ideal on a Polish space X such that every analytic I-positive set contains a Borel I-positive subset and the quotient poset P_I is proper. Let $\mathbf{E}$ be a class of Borel equivalence relations closed under Borel reducibility, containing the smooth equivalence relations. The following are equivalent:*

(i) I has total canonization for $\mathbf{E}$ and the selection property;
(ii) I has the Silver property for $\mathbf{E}$.

If the σ-ideal is $\mathbf{\Delta}^1_2$ on $\mathbf{\Sigma}^1_1$ and generated by Borel sets, then the conclusion of the proposition remains in force under the additional assumption that ω_1 is inaccessible to the reals.

Proof First, the (*i*)→(*ii*) direction. Let $E \in \mathbf{E}$ be an equivalence relation on X, let $A \subset X$ be an analytic set and consider the set $C = \{x \in A : [x]_E \cap A \notin I\}$; this set is again analytic by the definability assumption on I, or if the σ-ideal I is merely $\mathbf{\Delta}^1_2$ on $\mathbf{\Sigma}^1_1$, then the set C is $\mathbf{\Delta}^1_2$.

Case 1 There are only countably many equivalence classes in the set C. In this case, the set C is even relatively Borel in A and $A \setminus C$ is analytic. If $A \setminus C \in I$ then we are in the first alternative of the Silver property. On the other hand, if $A \setminus C \notin I$, then $A \setminus C$ contains an I-positive Borel set $B \subset X$ by the assumptions of the proposition. The equivalence E has I-small classes by the definition of the set C; the total canonization yields a Borel I-positive subset $B' \subset B$ consisting of pairwise F-inequivalent elements, witnessing the second alternative of the Silver property.

Case 2 If Case 1 fails, then Corollary 2.15 in the case of a $\mathbf{\Pi}^1_1$ on $\mathbf{\Sigma}^1_1$ σ-ideal, or Theorem 5.16 in the case of a $\mathbf{\Delta}^1_2$ on $\mathbf{\Sigma}^1_1$ σ-ideal, provides a perfect set $P \subset C$ of pairwise inequivalent points. Let $D \subset P \times X$ be the set defined by $\langle y, x\rangle \in D \leftrightarrow A_x \cap [y]_E \neq 0$; this is a Borel set with pairwise disjoint I-positive vertical sections. The selection property of the σ-ideal I yields a partial Borel function $f : P \to X$ such that $\forall y \in \mathrm{dom}(f)\ \langle y, f(y)\rangle \in D$ and $\mathrm{rng}(f) \notin I$. The analytic set $\mathrm{rng}(f)$ consists of pairwise E-unrelated points can be thinned out to a Borel I-positive set, witnessing the second alternative of the Silver property!

We are left with the (*ii*)→(*i*) implication. The Silver property clearly implies total canonization. If E is a Borel equivalence relation on X and $A \subset X$ is an I-positive analytic set, then either A decomposes into countably many equivalence classes and a set in I, in which case one of the equivalence classes in A must be I-positive, or A contains a Borel I-positive subset consisting of pairwise inequivalent elements. In both cases, the total canonization follows. The verification of the selection property is significantly harder. Let $A \subset 2^\omega \times X$ be an analytic set with I-positive vertical sections. Consider the set $B = \{x \in X :$ the horizontal section A^x is uncountable$\}$; this is an analytic set. There are two cases:

Case 1 Either, $B \in I$. By the first reflection theorem (Fact 2.2) in the $\mathbf{\Pi}^1_1$ on $\mathbf{\Sigma}^1_1$ case or the Borel basis assumption in the $\mathbf{\Delta}^1_2$ on $\mathbf{\Sigma}^1_1$ case, there is a Borel set $C \subset X$ such that $B \subset X \setminus C \in I$. By the first reflection theorem again, there is a Borel set $A' \supset A \cap (2^\omega \times C)$ with nonempty and countable horizontal sections. By the Novikov uniformization theorem (Fact 2.3), there are Borel functions $f_n : C \to 2^\omega$ for $n \in \omega$ such that the inverses of the graphs of f_n cover the set A'. Let F_n be the equivalence relation on $C \subset X$ defined by $x \; F_n \; y$ if $f_n(x) = f_n(y)$. Let $C_n = \{x \in C : \langle f_n(x), x\rangle \in A\}$. The Silver dichotomy for the σ-ideal I yields a split into two subcases:

Case 1a Either there is a number $n \in \omega$ and an I-positive Borel set $D \subset C_n$ such that D consists of F_n-inequivalent elements. In that subcase, the properness of the quotient forcing P_I can be used to thin out D if necessary to get $\mathrm{rng}(f_n \restriction D)$ to be a Borel set by Fact 2.51(*iii*), and the inverse of $f_n \restriction D$ will be the function required in the selection property.

Case 1b Or, for every number $n \in \omega$, the set C_n decomposes into countably many F_n-equivalence classes (sets $\{x \in C_n : f_n(x) = y_n^m\}$ for some points $y_n^m \in 2^\omega$) and an I-small set $D_n \subset \omega$. This subcase is impossible, however: let $z \in 2^\omega \setminus \{y_n^m : n, m \in \omega\}$ be an arbitrary point, and let $x \in A_z \cap C \setminus \bigcup_n D_n$ be another arbitrary point. The ordered pair $\langle z, x\rangle$ must then both belong to A and not belong to it, which yields a contradiction.

Case 2 Or, $B \notin I$. Thinning out the set B, we may assume that it is Borel. By a Shoenfield absoluteness argument, B forces in P_I the set $A^{\dot{x}_{\mathrm{gen}}}$ to be uncountable. By a result of Velickovic and Woodin (1998, theorem 3) in proper extensions adding new reals the set of ground model points does not contain a nonempty perfect subset. Therefore, the proper forcing P_I forces that the analytic set $A^{\dot{x}_{\mathrm{gen}}}$ contains a point $\dot{x}$ which is not in the ground model. Thinning out the set B further if necessary, we may assume that there is a Borel function $f : B \to X$ such that $B \Vdash \dot{x} = \dot{f}(\dot{x}_{\mathrm{gen}})$. Since $\dot{x}$ is forced not to belong to the ground model, preimages of singletons under the function f must be I-small, and another application of the Silver dichotomy for I shows that we may thin out B such that $f \restriction B$ is one-to-one. Thinning out further if necessary, we may assume that for every $x \in B$, $\langle f(x), x\rangle \in A$ and $\mathrm{rng}(f \restriction B)$ is Borel. It is clear that the inverse of $f \restriction B$ yields a function as required in the definition of the selection property. $\square$

6

The game ideals

6.1 σ-ideals σ-generated by closed sets

Following the exposition of Zapletal (2008), the σ-ideals that should be easiest to deal with are those σ-generated by closed sets. Regarding the canonization properties of σ-ideals in this class, on one hand there is the σ-ideal of countable sets which has the Silver property as an immediate corollary of Silver's theorem; on the other hand, there is the σ-ideal of meager sets, which has a very complicated spectrum. Thus, the treatment of this class must make finer distinctions between various σ-ideals. Still, there is a wealth of relevant general information available:

Fact 6.1 *Suppose that I is a σ-ideal on a Polish space X, σ-generated by closed sets. Then:*

(i) *every analytic I-positive set contains a G_δ I-positive subset;*
(ii) *I has the continuous reading of names;*
(iii) *the poset P_I is $< \omega_1$-proper and preserves Baire category;*
(iv) *every intermediate extension of the P_I extension is generated by a single Cohen real;*
(v) *the σ-ideal I is $\mathbf{\Pi}^1_1$ on $\mathbf{\Sigma}^1_1$ if and only if the collection of closed sets in I is coanalytic in the standard Borel space $F(X)$ if and only if I is σ-generated by a coanalytic collection of closed sets.*

Proof (*i*) is a result of Solecki (1994). (*ii*) and (*iii*) are in Zapletal (2008, theorem 4.1.2). (*iv*) was proved in Zapletal (2008, theorem 4.1.7) with some unnecessary assumptions; its present strong form is a special case of Corollary 6.46. For (*v*), it is clear that if I is $\mathbf{\Pi}^1_1$ on $\mathbf{\Sigma}^1_1$ then the set of closed sets in I is coanalytic. On the other hand, if I is σ-generated by a coanalytic collection of closed sets then it is $\mathbf{\Pi}^1_1$ on $\mathbf{\Sigma}^1_1$ by Kechris (1994, theorem 35.38). □

Proposition 6.2 *If I is a $\mathbf{\Pi}^1_1$ on $\mathbf{\Sigma}^1_1$ σ-ideal σ-generated by closed sets then I has the selection property.*

Proof The selection property follows from Theorems 5.4 and 5.20 if the σ-ideal is generated by a Borel collection of closed sets. However, there are important σ-ideals generated by closed sets without such a Borel basis, and the argument from Theorem 5.4 must be restated to work for them. Let $D \subset 2^\omega \times X$ be an analytic set all of whose vertical sections are I-positive. We must find a Borel partial uniformization of D with I-positive range.

Claim 6.3 *There is a perfect set $P \subset 2^\omega$ and a continuous function $g : P \times \omega^\omega \to X$ such that for every $z \in P$, the range of the function $g_z = g(z, \cdot) : \omega^\omega \to X$ is a subset of D_z, and the images of nonempty open sets under this function are I-positive.*

Proof Let $k : \omega^\omega \to 2^\omega \times X$ be a continuous function whose image is the analytic set D. For every $z \in 2^\omega$ let $A_z = f^{-1} D_z$; remove from A_z all the relatively open sets with I-small k-image, obtaining a closed set $\hat{A}_z = [T_z]$ for some pruned tree $T_z \subset \omega^{<\omega}$. Note that the f-images of nonempty relatively open subsets of $\hat{A}_z$ are I-positive, and for every $t \in \omega^{<\omega}$, the statement $t \in T_z$ is analytic since I is $\mathbf{\Pi}^1_1$ on $\mathbf{\Sigma}^1_1$. The map $z \mapsto T_z$ is $\sigma(\mathbf{\Sigma}^1_1)$-measurable, therefore Baire measurable, and so continuous on a dense G_δ-set. For every z there is a map $\pi : \omega^{<\omega} \to \omega^{<\omega}$ such that π preserves extension and length of sequences and for every $t \in \omega^{<\omega}$, $\pi''\{u : t \subset u\} = \{v \in T_z : \pi(t) \subset v\}$. Application of the Jankov–von Neumann uniformization theorem (Fact 2.6) yields a $\sigma(\mathbf{\Sigma}^1_1)$-measurable map $z \mapsto \pi_z$, which is then continuous on further dense G_δ subset. Let $P \subset 2^\omega$ be a perfect subset of this dense G_δ set, and let $g : P \times \omega^\omega \to X$ be the map defined by $g(z, y) = f(\bigcup_n \pi_z(y \restriction n))$. It is immediate to check that the function g has the desired properties. □

Choose an F_σ-universal F_σ set $F \subset P \times X$, and use the nonmeager section uniformization theorem (Fact 2.4) to find a Borel function $h : P \to \omega^\omega$ such that for every $z \in P$, if $g_z^{-1} F_z \subset \omega^\omega$ is meager then $h(z) \notin g_z^{-1} F_z$. Let $f : P \to X$ be the function defined by $f(z) = g(z, h(z))$. This function is certainly Borel, and for each $z \in P$ the value $f(z)$ belongs to D_z. We must show that $\text{rng}(f) \notin I$. Indeed, suppose that $C \in I$ is a set, and cover it by the union of closed sets in I, $C \subset \bigcup_n C_n$. There is $z \in Z$ such that $F_z = \bigcup_n C_n$. Note that the preimages $g_z^{-1} C_n \subset \omega^\omega$ are closed sets with empty interior by the choice of the function g, so they are meager and so is their union $g_z^{-1} F_z$. Then the point $f(z)$ is in $\text{rng}(f) \setminus \bigcup_n C_n$, showing that the set $\text{rng}(f)$ is I-positive. □

The classification of possible intermediate forcing extensions of the P_I extension yields an immediate attractive corollary:

Corollary 6.4 *If I is a σ-ideal σ-generated by closed sets, then the following are equivalent:*

(i) P_I does not add Cohen reals;
(ii) I has total canonization for equivalence relations classifiable by countable structures.

If, in addition, I is $\mathbf{\Pi}^1_1$ on $\mathbf{\Sigma}^1_1$ and (i) or (ii) hold then I has the Silver property for Borel equivalence relations classifiable by countable structures.

Proof We will first argue for an independently interesting fact:

Claim 6.5 *If I is σ-generated by closed sets and P_I is c.c.c. below some condition, then P_I is isomorphic to the Cohen forcing below some stronger condition.*

Proof Let $B \in P_I$ be such that $P_I \restriction B$ is c.c.c. Using Fact 6.1(*i*), thin down B to an I-positive G_δ-set whose intersections with open sets are either empty or I-positive. The closed sets in I are relatively meager in B; thus, writing J for the σ-ideal of meager-in-B subsets of B, we get that $I \subset J$. Note that $B \notin J$ since B is G_δ, therefore Polish and the Baire category theorem applies. We will now argue that there is a condition $C \in P_I$ below B such that $J \restriction C = I \restriction C$; clearly, $P_I \restriction C$ is then isomorphic to the Cohen forcing. Suppose for contradiction that such a set C does not exist. Then the set $D = J \setminus I$ is dense in the poset $P_I \restriction B$, and by c.c.c. of $P_I \restriction B$ it must contain a countable maximal antichain $A \subset D$. Since every set in A is meager in B, so is $\bigcup A$. The Borel set $B \setminus \bigcup A$ is then I-positive and incompatible with every element of A, contradicting the maximality of A. □

Towards the proof of the corollary, assume (*i*) holds. Now, if the poset P_I does not add Cohen reals, then it must be nowhere c.c.c. by Claim 6.5, and it must add a minimal forcing extension by Fact 6.1(*iv*). Corollary 4.10 then yields (*ii*). On the other hand, if (*i*) does not hold, then there is a Borel I-positive set $B \subset X$ and a Borel function $f : B \to 2^\omega$ representing the Cohen real; the f-preimages of meager sets must be I-small. Thus, f is a Katětov map between $I \restriction B$ and the meager ideal, showing that the spectrum of I includes the spectrum of the meager ideal, which is very rich by Theorem 6.23; in particular, it includes E_0, so (*ii*) must fail.

Finally, if the σ-ideal I is in addition $\mathbf{\Pi}^1_1$ on $\mathbf{\Sigma}^1_1$ then (*ii*) implies the Silver property for Borel equivalence relations classifiable by countable structures by Proposition 6.2 and Theorem 5.27. □

The corollary suggests an obvious question:

Question 6.6 Let I be a $\mathbf{\Pi}^1_1$ on $\mathbf{\Sigma}^1_1$ σ-ideal σ-generated by closed sets, such that the quotient poset P_I does not add Cohen reals. Does it follow that I has total canonization for analytic equivalence relations?

The work in this section should be viewed as an effort to give a positive answer to this question by considering certain special classes of σ-ideals first. We will show that calibrated σ-ideals and certain other σ-ideals generalizing the σ-ideal generated compact subsets of the Baire space have the free set property, and therefore do have total canonization for analytic equivalence relations. Furthermore, Corollary 6.36 in Section 6.2 shows that if a σ-ideal σ-generated by closed sets does not add an infinitely equal real (a condition slightly stronger than not adding a Cohen real) then it has total canonization and Silver property for the equivalence relations reducible to E_{K_σ} or $=_J$ for an analytic P-ideal J. This still leaves a good number of interesting open special cases.

6.1.1 Calibrated ideals

In this section, we will prove a general canonization theorem for a class of σ-ideals widely encountered in abstract analysis:

Definition 6.7 (Kechris *et al.* 1987) A σ-ideal on a Polish space X is *calibrated* if for every closed I-positive set $C \subset X$ and countably many closed sets $\{D_n : n \in \omega\}$ in the ideal I there is a closed set $C' \subset C$ disjoint from all the sets $\{D_n : n \in \omega\}$.

Theorem 6.8 *Let I be a $\mathbf{\Pi}^1_1$ on $\mathbf{\Sigma}^1_1$ σ-ideal on a compact metrizable space X, σ-generated by compact sets. If I is calibrated, then I has the free set property.*

Corollary 6.9 *For such a calibrated σ-ideal I, the total canonization for analytic equivalence relations and the Silver property for Borel equivalence relations both hold.*

To illustrate the strength of the theorem, we first provide a list of examples of calibrated σ-ideals on compact spaces. One wide class of calibrated σ-ideals generated by compact sets consists of those σ-ideals with the *covering property*, i. e., those where every I-positive analytic set contains a compact I-positive subset. This includes the σ-ideal of countable sets; the σ-ideal generated by sets of finite packing measure of some fixed dimension (Zapletal 2008, section 4.1.6); the σ-ideal generated by the sets of monotonicity of a given continuous function (Zapletal 2004, lemma 2.3.7); the σ-ideal generated by H-sets (Rajchman 1922; Zapletal 2011); or the σ-ideal generated by compact sets of extended uniqueness (Kechris and Louveau 1989).

Another class of calibrated σ-ideals consists of those σ-ideals σ-generated by all closed sets in a fixed larger σ-ideal with the covering property. This includes the σ-ideal σ-generated by closed Lebesgue measure zero sets; the σ-ideal σ-generated by the compact sets on which E_0 is smooth; the σ-ideal σ-generated by closed σ-porous sets; or the σ-ideal σ-generated by closed sets of capacity 0 for a fixed subadditive capacity.

Other σ-ideals σ-generated by closed sets may satisfy calibration for more sophisticated reasons, such as the σ-ideal σ-generated by closed sets of uniqueness (Debs and Saint-Raymond 1987). Some σ-ideals may fail calibration but satisfy some weakenings of it which still allow our argument to go through without significant changes. Such is the case of stratified calibration and the σ-ideal σ-generated by closed sets of finite dimension in the Hilbert cube (Pol and Zakrzewski 2012).

A simple example of a σ-ideal which is not calibrated is the meager ideal. A more sophisticated class of non-calibrated σ-ideals is discussed in the next section. These σ-ideals in fact remain non-calibrated no matter what the choice of Polish topology on the underlying space as long as the Borel structure is preserved.

All the σ-ideals discussed above have a coanalytic set of generators and therefore are $\mathbf{\Pi}^1_1$ on $\mathbf{\Sigma}^1_1$.

Proof of Theorem 6.8 Let $B \subset X$ be a Borel I-positive set and let $D \subset B \times B$ be an analytic set with I-small vertical sections. We must find a Borel I-positive set $C \subset B$ such that $D \cap (C \times C) \subset \text{id}$.

Use the Burgess–Hillard theorem (Kechris 1994, theorem 35.43) to cover the set D with countably many Borel sets with I-small compact vertical sections. The functions $f_n : B \to K(X)$ for $n \in \omega$ assigning to each point its corresponding vertical section are Borel by Kechris (1994, theorem 28.8) – so we have that for every $x \in B$, $f_n(x) \in I$ and $D_x \subset \bigcup_n f_n(x)$. Thinning out the set B if necessary, we may assume that the functions f_n are all continuous – Fact 6.1(*ii*). Further thinning out the set B, we may assume that it is a G_δ set such that intersections with open sets are either I-positive or empty by Solecki's theorem (Solecki 1994). Let $B = \bigcap_n O_n$ for some open sets $O_n \subset X$.

Construct a suitable tree representation of the set B. By tree induction on $t \in \omega^{<\omega}$, build open sets $O_t \subset X$ such that $s \subset t$ implies $\bar{O}_t \subset O_s$, the diameters of the sets O_t tend to zero in some fixed complete compatible metric for the space X, $O_t \subset O_{|t|}$, $O_t \cap B \neq 0$, and for every neighborhood O with nonempty intersection with $O_t \cap B$ there is $m \in \omega$ such that $O_{t^\frown m} \subset O$. For every point $y \in \omega^\omega$, the intersection $\bigcap_n O_{y \restriction n}$ is a singleton and its single element $g(y) \in X$ belongs to the set B. The function $g : \omega^\omega \to X$ is easily checked to be continuous.

Pick a countable dense set $a \subset \omega^\omega$. Use the calibration assumption to find compact sets K_t included in both O_t and the closure of B that are I-positive and disjoint from all sets $\{f_n(g(y)) : n \in \omega, y \in a\}$. We will find a certain tree $T \subset \omega^{<\omega}$ which splits densely often and

($\star$) for every splitnode $t \in T$, every $x \in K_t$ and every neighborhood P of x there is m such that $t^\frown m \in T$ and $O_{t^\frown m} \subset P$.

Claim 6.10 *Whenever T satisfies ($\star$) then $g''[T]$ is an I-positive set.*

Proof Let $\{H_n : n \in \omega\}$ be compact subsets of X in the σ-ideal I; we must produce a point $y \in \omega^\omega$ such that $g(y) \notin \bigcup_n H_n$. To do that, by induction on $n \in \omega$ build an increasing sequence $\langle t_n : n \in \omega\rangle$ of splitnodes t_n such that $O_{t_{n+1}} \cap H_n = 0$; then $y = \bigcup_n t_n$ will be as required. Now suppose that t_n has been constructed. Since K_{t_n} is not a subset of H_n, there must be a point $x \in K_{t_n} \setminus H_n$. Use ($\star$) to find a number m such that $O_{t^\frown m} \subset X \setminus H$ and let t_{n+1} be some splitnode of T extending $t_n^\frown m$. The induction can proceed. □

Of course, the point of the construction of T will be that the set $g''[T]$ consists of pairwise E-independent elements. Then, it is an I-positive analytic set which has an I-positive G_δ-subset C by a theorem of Solecki (1994). The set C will be as required.

Let $\langle\langle t_i, P_i\rangle : i \in \omega\rangle$ be an enumeration of the product $\omega^{<\omega} \times \mathcal{O}$, where $\mathcal{O}$ is some fixed countable basis of the space X, with infinite repetitions. By induction on $i \in \omega$ build finite sets $T_i \subset \omega^{<\omega}$ and open sets $U_i \subset \omega^\omega$ so that:

I1. $T_0 \subset T_1 \subset \cdots$ and $U_0 \supset U_1 \supset \cdots$. Each set T_i is closed under the operation of taking the longest common initial segment. Moreover, T_{i+1} is an end-extension of T_i, and for every $t \in T_{i+1} \setminus T_i$ it is the case that every infinite extension of t in T belongs to U_i;

I2. for every $t \in T_i$, for every neighborhood P with nonempty intersection with K_t there is $m \in \omega$ and $y \in U_i$ such that $t^\frown m \subset y$ and $O_{t^\frown m} \subset P$;

I3. if (a) $t_i \in T_i$ and P_i has nonempty intersection with K_{t_i} then $T_{i+1} = T_i \cup \{s\}$ for some node $s \in \omega^{<\omega}$ such that some $m \in \omega$, $t_i^\frown m \subset s$ and $O_{t^\frown m} \subset P_i$. Otherwise (b) $T_{i+1} = T_i$;

I4. if I3(a) holds at i, then if $y, z \in U_{i+1}$ and $s \in \omega^{<\omega}$ is a sequence with $s \subset y$ and $s \not\subset z$ then $g(z) \notin \bigcup_{n \in i} f_n(g(y))$.

After the induction has been performed, let S be the downward closure of $\bigcup_i T_i$. The first item shows that $[S] \subseteq \bigcap_i U_i$; in fact, an equality can be arranged here. It also implies that $\bigcup_i T_i$ is exactly the set of splitnodes of the tree S, and the third item implies that S satisfies $(\star)$. Now suppose that y, z are two distinct branches through S and $n \in \omega$, and argue that $g(z) \notin f_n(g(y))$. There has to be $i \in \omega$ larger than n and such that the longest initial segment of y in T_i is t_i, I3(a) holds, the node $s \in T_{i+1} \setminus T_i$ is still an initial segment of y, and moreover t_i is not an initial segment of z. But then, I4 shows that $g(z) \notin f_n(g(y))$ as desired!

The induction is started with $T_0 = \{0\}$ and $U_0 = [T]$. Suppose T_i, U_i have been obtained. If I3(a) fails then let $T_{i+1} = T_i$ and $U_{i+1} = U_i$. If I3(a) occurs, first use I2 to find a point $y \in U_i$ such that $t_i^\frown m \subset y$ for a number m which is not used on any of the sequences in the set T_i and such that $O_{t^\frown m} \subset P_i$. Adjusting y slightly, we may find it in the dense set a. The compact set $\bigcup_{n \in i} f_n(g(y))$ is disjoint from all the compact sets K_t for $t \in T_i$. By the construction of sets K_t, this compact set has an open neighborhood P such that its closure is still disjoint from all of them. Use the continuity of the functions f_n and g to find an initial segment s of y such that every infinite extension y' of s belongs to the set U_i, and $\bigcup_{n \in i} f_n(g(y')) \subset P$. Let $T_{i+1} = T_i \cup \{s\}$ and $U_{i+1} = \{z \in U_i : s \subset z$ or $g(z)$ is not in the closure of $P\}$. It is not difficult to verify that the induction hypotheses continue to hold. □

The σ-ideals with the covering property form a very special subclass of the calibrated ideals. For them, we can even prove the mutual genericity property as well as the rectangular Ramsey property, and this is utilized in Section 9.1 dealing with their products. The total canonization of analytic equivalence relations could be proved in this subclass simply by quoting Theorem 3.5 and observing that the rectangular Ramsey property prevents the nontrivial ergodicity in Case 1 of that theorem from ever occurring. The general calibrated σ-ideals have much more complicated behavior, and typically both the mutual genericity and the rectangular property will fail. In one especially common case, we can find a Borel function on pairs that on any Borel positive rectangle covers a large set.

Theorem 6.11 *Let ν be a diffuse Borel probability measure on a compact metric space X, and let I_ν be the σ-ideal σ-generated by the closed sets of ν-mass 0. There is a rectangular coding function for I_ν defined on X^2.*

The proof uses a measure theoretic proposition of independent interest:

Proposition 6.12 *Let X be a compact metric space with a diffuse Borel probability measure ν on it. Let μ be the usual Borel probability measure* *on 2^ω. Let I_ν be the σ-ideal σ-generated by closed ν-mass zero sets. Then, $I_\nu \geq_K I_\mu$.*

Proof For the duration of this argument the word "mass" refers to a ν-mass of a Borel subset of X. The following simple claim will be instrumental:

Claim 6.13 *Let $O \subset X$ be an open set of positive mass, whose boundary has zero mass. Let $\varepsilon > 0$ be a real number. There is an open subset $P \subset O$ whose boundary has zero mass and $\frac{1}{2}\nu(O) < \nu(P) < (1+\varepsilon)\frac{1}{2}\nu(O)$.*

Proof First note that for every point $x \in X$ there is an open neighborhood P_x of x with boundary of zero mass, such that $\nu(P_x) < \frac{\varepsilon}{2}\nu(O)$. To find the neighborhood, fix a compatible metric for the space X and for every real number $\delta > 0$ write $B(x,\delta)$ for the open ball around x of radius δ. The function $\delta \mapsto \nu(B(x,\delta))$ is nondecreasing, and tends to zero as δ tends to zero. Therefore, there has to be a point $\delta > 0$ of continuity of this function such that $B(x,\delta) < \frac{\varepsilon}{2}\nu(O)$. The set $P_x = B(x,\delta)$ works.

Now, use the compactness of the space X to find a finite collection $\{x_i : i \in n\}$ of points in X such that $X = \bigcup_{m<n} P_{x_m}$. Let $a \subset n$ be an inclusion-minimal set such that the set $P = \bigcup_{m\in a} P_{x_m} \cap O$ has mass larger than $\frac{1}{2}\nu(O)$. The set P works as required in the claim.

To see why this is true, note that since the masses of the sets P_{x_m} are small, the mass of P cannot be bigger than $(1+\varepsilon)\frac{1}{2}\nu(O)$. Moreover, the boundary of the set P is included in the union of boundaries of P_{x_m} for $m \in a$ with the boundary of O, and therefore has zero mass. □

For $n \in \omega$, find real numbers $\varepsilon_n > 0$ such that $\prod_n(1+\varepsilon_n) < 2$. By tree induction on $2^{<\omega}$ build sets $O_t \subset X$ such that:

I1. $O_0 = X$ and each set O_t is open with boundary of zero mass;
I2. $s \subset t$ implies $O_t \subset O_s$;
I3. for every $t \in 2^{<\omega}$, the set O_t is the disjoint union of $O_{t^\frown 0}$, $O_{t^\frown 1}$, and some part of their joint boundary, and $\frac{1}{2}\nu(O_t) < \nu(O_{t^\frown 0}) < (1+\varepsilon_n)\frac{1}{2}\nu(O_t)$.

This is easy to do using Claim 6.13 repeatedly. Note that for every sequence $t \in 2^{<\omega}$, $\nu(O_t) < \prod_n(1+\varepsilon_n)2^{-|t|} < 2 \cdot 2^{-|t|}$. In the end, set B to be the union of the boundaries of the sets O_t for all $t \in 2^{<\omega}$; thus, $B \in I_\nu$. For every point $x \in X \setminus B$, there is a unique infinite binary sequence $g(x) \in 2^\omega$ such that $\forall n\ x \in O_{g(x)\restriction n}$. For $x \in B$ set $g(x) = 0$. Note that the function $g \restriction X \setminus B$ is continuous, as for every sequence $t \in \omega^{<\omega}$, $g^{-1}[t] \setminus B = O_t \setminus B$. We claim that the Borel function g is the desired Katětov reduction: g-preimages of I_μ-small sets are in I_ν.

To verify this, let $C \subset 2^\omega$ be a closed set of μ-mass zero; we will show that $g^{-1}C$ is covered by B and a closed ν-null set, which will complete the proof as $B \in I$. Write $D = g^{-1}C \setminus B \subset X$; it will be enough to show that $\nu(\overline{D}) > 0$. Now, as the function $g \restriction X \setminus B$ is continuous, the set D is relatively closed in $X \setminus B$, and $\overline{D} \subset D \cup B$. Since $\nu(B) = 0$, it is enough to prove that $\nu(D) = 0$. Let $\varepsilon > 0$ be any real number, and cover the set C with some union $\bigcup_n [t_n]$ for some collection $\{t_n : n \in \omega\}$ of finite binary sequences such that $\Sigma_n 2^{-|t_n|} < \varepsilon$. The definition of the function g shows that the set D is covered by the union $\bigcup_n O_{t_n}$, and $\Sigma_n \nu(O_n) < 2\varepsilon$. Since ε was arbitrary, this shows that $\nu(D) = 0$ as required. □

Proof of Theorem 6.11 We will first perform the argument for the usual Borel probability measure μ on $X = 2^\omega$. The theorem then immediately follows from the treatment of I_μ and Proposition 2.69. It will be enough to find a continuous function $f : 2^\omega \times 2^\omega \to 2^\omega$ such that for any two Borel I_μ-positive sets $B, C \subset 2^\omega$, the set $f''(B \times C)$ has a nonempty interior. Proposition 2.67 will then provide the desired rectangular coding function for the ideal I_μ.

Use the concentration of measure, Corollary 2.66, to find numbers $\{n_i : i \in \omega\}$ such that for every $i \in \omega$, $n_i < n_{i+1}$ and, considering the Hamming cube $2^{n_{i+1} \setminus n_i}$ with the normalized counting measure μ_i and the normalized Hamming distance d_i, the d_i-1/8-neighborhood of any set of μ_i-mass $> 2^{-i-1}$ has μ_i-mass greater than $1/2$. Extend the definition of the distance d_i to 2^ω and finite binary sequences of length greater than n_{i+1} by setting $d_i(s, t) = d_i(s \restriction [n_i, n_{i+1}), t \restriction [n_i, n_{i+1}))$. For points $x, y \in 2^\omega$, let $\langle i_j(x, y) : j \in \omega \rangle$ be an increasing enumeration of the numbers i such that x is d_i-1/4-close to y. For $x, y \in 2^\omega$ define $f(x, y)(j) = 0$ if $i_{2j+1}(x, y) = i_{2j}(x, y) + 1$. We claim that if $B, C \subset 2^\omega$ are analytic I_μ-positive sets, then the image $f''(B \times C)$ has a nonempty interior. In order to show this, the following claim will be useful:

Claim 6.14 *Let $i \in \omega$, let $s \in 2^{n_i}$ be a sequence and let $D \subset 2^\omega$ be a Borel set of relative mass $> 2^{-i}$ in $[s]$. Then:*

(i) if $j > i$ and $u \in 2^{n_j \setminus n_i}$ is a binary sequence, there is an extension $s' \supset s$ such that $s' \in 2^{n_j}$, D has relative mass at least 2^{-j} inside $[s']$, and for every $i \leq k < j$, $d_k(s', u) > 1/4$;

(ii) if $t \in 2^{n_i}$ is another sequence and $C \subset 2^\omega$ is a Borel set with relative mass $> 2^{-i}$ in $[t]$, $j > i$ and $z \in 2^{j \setminus i}$ is a binary sequence, there are extensions $s' \supset s$ and $t' \supset t$ of length n_j such that D has relative mass at least 2^{-j} inside $[s']$, C has relative mass at least 2^{-j} inside $[t']$, for every $i \leq k < j$, $d_k(s', t') < 1/4$ if and only if $z(k) = 0$.

Proof Both parts of the claim are obtained by induction on $j - i$, and it is clearly enough to resolve the case $j = i + 1$. For (*i*), note that the set $b = \{v \in 2^{n_{i+1} \setminus n_i} : D$ has relative mass $> 2^{-i}$ in $[s^\frown v]\}$ has μ_{i+1}-mass at least 2^{-i} by the Fubini theorem. By the choice of n_{i+1}, there are points $v_0, v_1 \in b$ such that v_0 is closer than $1/4$ from the flip of v_1. Then v_0, v_1 must have distance at least $1/2$, so at least one of them is d_i-farther than $1/4$ from the sequence u; say it is v_0. The sequence $s' = s^\frown v_0 \in 2^{n_{i+1}}$ is as required.

For (*ii*), consider the sets $b = \{v \in 2^{n_{i+1} \setminus n_i} : D$ has relative mass $> 2^{-i-1}$ in $[s^\frown v]\}$ and $c = \{v \in 2^{n_{i+1} \setminus n_i} : C$ has relative mass $> 2^{-i-1}$ in $[s^\frown v]\}$; both of them have μ_{i+1}-mass at least 2^{-i-1} by the Fubini theorem. By the choice of n_{i+1}, there are sequences $v_0 \in b$ and $v_1 \in c$ of d_i-distance $> 1/4$, proved just as in the previous paragraph. There are also sequences $v_0 \in b$ and $v_1 \in c$ of d_i-distance $< 1/4$. The conclusion of item (*ii*) immediately follows. □

Now, let $B, C \subset 2^\omega$ be analytic I-positive sets. Thinning out the sets B, C if necessary, we may assume that both are G_δ and their intersections with open sets are either empty or I-positive; write $B = \bigcap_n O_n$ and $C = \bigcap_n P_n$, intersections of open sets. The closures of B and C have positive μ-mass, and by the Lebesgue density theorem, there must be finite binary sequences s_0, t_0 of equal length such that the closures of the sets B and C have relative masses $> 1/2$ in the respective open neighborhoods $[s_0], [t_0]$. Let $\langle i_j : j \in j_0 \rangle$ be an increasing enumeration of the numbers i such that t_0 is d_i-$1/4$-close to s_0. To simplify the notation assume that j_0 is an even number. Let $u_0 : j_0/2 \to 2$ be the function defined by $u(j) = 0$ if $i_{2j+1} = i_{2j} + 1$. We will show that for every $z \in 2^\omega$ extending u there is $x \in B \cap [s_0]$ and $y \in C \cap [t_0]$ such that $f(x, y) = z$.

By induction on $k \in \omega$ build finite binary sequences s_k, t_k so that:

I1. $s_0 \subset s_1 \subset \cdots$ and similarly on the t side. Moreover, $[s_{k+1}] \subset O_k$ and $[t_{k+1}] \subset P_k$;

I2. there are numbers l_k such that s_k, t_k are both of length l_k, and the closures of the sets B and C are sets of relative mass $> 2^{-j_0-k}$ in the respective open neigborhoods $[s_k], [t_k]$;

I3. there are numbers $i_{2(j+k)}$ and $i_{2(j+k)+1}$ such that they are the only numbers i between l_k and l_{k+1} such that s_{k+1} is d_i-$1/4$-close to t_{k+1}, and $z(j_0+k) = 0$ if and only if these numbers are successive.

In the end, we will set $x = \bigcup_k s_k$ and $y = \bigcup_k t_k$. I1 will imply that $x \in B$ and $y \in C$, and I3 will imply that $f(x, y) = z$ as required.

To perform the induction, suppose that s_k, t_k have been found. First, use the Lebesgue density theorem to find $l' > l_k$ and an extension $s' \supset s_k$ in $2^{n_{l'}}$ such that $[s'] \subset O_k$ and $\overline{B}$ has relative mass in $[s']$ larger than $2^{-l'}$. Use Claim 6.14 (*i*) to find an extension $t' \supset t_k$ in $2^{n_{l'}}$ such that $\overline{C} \cap [t']$ has relative mass $> 2^{-l'}$ and for every $l \leq i < l'$ it is the case that $d_i(s', t') > 1/4$. Now switch the roles of s, t: find a number $l'' > l'$ and extensions $t'' \supset t'$ and $s'' \supset s'$ in $2^{n_{l''}}$ such that both sets B, C have relative mass $> 2^{-l''}$ in $[s''], [t'']$ respectively, $[t''] \subset P_k$, and for every $l_k \leq i < l''$ it is the case that $d_i(s'', t'') > 1/4$. Finally, use Claim 6.14 (*ii*) on s'', t'' to find their extensions s_{k+1}, t_{k+1} that satisfy the induction hypothesis at $k+1$. This concludes the induction step and the proof of the theorem. □

6.1.2 The σ-compact ideal and generalizations

The calibration feature of a σ-ideal I is highly dependent on the choice of the topology for the underlying Polish space X. Moreover, passing to an I-positive Borel set $B \subset X$, it may be difficult to recognize the calibration just from the structure of the σ-ideal below B–even though the total canonization or Silver property certainly survive such a restriction. In this section, we will deal with a class of σ-ideals σ-generated by closed sets that cannot be presented as calibrated σ-ideals restricted to a condition, no matter what the choice of the topology or the Borel positive set. The most prominent representative of this class is the σ-ideal I on ω^ω σ-generated by compact sets. Its dichotomy properties are well known.

Fact 6.15 (Hurewicz (Kechris 1977)) *Whenever $A \subset \omega^\omega$ is an analytic set, exactly one of the following is true: either $A \in I$ or A contains all branches of a superperfect tree. The ideal I is $\mathbf{\Pi}^1_1$ on $\mathbf{\Sigma}^1_1$.*

(Solecki–Spinas (Solecki and Spinas 1999)) *The ideal I has the free set property.*

As every $\mathbf{\Pi}^1_1$ on $\mathbf{\Sigma}^1_1$ σ-ideal σ-generated by closed sets, the ideal I has the selection property by Proposition 6.2, its free set property implies total canonization, and so in conjunction with Theorem 5.27 this yields the following:

Corollary 6.16 *The ideal I has the Silver property. Restated, for every Borel equivalence relation E on ω^ω, either the space ω^ω is covered by countably many E-equivalence classes and countably many compact sets, or it has a closed subset homeomorphic to ω^ω consisting of E-inequivalent elements.*

We will now generalize the Solecki–Spinas result to partially ordered sets of infinitely branching trees with various measures of the size of branching. For an ideal K on a countable set a let $P(K)$ be the poset of all trees $T \subset a^{<\omega}$ such that every node of T extends to a splitnode of T, and for every splitnode

$t \in T$, the set $\{i \in a : t^\frown i \in T\}$ is not in the ideal K. Thus, for example, the Miller forcing is P(Frechet ideal on ω). The computation of the ideal $I(K)$ associated with the forcing gives complete information. Let $X = a^\omega$ with the product topology, where a is taken with the discrete topology. For every function $g : a^{<\omega} \to K$ consider the closed set $A_g = \{x \in a^\omega : \forall n\ x(n) \in g(x \restriction n)\} \subset X$, and the σ-ideal $I(K)$ on the space X σ-generated by all closed sets of this form. The following is obtained by a straightforward generalization of the proof of Fact 6.15:

Fact 6.17 *Let $A \subset X$ be an analytic set. Exactly one of the following happens:*

- *$A \in I(K)$;*
- *there is a tree $T \in P(K)$ such that $[T] \subset A$.*

Moreover, if the ideal K is coanalytic then the σ-ideal $I(K)$ is $\mathbf{\Pi}^1_1$ on $\mathbf{\Sigma}^1_1$.

Thus the map $T \mapsto [T]$ is a dense embedding of the poset $P(K)$ into $P_{I(K)}$. It is clear now that the poset $P(K)$ is proper and preserves Baire category since the corresponding ideal is generated by closed sets (Zapletal 2008, theorem 4.1.2). If the ideal K is nonprincipal then the forcing adds an unbounded real. Other forcing properties depend very closely on the position of the ideal K in the Katětov ordering. The poset may have the Laver property (such as with K equal to the Frechet ideal), or it may add a Cohen real (such as in K equal to the nowhere dense ideal on $2^{<\omega}$). These issues are addressed in Sabok and Zapletal (2011, section 3). We first look at a very well-behaved special case, generalizing the result of Solecki and Spinas (1999).

Theorem 6.18 *Suppose that K is an F_σ ideal on a countable set. Then $I(K)$ has the free set property.*

Corollary 6.19 *For F_σ ideals K, the ideal $I(K)$ has the Silver property.*

Proof of Theorem 6.18 Fix an F_σ-ideal K on ω and use Mazur's theorem (Mazur 1991) to find a lower continuous submeasure ϕ on $\mathcal{P}(\omega)$ such that $K = \{a \subset \omega : \phi(a) < \infty\}$. The argument now proceeds with a series of claims. Recall that a forcing P has the *weak Laver property* if for every condition $p \in P$, every name $\dot{f}$ for a function in ω^ω forced to be dominated by a ground model function there is a condition $q \leq p$, an infinite set $b \subset \omega$, and sets $\{c_n : n \in b\}$ of the respective size n such that $q \Vdash \forall n \in \check{b}\ \dot{f}(n) \in c_n$. This is a forcing property appearing prominently in connection with P-point preservation (Zapletal 2009).

Claim 6.20 *The forcing $P(K)$ has the weak Laver property.*

Proof Let $p \in P(K)$ be a condition, $g \in \omega^\omega$ a function and $\dot{f}$ a $P(K)$-name for a function in ω^ω pointwise dominated by g. By induction on $n \in \omega$ construct trees $U_n \in P(K)$ and their finite subsets $u_n \subset U_n$, numbers m_n, and finite sets $c_n \subset \omega$ so that:

I1. $T = U_0 \supset U_1 \supset \cdots$ and $\{t\} = u_0 \subset u_1 \subset \cdots$ where t is the shortest splitnode of T;

I2. u_n is an inclusion initial set of splitnodes of U_n and for every node $t \in u_n$, the set of its immediate successors that have some successor in u_{n+1} has ϕ-mass at least n;

I3. $U_{n+1} \Vdash \dot{f}(m_n) \in \check{c}_n$ and $|c_n| \leq m_n$.

The induction is not difficult to perform. Given U_n, u_n, let $m_n = |u_n|$ and for every splitnode $t \in u_n$ and every immediate successor s of t that has no successor in u_n, thin out the tree $U_n \restriction s$ to decide the value of $\dot{f}(m_n)$. Since this value must be smaller than $g(m_n)$, there are only finitely many possible values for this decision and so, thinning out the set of immediate successors further if necessary, the decision can be assumed to be the same for all such immediate successors of t, yielding a number $k_t \in \omega$. Let $c_n = \{k_t : t \in u_n\}$ and let U_{n+1} be the thinned out tree. Finally, find $u_{n+1} \subset U_{n+1}$ satisfying I2.

In the end, $U = \bigcap_n U_n$ is a tree in the poset $P(K)$ forcing $\forall n\ \dot{f}(m_n) \in \check{c}(n)$ as desired. □

Claim 6.21 *Whenever $a \subset \omega$ is a K-positive subset and $T \in P(K)$ forces $\dot{b} \subset \omega$ is an element of $\dot{K}$, then there is a condition $S \leq T$ and a K-positive ground model set $c \subset a$ such that $S \Vdash \check{c} \cap \dot{b} = 0$.*

Proof This is in fact true for any forcing with the weak Laver property in place of $P(K)$. Thinning out the condition T if necessary we may find a number $m \in \omega$ such that $T \Vdash \phi(\dot{b}) < \check{m}$. Find disjoint finite sets $a_n \subset a$ with $\phi(a_n) > 2mn$, and use the weak Laver property to find a condition $S \leq T$, an infinite set $b \subset \omega$ and sets $B_n \subset \mathcal{P}(a_n)$ for every $n \in b$ such that all sets in B_n have ϕ-mass $< m$, $|B_n| < n$, and $S \Vdash \dot{b} \cap \check{a}_n \in \check{B}_n$. Then for every number $n \in b$, the set $c_n = a_n \setminus \bigcup B_n$ has mass at least nm, and it is forced by S to be disjoint from $\dot{b}$. Thus the set $c = \bigcup_{n \in b} c_n$ works as desired. □

Now, suppose that $D \subset \omega^\omega \times \omega^\omega$ is a Borel set with $I(K)$-small vertical sections, and $T \in P(K)$ is a tree. We must find a tree $S \subset T$ in $P(K)$ such that $[S] \times [S] \cap D \subset \text{id}$. Thinning out the tree T if necessary, by Fact 6.1(*ii*) we may assume that there is a continuous function $f : [T] \times \omega^{<\omega} \to K$ such that for every branch $x \in [T]$ the section D_x is included in the $I(K)$-small set $\{y \in \omega^\omega : \forall^\infty n\ y(n) \in f(x)(y \restriction n)\}$. Let $\{t_n : n \in \omega\}$ be an enumeration of

$\omega^{<\omega}$ with infinite repetitions, and by induction construct trees $U_n \in P(K)$ and their finite subsets $u_n \subset U_n$ so that:

I1. $T = U_0 \supset U_1 \supset \cdots$ and $\{t\} = u_0 \subset u_1 \subset \cdots$ where t is the shortest splitnode of T;

I2. u_n is an inclusion initial set of splitnodes of U_n;

I3. if $t_n \in u_n$ then $u_{n+1} = u_n$ together with some set of splitnodes extending t_n. All the new splitnodes differ from all the old ones and among each other already at their $|t_n|$-th entry. Moreover the set $\{i \in \omega : \exists t \in u_{n+1}\ t_n{}^\frown i \subset t\}$ has ϕ-mass at least n. If $t_n \notin u_n$ then $u_{n+1} = u_n$;

I4. if $t_n \in u_n$ and $t \neq t_n$ is another node in u_n, then for every node $s \in u_{n+1} \setminus u$ and every $x \in [U_{n+1} \restriction s]$ it is the case that $f(x)(t)$ is disjoint from the set $\{i \in \omega : t^\frown i \in U_{n+1}$ but no node of u_n extends $t^\frown i\}$.

This is not difficult to do using Claim 6.21. In the end, the tree $S = \bigcap_n U_n$ belongs to $P(K)$, and its set of splitnodes is exactly $\bigcup_n u_n$. We must show that $[S] \times [S] \cap D \subset \mathrm{id}$. To verify the inclusion, suppose for contradiction that $x \neq y \in [S]$ are two branches and $y \in D_x$. This means that there is a number m_0 such that for every $m > m_0$, $y(m) \in f(x)(y \restriction m)$. Find a number $n \in \omega$ such that $t_n \in u_n$ is an initial segment of x, $u_{n+1} \setminus u_n$ contains a still longer initial segment of x, and the longest node $t \in u_n$ which is an initial segment of y is of length greater than m_0. Then I4 shows that $y(|t|) \notin f(x)(t)$, contradicting the choice of m_0! □

The previous result cannot be generalized to much more general ideals. A more or less canonical ideal on a countable set which is not a subset of an F_σ-ideal is $K = \mathrm{Fin} \times \mathrm{Fin}$, the ideal on $\omega \times \omega$ generated by vertical sections and sets with all vertical sections finite. To simplify notation, restrict the σ-ideal $I(K)$ to the closed subset $X \subset (\omega \times \omega)^\omega$ of those functions x such that, for every $n < m$, the numbers in the pair $x(n)$ are smaller than both numbers in the pair $x(m)$.

Theorem 6.22 *The ideal $I(K)$ does not have the free set property. There is a rectangular coding function for $I(K)$ defined on the whole space X.*

This theorem would not be particularly exciting if the poset $P_{I(K)}$ added a Cohen real. However, it is not difficult to show that it preserves the outer Lebesgue measure. Corollary 6.37 then implies that the σ-ideal $I(K)$ has the Silver property for equivalences that are Borel reducible to E_{K_σ}, $=_J$ for an analytic P-ideal J, or for Borel equivalences classifiable by countable structures.

Proof For every point $x \in X$ and every number $n \in \omega$, the value $x(n)$ is an ordered pair of natural numbers denoted as $\langle x(n)(0), x(n)(1)\rangle$. Say that points $x, y \in X$ are *entangled* at n, m if the pairs $x(n)$, $y(m)$ are not comparable in the natural coordinatewise ordering of pairs of natural numbers. Note that if x, y are entangled at n_0, m_0 and at n_1, m_1, then either $n_0 \in n_1, m_0 \in m_1$ or $n_1 \in n_0, m_1 \in m_0$; so the locations where the pair x, y is entangled are naturally linearly ordered. Now let $f(x, y)(i) = 0$ if at the i-th location n, m where x, y are entangled, $x(n)(0) > y(m)(0)$, and let $f(x, y) = 1$ otherwise. We will show that for any two Borel $I(K)$-positive sets $B, C \subset X$, the image $f''(B \times C)$ has nonempty interior. Proposition 2.67 will then provide the desired rectangular coding function for the ideal $I(K)$.

Let $B, C \subset X$ be $I(K)$-positive Borel sets, and let $S, T \subset (\omega \times \omega)^{<\omega}$ be trees such that $[S] \subset B$, $[T] \subset C$, and K-positively splitting nodes are cofinal in both S and T. Let s_0, t_0 be the first K-positively splitting nodes in S, T respectively, let a be the finite set of positions at which they are entangled, and let $u : |a| \to 2$ be the binary sequence defined by $u(j) = 0$ if at the j-th position (n, m) of entanglement of s_0, t_0, it is the case that $s_0(n)(0) > t_0(m)(0)$. We will show that whenever $v \in 2^\omega$ extends u then there are points $x \in [S]$, $y \in [T]$ such that $f(x, y) = v$.

By induction on $i \in \omega$ build splitnodes s_i, t_i of S, T respectively so that:

I1. $s_0 \subset s_1 \subset \cdots$ and $t_0 \subset t_1 \subset \cdots$;
I2. s_i, t_i are entangled exactly at pairs in a and the pairs $(|s_j|, |t_j|)$ for $j \in i$;
I3. $v(|a| + i) = 0$ if and only if $s_{i+1}(|s_i|)(0) > t_{i+1}(|t_i|)(0)$.

If this succeeds, the points $x = \bigcup_i s_i \in [S]$ and $y = \bigcup_i t_i \in [T]$ will have $f(x, y) = v$ as required. The induction itself is easy. Suppose that s_i, t_i have been constructed, and for simplicity assume that $v(|a| + i) = 0$. Use the fact that t_i is K-positively splitting node of the tree T to find a number $k \in \omega$ such that there are infinitely many l such that $t_i^\frown\langle k, l\rangle \in T$. Let s_{i+1} be any splitnode of S extending s_i such that $s_{i+1}(|s_i|)(0) > k$. Let t_{i+1} be any splitnode of the tree T extending t_i such that $t_{i+1}(|t_i|)(0) = k$ and $t_{i+1}(|t_i|)(1)$ is greater than all numbers appearing in the pairs in $\text{rng}(s_{i+1})$. This concludes the induction step.

For the failure of the free set property, let $D = \{\langle x, y\rangle : f(x, y)$ ends with a tail of 1's$\}$. The vertical sections of D are I-small: whenever $x \in X$ is a point, let $A \subset \omega \times \omega$ be the set of all pairs $\langle k, l\rangle$ such that $\langle n, m\rangle \in \text{rng}(x)$ and $n > k$ implies $m > l$. The set A has finite vertical sections and as such is in the ideal K. The vertical section D_x consists of those $y \in Y$ such that for all but finitely many $i \in \omega$, $y(i) \in A$, and therefore $D_x \in I(K)$. No $I(K)$-positive Borel set $B \subset X$ can be free for D, since the set $f''(B \times B)$ has a nonempty interior, and as such must contain a pair of points $\langle x, y\rangle$ such that $f(x, y)$ ends with a tail of 1's. □

6.1.3 The meager ideal

The spectrum of the meager ideal is very complex.

Theorem 6.23 *The equivalence relations E_0, E_2, E_{K_σ}, and F_2 belong to the spectrum of the meager ideal. The spectrum is cofinal in the reducibility of Borel equivalence relations.*

In the canonization direction, note that no countable Borel equivalence relations belong to the spectrum as by Theorem 4.7, essentially countable$\rightarrow_I$reducible to E_0. In the realm of non-countable Borel equivalence relations, we can prove the following:

Theorem 6.24 *Let E be an analytic equivalence relation on $X = (2^\omega)^\omega$ such that $F_2 \subset E$. Then:*

(i) either E has a comeager equivalence class;
(ii) or $F_2 \leq_B E \restriction C$ for every Borel nonmeager set $C \subset X$.

In other words, for every analytic equivalence relation E, exactly one of the following holds: either $F_2 \leq_B E$ or F_2 is generically E-ergodic.

The proof of Theorem 6.24 requires several preliminary remarks of independent interest. Let $X = (2^\omega)^\omega$ and let I be the meager ideal on X. It is well known that the poset $P = P_I$ is in the forcing sense equivalent to the finite support product of countably many Cohen forcings, adding a point $\dot{x} \in X$. For natural number $n \in \omega$, the symbol P^n denotes the n-fold product of the poset P, adding an n-tuple $\langle \dot{x}_i : i \in n\rangle$ of points in X.

Definition 6.25 Let M be a transitive model of set theory and $a \subset 2^\omega$ be a countable set. We will say that a is a *Cohen-symmetric* over M if there is a point $x \in X$ P-generic over M such that $a = \text{rng}(x)$. A finite collection $\{a_i : i \in n\}$ of countable subsets of 2^ω is *mutually Cohen-symmetric* over M if there is an n-tuple $\langle x_i : i \in n\rangle$ of points in X, P^n-generic over M, such that for every $i \in n$, $a_i = \text{rng}(x_i)$ holds.

The following claim records the basic properties of the posets P^n. Define the *shift* on X^n by identifying X with $((2^\omega)^{\mathbb{Z}}$, letting $s : X \to X$ be the function $s(x)(m) = x(m+1)$ and letting $s(\langle x_i : i \in n\rangle) = \langle s(x_i) : i \in n\rangle$ for every $n \in \omega$.

Claim 6.26 *Let M be transitive model of set theory, let $n \in \omega$ and let $\langle x_i : i \in n\rangle \in X^n$ by a P^n-generic point over M.*

(i) Permuting finitely many points in $\text{rng}(x_i)$ *for each $i \in n$ yields another P^n-generic point over M.*
(ii) If $\{b_j : j \in m\}$ are pairwise disjoint nonempty subsets of n then there is a P^m-generic point $\langle y_j : j \in m\rangle$ over M such that $\text{rng}(y_j) = \bigcup_{i \in b_j} \text{rng}(x_i)$.
(iii) If $C \in M$ is a Borel nonmeager set then some shift of $\langle x_i : i \in n\rangle$ is a P^n-generic point over M belonging to C.

Proof (*i*) is standard. For (*ii*) choose any bijections $\pi_j : b_j \times \omega \to \omega$ and define $f : X^{\bigcup_j b_j} \to X^m$ by $f(\langle z_k : k \in \bigcup_j b_j\rangle) = \langle y_j : j \in m\rangle$ if $y_j(\pi_j(k,l)) = z_k(l)$ for every $j \in m$, $k \in b_j$ and $l \in \omega$. The map f is a homeomorphism and so (*ii*) is witnessed by $\langle y_j : yj \in m\rangle = f(\langle x_i : i \in \bigcup_j b_j\rangle)$. For the last item, the shift preserves Baire category and every Borel subset of X^n invariant under the shift is meager or comeager. Thus, the closure of the nonmeager set C under the shift is comeager, and so it must contain the generic point $\langle x_i : i \in n\rangle$. In other words, some shift of the generic point belongs to the set C. □

Corollary 6.27 *Sentences of the P^n-forcing language using only canonical names for elements of the ground model and names for* $\text{rng}(\dot{x}_i)$ *for $i \in n$ are decided by the largest condition in P^n.*

Proof Suppose that ϕ is such a sentence and for contradiction assume that $C \Vdash \phi$ and $D \Vdash \neg\phi$ for some conditions $C, D \in P^n$. Let $\langle x_i : i \in n\rangle \in X^n$ be a P^n-generic point over V in the set D, and use Claim 6.26(*iii*) to shift it to some $\langle y_i : i \in n\rangle \in C$. The models $V[x_i : i \in n]$ and $V[y_i : i \in n]$ are the same and also $\text{rng}(x_i) = \text{rng}(y_i)$ for all $i \in n$. Thus the models have to either both satisfy ϕ or both satisfy $\neg\phi$, contradicting the forcing theorem. □

Proof of Theorem 6.24 There are four distinct cases. In the first three we end up in the first clause of the theorem, and only in the last one will the second clause prevail. Since $F_2 \subset E$, we will abuse the notation and extend the definition of E to countable subsets of 2^ω. Note that $F_2 \subset E$ remains true also in every forcing extension by the Shoenfield absoluteness.

Case 1 Suppose first $P^2 \Vdash \dot{x}_0 \; E \; \dot{x}_1$. In such a case, let M be a countable elementary submodel of a large structure and let $B = \{x \in X : x \text{ is } M\text{-generic for } P\}$. This is a Borel comeager subset of X. Any two elements of B must be E-related. To prove this, for points $x_0, x_1 \in B$ pick a point $y \in X$ such that y

is P-generic over both models $M[x_0]$ and $M[x_1]$. Use the forcing theorem to argue that $M[x_0, y] \models x_0\ E\ y$ and $M[x_1, y] \models x_1\ E\ y$. Use the Mostowski absoluteness for wellfounded models to conclude that $x_0\ E\ y\ E\ x_1$, and finally argue that $x_0\ E\ x_1$ by transitivity of E. The first clause of the conclusion of the theorem has just been verified.

Case 2 Suppose instead $P^3 \Vdash \mathrm{rng}(x_0) \cup \mathrm{rng}(\dot{x}_1)\ E\ \mathrm{rng}(\dot{x}_0) \cup \mathrm{rng}(\dot{x}_2)$. Again, let M be a countable elementary submodel of a large structure, let $B = \{x \in X : x$ is M-generic for $P\}$ and argue that any two elements of B must be E-related. Let $x_0, x_1 \in B$ and pick a point $y \in X$ such that y is is P-generic over both models $M[x_0]$ and $M[x_1]$. As in Case 1, we will argue that $x_0\ E\ y\ E\ x_1$ and conclude the argument by transitivity of E. To see that $x_0\ E\ y$, define points $z_0, z_1, z_2, z_3 \in X$ by $z_0(i) = x_0(2i)$, $z_1(i) = x(2i + 1)$, $z_2(i) = y(2i)$ and $z_3(i) = y(2i + 1)$, and observe that the tuple $\langle z_0, z_1, z_2, z_3\rangle$ is P^4-generic over the model M. Use the case assumption and the forcing theorem to argue that $\mathrm{rng}(x_0) = \mathrm{rng}(z_0) \cup \mathrm{rng}(z_1)$ is E-related to $\mathrm{rng}(z_0) \cup \mathrm{rng}(z_3)$ and then to $\mathrm{rng}(z_2) \cup \mathrm{rng}(z_3) = \mathrm{rng}(y)$. The first clause of the conclusion of the theorem is satisfied.

Case 3 As another case, suppose that $P^2 \Vdash \mathrm{rng}(x_0)\ E\ \mathrm{rng}(\dot{x}_0) \cup \mathrm{rng}(\dot{x}_1)$. In this case again, we get an E-comeager equivalence class similarly to the previous argument.

Case 4 Now assume that all of the previous cases have failed; we will argue that the second clause of the theorem must hold. Let $C \subset X$ be a Borel non-meager set, and let M be a countable elementary submodel of a large structure containing the set C.

Claim 6.28 *There is a continuous map $h : 2^\omega \to X$ such that for each $n \in \omega$ and each n-tuple $\langle y_i : i \in n\rangle$ of pairwise distinct points of 2^ω, the tuple $\langle h(y_i) : i \in n\rangle$ is P^n-generic over the model M.*

Proof Let $\langle D_m : m \in \omega\rangle$ enumerate all dense open subsets of all the posets P^n for all $n \in \omega$ that belong to the model M, with infinite repetitions. By induction on $m \in \omega$ build maps $h_m : 2^m \to P$ so that:

I1. if $t \in 2^m$ and $s \in 2^{m+1}$ are such that $t \subset s$, then $h_m(s) \leq h_m(t)$;
I2. if D_m is an open dense set in P^n for some $n < 2^m$, then for every n-tuple $\langle t_i : i \in n\rangle$ of pairwise distinct elements of 2^m it is the case that $\langle h_m(t_i) : i \in n\rangle$ belongs to D_m.

The induction is trivial, at each stage strengthening the sequences $h_m(t)$ for $t \in 2^m$ repeatedly so as to satisfy I2 for all the finitely many choices of the tuples $\langle t_i : i \in n \rangle$. In the end, for every $y \in 2^\omega$ the filter on P generated by the set $\{h_m(y \restriction m) : m \in \omega\}$ is generic over the model M by I2, and so it produces a point $h(y) \in X$. The function h is as required. □

Now, fix a bijection $\pi : \omega \times \omega \to \omega$ and for every $z \in X$ let $k(z) \in X$ be the function defined by $k(z)(\pi(k, l)) = h(z(k))(l)$; k is a Borel function. We will show that for every $z \in X$, $\mathrm{rng}(k(z))$ is a Cohen-symmetric set over the model M. Even more is true:

Claim 6.29 *Let $n \in \omega$ and $\langle z_i : i \in n \rangle$ be points in X with pairwise disjoint ranges. The sets $a_i = \mathrm{rng}(k(z_i))$ for $i \in n$ are mutually Cohen-symmetric over M.*

Proof Let $\langle D_m : m \in \omega \rangle$ enumerate all open dense subsets of P^n in the model M. By induction on $m \in \omega$ build conditions $p_m \in P^n$, numbers $l_m \in \omega$ and points $x(i, l) \in 2^\omega$ for $i \in n$ and $l \in l_m$ so that:

I1. $p_0 \geq p_1 \geq \cdots, 0 = l_0 < l_1 < \cdots$;
I2. for every $i \in n$, the points $\{x(i, l) : l \in l_{m+1}\}$ all appear in $\mathrm{rng}(k(z_i))$ and include the m-th point of $\mathrm{rng}(k(z_i))$ under some fixed enumeration;
I3. $p_{m+1} \in D_m$ and (interpreting p_m as an open subset of X^n) whenever $\langle x_i : i \in n \rangle \in X^n$ is a tuple such that for all $i \in n$ and all $l \in l_{m+1}$ $x_i(l) = x(i, l)$ then $\langle x_i : i \in n \rangle \in p_m$.

Once the induction is performed, let g be the filter on P generated by the conditions $\{p_m : m \in \omega\}$, and for every $i \in n$ let $x_i \in X$ be the point defined by $x_i(l) = x(i, l)$ for every $l \in \omega$. Item I3 shows that g is a filter generic over M and $\langle x_i : i \in n \rangle$ is the generic tuple associated with the filter g, and I2 shows that for every $i \in n$, $\mathrm{rng}(x_i) = \mathrm{rng}(k(z_i))$ holds. This will complete the proof of the claim.

The induction starts with letting p_0 be the largest condition in P^n and $l_0 = 0$. Suppose that p_m and $x(i, l)$ for $i \in n, l \in l_m$ have been obtained. Let $\{b_i : i \in n\}$ be finite subsets of 2^ω such that $b_i \subset \mathrm{rng}(z_i)$ and the set $\bigcup\{\mathrm{rng}(h(y)) : y \in b_i\}$ includes all the points $x(i, l)$ for $l \in l_m$ as well as the m-th element of $\mathrm{rng}(k(z_i))$ under some fixed enumeration. Now, the points $\langle h(y) : y \in \bigcup_i b_i \rangle$ are mutually generic over M by the choice of the function h. By Claim 6.26(*ii*), there is an n-tuple $\langle x_i^m : i \in n \rangle$ P^n-generic over M such that $\mathrm{rng}(x_i^m) = \bigcup_{y \in b_i} \mathrm{rng}(h(y))$. Permuting the range of x_i^m as in Claim 6.26(*ii*) if necessary, we can obtain $x_i^m(l) = x(i, l)$ for all $l \in l_m$. By I3, the filter $g \subset P^n$ generic over M associated with the tuple $\langle x_i^m : i \in \omega \rangle$ contains the condition p_m, and by genericity also contains a condition $p_{m+1} \leq p_m$ in the open dense set D_m.

Choose a large enough number $l_{m+1} \in \omega$ and let $x(i,l) = x_i^m(l)$ for all $i \in n$ and $l \in l_{m+1}$. This completes the induction step. $\square$

Theorem 2.29 shows that the set $C' = \{y \in X : \exists x \in C$ P-generic over M such that $x \; F_2 \; y\}$ is Borel and there is a Borel function $g : C' \to C$ which for every $y \in C'$ produces a P-generic point over M such that $x \; F_2 \; y$. Claim 6.29 for $n = 1$ shows that $\mathrm{rng}(k) \subset C'$. We claim that the function $f = g \circ k$ is a Borel reduction from F_2 to $E \restriction C$. It is clear that the function f is an F_2-homomorphism since k is, and so F_2-equivalent points are sent to E-equivalent points. We must prove that if $z_0, z_1 \in X$ are not F_2-equivalent, then their f-images are not E-equivalent. There are several cases depending on the inclusion comparison of $\mathrm{rng}(z_0)$ and $\mathrm{rng}(z_1)$; we will handle the case when the sets $\mathrm{rng}(z_0) \cap \mathrm{rng}(z_1)$, $\mathrm{rng}(z_0) \setminus \mathrm{rng}(z_1)$, $\mathrm{rng}(z_1) \setminus \mathrm{rng}(z_0)$ are all nonempty. In such a case, let $z'_0, z'_1, z'_2 \in X$ enumerate these three respective pairwise disjoint sets and use Claim 6.29 to find points $\langle x_i : i \in 3\rangle$ P^3-generic over M such that $\mathrm{rng}(x_i)$ enumerates $\mathrm{rng}(k(z'_i))$ for every $i \in 3$. By the failure of Case 2 and Corollary 6.70 we see that $\mathrm{rng}(x_0) \cup \mathrm{rng}(x_2) \; E \; \mathrm{rng}(x_0) \cup \mathrm{rng}(x_2)$ fails, in other words $f(z_0) \; E \; f(z_1)$ must fail. The other cases are dealt with similarly. $\square$

Corollary 6.30 *Whenever $\{G_n : n \in \omega\}$ is a collection of analytic equivalences on a fixed Polish space Y such that F_2 is not Borel reducible to any of them, then F_2 is not reducible to $\bigcap_n G_n$ either.*

This should be compared with the case of $E_3 = \bigcap_n E_0^n$, where E_0^n is the equivalence on $(2^\omega)^\omega$ connecting two points if their entries on the first n coordinates are E_0 connected. The equivalences forming the intersection are each reducible to E_0 while E_3 is known not to be even essentially countable (Kanovei 2008, chapter 14).

Proof Suppose for contradiction that $f : (2^\omega)^\omega \to Y$ is a Borel function reducing F_2 to $\bigcap_n G_n$. Write $\bar{G}_n$ for the pullback of G_n by the function f and observe that these are all analytic equivalence relations containing F_2 as a subset, to which F_2 is not reducible. Thus, each of the equivalence relations $\bar{G}_n$ has a comeager equivalence class $C_n \subset (2^\omega)^\omega$. The set $C = \bigcap_n C_n$ is still comeager, and as F_2-classes are meager, it contains non-F_2-equivalent elements x_0, x_1. Now, the points $f(x_0), f(x_1)$ are G_n-related for every n, and so $\bigcap_n G_n$-related. This contradicts the assumption that f was a reduction of F_2 to $\bigcap_n G_n$. $\square$

Proof of Theorem 6.23 First, let I be the meager ideal on the space 2^ω. To show that E_0 is in the spectrum of the meager ideal, we will argue that E_0

retains its complexity on every Borel nonmeager set. Consider the E_0 σ-ideal J on 2^ω introduced in Section 7.1; it is generated by Borel E_0 selectors as indicated by Fact 7.2. The σ-ideal consists of meager sets, since its generators are meager by the Pettis theorem (Kechris 1994). Thus, whenever $B \subset 2^\omega$ is a nonmeager Borel set, it is J-positive, and then Fact 7.4 provides a continuous reduction of E_0 to $E_0 \restriction B$.

To show that E_2 is in the spectrum of the meager ideal, we will argue that E_2 retains its complexity on every Borel nonmeager set. Consider the σ-ideal J σ-generated by Borel grainy subsets of 2^ω of Definition 7.41. The σ-ideal J consists of meager sets only as shown in Claim 7.46. Thus, whenever $B \subset 2^\omega$ is a Borel nonmeager set then $B \notin J$ and therefore E_2 is Borel reducible to $E_2 \restriction B$ by Fact 7.38.

To show that the spectrum of Cohen forcing is cofinal in $\leq_B$ and it includes E_{K_σ}, consider the space $X = (2^\omega)^\omega$ instead, with its associated meager ideal I. Consider the σ-ideal J consisting of Borel subsets of X containing no infinite product of perfect subsets of 2^ω as defined in Section 9.3. Then $J \subset I$ by Proposition 9.31, therefore the identity is a Katětov map from I to J, and the spectrum of J is included in the spectrum of I by Proposition 2.70. The spectrum of J is cofinal in the reducibility order of Borel equivalence relations and includes E_{K_σ} as desired by Theorem 9.29.

To show that F_2 is in the spectrum, argue that F_2 retains its complexity of every nonmeager Borel subset of X. To this end, apply Theorem 6.24 to $E = F_2$. The F_2 equivalence classes are meager, and so the second clause of the theorem has to hold, as required. □

6.2 Porosity ideals

A great number of game σ-ideals falls into the following general scheme, isolated in Zapletal (2008, section 4.2). The scheme was originally motivated by the metric porosity σ-ideals studied in connection with differentiability of functions on Banach spaces (Lindenstrauss *et al.* 2012). It turns out though that it encompasses a wide spectrum of other examples, including the σ-ideal generated by sets of continuity of a fixed Borel function as well as all the σ-ideals σ-generated by closed sets.

Definition 6.31 Let X be a Polish space, and let U be a countable collection of its Borel subsets. An *abstract porosity* is a function por with domain $\mathcal{P}(U)$ and range consisting of Borel subsets of X, such that $a \subset b$ implies $\text{por}(a) \subset \text{por}(b)$. Given an abstract porosity por, a set $A \subset X$ will be called *porous* if it is covered by a set of the form $\mathrm{por}(a) \setminus \bigcup a$ for some $a \subset U$. The associated *porosity σ-ideal* is the σ-ideal on X σ-generated by porous sets.

While all of our specific examples come from quite simple porosities, the statements of general theorems require complexity specifications. Equip the underlying countable set U with the discrete topology, and the set $\mathcal{P}(U)$ with the usual product topology. We will say that the porosity por is *Borel* (or *coanalytic*) if the set $\{\langle a, x\rangle \in \mathcal{P}(U) \times X : x \in \mathrm{por}(a)\} \subset \mathcal{P}(U) \times X$ is Borel, (or coanalytic, respectively). The following fact records the general properties of porosity σ-ideals proved in Zapletal (2008, section 4.2).

Fact 6.32 *Let X be a Polish space and I be a porosity σ-ideal. Then:*

(i) the quotient poset P_I is proper and preserves Baire category;
(ii) every analytic I-positive set contains a Borel I-positive subset;
(iii) if the porosity is coanalytic then I is $\mathbf{\Pi}^1_1$ on $\mathbf{\Sigma}^1_1$;
(iv) if in addition the porosity is Borel, then I is a game ideal.

Proof (*i*) is proved in Zapletal (2008, theorem 4.2.2 and theorem 4.2.7), (*ii*) comes from Zapletal (2008, theorem 4.2.3), (*iii*) comes from Zapletal (2008, theorem 3.8.7). For the last item, the Borel basis for I is naturally indexed by the space $\mathcal{P}(U)^\omega$. The scrambling tool of Definition 5.2 is obtained either abstractly, by an application of Theorem 5.4, or by a direct construction of the integer game, as is done in the proof of Zapletal (2008, theorem 4.2.3(3)). □

It is now time to state our main canonization results for porosity σ-ideals.

Theorem 6.33 *Let I be a coanalytic porosity σ-ideal on a Polish space X such that the poset P_I is bounding. Then I has total canonization for analytic equivalence relations. If the porosity is in addition Borel, then I has the Silver property for Borel equivalence relations.*

Theorem 6.34 *Let I be a porosity σ-ideal on a Polish space X such that the poset P_I does not add an infinitely equal real. Then I has total canonization for equivalence relations Borel reducible to E_{K_σ}, or $=_J$ for an analytic P-ideal J, and for equivalence relations classifiable by countable structures. If the porosity is in addition Borel, then I has the Silver property for Borel equivalence relations in these classes.*

These theorems will be proved in Section 6.2.2. Here, we formulate and prove several corollaries pertaining to specific σ-ideals.

Corollary 6.35 *The σ-ideal I of metrically σ-porous sets on a compact metric space X has total canonization for analytic equivalence relations, and the Silver property for all Borel equivalence relations.*

Corollary 6.36 *If $f : X \to Y$ is a Borel function between Polish spaces and I is the σ-ideal on X σ-generated by the sets on which f is continuous, then I has the Silver property for Borel equivalence relations reducible to E_{K_σ} or to $=_J$ for an analytic P-ideal J on ω, or classifiable by countable structures.*

The final corollary speaks about σ-ideals generated by closed sets, and it gives a canonization result that is impossible to obtain by the methods of Section 6.1:

Corollary 6.37 *If I is a σ-ideal on a Polish space σ-generated by a coanalytic collection of closed sets, and the quotient poset P_I adds no infinitely equal real, then I has the Silver property for Borel equivalence relations reducible to E_{K_σ} or to $=_J$ for an analytic P-ideal J on ω, or classifiable by countable structures.*

We will first pause and elaborate on the assumption of Theorem 6.34. An *infinitely equal real* is a point of ω^ω in the generic extension which has infinite intersection with every point in the ground model. Certainly, a Cohen real is an infinitely equal real; thus, the assumption implies (and is in fact strictly stronger than) not adding Cohen reals. Not adding an infinitely equal real follows from a number of more familiar and quite common forcing preservation properties:

Proposition 6.38 *Suppose that P is a partial order such that one of the following holds:*

(i) P is bounding;
(ii) P preserves outer Lebesgue measure;
(iii) P has the Laver property.

Then P adds no infinitely equal real.

Proof For (*i*), let $y \in \omega^\omega$ be a point in the generic extension. Since P is bounding, there is a ground model point $z \in \omega^\omega$ such that for every $n \in \omega$, $y(n) < z(n)$. Certainly, $y \cap z = 0$ and so y is not an infinitely equal real.

For (*ii*), let μ be the product measure on $\prod_n 2^{n+2}$, where the number 2^{n+2} is equipped with the normalized counting measure. Let $y \in \omega^\omega$ be a point in the forcing extension, and let $O = \{z \in \Pi_n 2^{n+1} : \exists n\ z(n) = y(n)\}$. It is immediate that $\mu(O) \leq \Sigma_n 2^{-n-2} = 1/2$. Since P preserves outer Lebesgue measure, there must be a ground model point $z \in \prod_n 2^{n+2}$ such that $z \notin O$. Then, $z \cap y = 0$ and so y is not an infinitely equal real.

For (*iii*), let $y \in \omega^\omega$ be a point in the generic extension, and let $y' \in \omega^\omega$ be defined as $y'(n) = \min\{2^n, y(n)\}$. The function $y' : \omega \to \omega$ is bounded by the exponential function, and so by the Laver property of the poset P, there is a ground model function $t : \omega \to \mathcal{P}(\omega)$ such that $\forall n\ |t(n)| \leq n+1 \wedge y'(n) \in t(n)$. Let $z \in \omega^\omega$ be a ground model function below the exponential such that $\forall n\ z(n) \notin t(n)$. Clearly, $y \cap z = 0$ and so y is not an infinitely equal real. □

It is now time to discuss examples of porosity ideals.

Example 6.39 Let X, d be a Polish space with a complete metric on it, and let I be the σ-ideal of metrically σ-porous sets on X. Then I is a porosity σ-ideal. The generating porosity is Borel.

Proof Let us first recall the definition of metrically σ-porous sets from Zajíček's expository article (Zajíček 1987/88). While there are many variations in the definition (Lindenstrauss *et al.* 2012), all of them yield abstract porosity ideals; we will choose one typical case. For a set $A \subset X$, a point $x \in X$ and $\delta > 0$ write $r(A, x, \delta)$ for the supremum of all radii of open balls that are included in the open ball centered at x of radius δ, that are disjoint from the set A. Define the porosity of A at x as the $\limsup_{\delta \to 0+} r(A, x, \delta)/\delta$. The set A is *metrically porous* if it has nonzero porosity at all of its points. Let I be the σ-ideal σ-generated by metrically porous sets.

To see that this σ-ideal fits into our framework, just let the set U consist of all open balls of rational radius centered at a point in a fixed countable dense subset of X. For $a \subset U$, $x \in X$ and $\delta > 0$ write $s(a, x, \delta)$ for the supremum of the radii of balls in a which are subsets of the ball of radius δ around the point x. Let $\mathrm{por}(a)$ be the set of all points $x \in X$ where $\limsup_{\delta \to 0+} s(a, x, \delta)/\delta > 0$. This is a Borel abstract porosity generating the σ-ideal I. □

Example 6.40 Whenever X, Y are Polish spaces and $f : X \to Y$ is a Borel function, then the σ-ideal I on X σ-generated by sets of continuity of f is a porosity σ-ideal. The generating porosity is Borel.

The quotient forcing P_I was investigated in its combinatorial presentation in Steprans (1993) and Sabok (2009).

Proof Fix countable bases B_X and B_Y for the two Polish topologies on X and Y and for open sets $O \in B_X$ and $P \in B_Y$ write $u(O, P) = O \setminus f^{-1}P$. Let $U = \{u(O, P) : O \in B_X, P \in B_Y\}$ and $\mathrm{por}(a) = \{x \in X : \forall P \in B_Y\ f(x) \in P \to \exists O \in B_X\ x \in O \wedge u(O, P) \in a\}$. This is clearly an inclusion preserving function from $\mathcal{P}(U)$ to $\mathcal{P}(X)$. We will show that for every set $A \subset X$, $f \restriction A$ is continuous if and only if there is an $a \subset U$ such that $A \subset \mathrm{por}(a) \setminus \bigcup a$; this proves that I is the porosity σ-ideal generated by por. The porosity is clearly Borel.

Suppose on the one hand that $f \restriction A$ is continuous, and let $a \subset U$ be the set $\{u(O, P) : A \cap O \subset f^{-1}P\}$; we claim that for every $x \in A$, $x \in \mathrm{por}(a) \setminus \bigcup a$

holds. First, $x \in \mathrm{por}(a)$ since for every $P \in B_Y$ such that $f(x) \in P$ there is $O \in B_X$ such that $x \in O$ and $A \cap O \subset f^{-1}P$ by the continuity of the function f at x. Second, $x \notin \bigcup a$ since for every $u(O, P) \in a$, it is the case that $x \notin u(O, P)$ by the definition of the set a.

Suppose on the other hand that $a \subset U$ is an arbitrary set; we claim that $f \restriction \mathrm{por}(a) \setminus \bigcup a$ is continuous. For this, suppose that $x \in \mathrm{por}(a) \setminus \bigcup a$ is a point and $P \in B_Y$ is an open neighborhood of $f(x)$. Since $x \in \mathrm{por}(a)$, there must be a set $O \in B_X$ such that $x \in O$ and $u(O, P) \in a$. The definition of the set $u(O, P)$ shows that for every $y \in O \setminus u(O, P)$, $f(y) \in P$ holds, and so $O \cap A \subset f^{-1}P$ as desired. □

Example 6.41 Every σ-ideal I on a Polish space X which is σ-generated by closed sets is a porosity σ-ideal. If the σ-ideal is σ-generated by a coanalytic collection of closed sets then the generating porosity is coanalytic.

Proof Let U be a basis for X, and for $a \subset U$, let $\mathrm{por}(a) = X$ if $X \setminus \bigcup a \in I$, and let $\mathrm{por}(a) = 0$ otherwise. It is immediate that por preserves inclusion. In addition, every set of the form $\mathrm{por}(a) \setminus \bigcup a$ is a closed set in I, and every closed set $C \in I$ is equal to $\mathrm{por}(a) \setminus \bigcup a$ where $a \subset U$ is such that $X \setminus \bigcup a = C$. Thus the σ-ideal I is exactly the σ-ideal generated by porous sets for the porosity por. If the σ-ideal I is σ-generated by a coanalytic collection of closed sets, then it is $\mathbf{\Pi}^1_1$ on $\mathbf{\Sigma}^1_1$, and the porosity por is coanalytic by its definition. □

Proof of Corollary 6.35 Example 6.39 shows that this σ-ideal is generated by a Borel notion of abstract porosity. It is necessary to verify the bounding property for the σ-ideal I; after that, Theorem 6.33 gives the desired conclusion.

Pelant and Zelený (2004) showed that every analytic I-positive set contains a compact I-positive subset; for the case of zero-dimensional spaces there is even a simpler determinacy argument of Rojas (2006). To get the bounding property, by Fact 2.52 we must in addition prove the continuous reading of names: every Borel function $f : B \to 2^\omega$ on an I-positive Borel set can be restricted to an I-positive compact set on which it is continuous. This was proved by Zajíček and Zelený (2005), even though there is no statement of this exact result in that paper. In order to extract it, consider the σ-ideal $\hat{I}$ of the subsets of $X \times 2^\omega$ whose projection into the first coordinate is in I. Zajíček and Zelený (2005, lemma 4.1) show that $\hat{I}$ is a generalized porosity ideal in their sense and Zajíček and Zelený (2005, theorem 4.5) then show that every analytic $\hat{I}$-positive set contains a compact $\hat{I}$-positive subset. In particular, the Borel function $f \subset B \times 2^\omega$ contains a compact subset with (compact) I-positive projection; on that projection, it is clearly continuous. □

Proof of Corollary 6.36 Example 6.40 shows that the σ-ideal is associated with a Borel notion of abstract porosity. The poset P_I does not add infinitely equal real, since it preserves outer Lebesgue measure by Zapletal (2008, section 4.2.3). Thus, the corollary follows from Theorem 6.34. □

Proof of Corollary 6.37 Example 6.41 shows that the σ-ideal is associated with a coanalytic notion of abstract porosity. Thus, the σ-ideal has total canonization for the equivalence relations in the classes indicated. To upgrade to the Silver property, we need the selection property of I proved in Proposition 6.2. The Silver property follows from Theorem 5.27. □

6.2.1 The refinement game

The whole treatment of the porosity σ-ideals in this section relies on a Boolean game.

Definition 6.42 Let P be a partial ordering. The *refinement game* G (or $G(P)$ if confusion is possible) between Players I and II proceeds in the following way. In the beginning, Player II indicates an initial condition $p_{\text{ini}} \in P$. After that, the players alternate for ω many rounds, in round n Player I plays a collection $\{p_i^n : i \in \omega\}$ of countably many conditions in P, and Player II responds with a collection $\{q_i^n : i \in \omega\}$ of conditions such that $q_i^n \leq p_i$ for every $i, n \in \omega$. Player I wins if, in the end, the *result* of the play is that the Boolean expression $p_{\text{ini}} \wedge \bigwedge_n \bigvee_i q_i^n$ is nonzero in the Boolean completion of P.

Note that if I is a σ-ideal on a Polish space X, the moves of the game $G(P_I)$ are I-positive Borel sets. In this case, we will use the Borel nomenclature. Player II starts out with some $B_{\text{ini}} \subset X$, and then in round n Player I plays Borel sets B_i^n, while Player II answers with Borel sets $C_i^n \subset B_i^n$. Player I wins if the result of the play, the set $B_{\text{ini}} \cap \bigcap_n \bigcup_i C_i^n$, is I-positive.

The refinement game seems to be a close relative of the Boolean games of Zapletal (2008, section 3.10). In contradistinction to the games described there, this game does not seem to have a classical forcing equivalent. It is nevertheless very useful for the present task. In this section, we will prove the following:

Theorem 6.43 *Let P be a partial ordering.*

(i) If Player I has a winning strategy in the refinement game then the poset P is proper.
(ii) If Player II has no winning strategy, then every intermediate extension of the P-extension either is given by a single Cohen real or else contains all reals of the P-extension.

It is most interesting to investigate the partial orders of the form $P = P_I$. We provide a basic determinacy result:

Theorem 6.44 *If I is a $\mathbf{\Pi}^1_1$ on $\mathbf{\Sigma}^1_1$ σ-ideal on a Polish space X, then the refinement game $G(P_I)$ is determined.*

In the case of porosity ideals, we can directly construct a winning strategy for Player I with no reference to determinacy:

Theorem 6.45 *If I is a a porosity σ-ideal on a Polish space X, then Player I has a winning strategy in the refinement game $G(P_I)$.*

The previous theorems yield an immediate corollary which greatly improves the similar (Zapletal 2008, theorem 4.2.7):

Corollary 6.46 *If I is a porosity σ-ideal then every intermediate extension of the P_I extension is either equal to the ground model, or to the whole extension, or else it is given by a single Cohen real.*

Proof of Theorem 6.43 For (*i*), suppose that Player I has a winning strategy σ in the game G, $p \in P$ is an arbitrary condition and M is a countable elementary submodel of a large structure containing p and σ. Consider a play against the strategy σ in which all moves of Player II are in the model M, $p_{\text{ini}} = p$ and the conditions q^n_i belong to the n-th open dense subset of P_I in M in some fixed enumeration. The result of the play is a master condition for the model M.

For (*ii*), since every intermediate extension is given by a complete subalgebra of the completion of P and the Cohen algebra is up to isomorphism the unique nonatomic algebra with a countable dense subset, it is enough to do the following. Given a complete subalgebra A of the completion of the poset P which does not have a countable dense subset below any condition, we must show that the model $V[H \cap A]$ contains all reals of $V[H]$ where H is any generic filter on (the completion of) P. Let $p \in P$ be an arbitrary condition and $\dot{x}$ a P-name for an element of ω^ω. Consider the strategy σ in the game G for Player II with the following description. His first move is $p_{\text{ini}} = p$. At round n, after Player I chooses his list $\{p^n_i : n \in \omega\}$, consider the projections $\{a^n_i : i \in \omega\}$ of the sets on this list to the complete Boolean algebra A. Since the algebra A has no countable somewhere dense subsets, this collection has a disjoint refinement $\{b^n_i : n \in \omega\}$ by a result of Balcar (Balcar and Simon 1989). Let $q^n_i \leq p^n_i$ be any condition forcing b^n_i into the generic filter, which also decides the value of $\dot{x}(n)$. Note that the conditions $\{q^n_i : i \in \omega\}$ must be pairwise incompatible as $\{b^n_i : i \in \omega\}$ are. The strategy σ given by this description cannot be winning for Player II, and so there is a counterplay against it that Player I wins. The outcome of the play forces the following: for every $n \in \omega$

there is exactly one $i \in \omega$ such that $q_i^n \in \dot{H}$, and this is exactly the only $i \in \omega$ such that $b_i^n \in \dot{H}$. Therefore, the collection $\{q_i^n : i, n \in \omega\} \cap H$ is in the model $V[H \cap A]$. The point $\dot{x}/H$ can be reconstructed from it as the function which assigns to each n the unique m such that for the unique $i \in \omega$ with $q_i^n \in H$, $q_i^n \Vdash \dot{x}(n) = \check{m}$. Thus, $\dot{x}/H \in V[H \cap A]$ as desired. □

Proof of Theorem 6.44 We will unravel the refinement game to a game on a wide tree whose payoff set is G_δ, and then use Borel determinacy. To this end, let $A \subset 2^\omega \times X$ be an analytic set universal for analytic subsets of X, and let $D \subset (2^\omega)^\omega \times \omega^\omega$ be a closed set such that its projection is the analytic set $\{z \in (2^\omega)^\omega : \bigcap_n A_{z(n)} \notin I\}$. Now define the unraveled version G_u of the refinement game G: Player II begins with a point $z_{\text{ini}} \in 2^\omega$ such that $A_{z_{\text{ini}}}$ is Borel and I-positive. After that, each round n will consist of three moves: first, Player I, then Player II, and then again Player I will make a move in their turn. Player I indicates a countable list $\{B_i^n : i \in \omega\}$ of Borel I-positive sets, Player II responds with a list $\{C_i^n : i \in \omega\}$ of their Borel I-positive subsets, and Player I reacts with a point $z_n \in 2^\omega$ such that the set $A_{z(n)}$ is Borel, I-positive and a subset of $\bigcup_i C_i^n$. At some rounds of his choice, Player I in addition indicates a natural number $m(n)$. In the end, Player I wins if he produced infinitely many of the additional numbers, and the point $\langle\langle z_{\text{ini}}, z(n) : n \in \omega\rangle, \langle m(n) : n \in \omega\rangle\rangle$ belongs to D.

The game G_u has a payoff set for Player I which is G_δ in the wide tree of all possible plays: a point in the space $(2^\omega)^\omega \times \omega^\omega$ is continuously produced in the play and Player I must make sure that the production process will be complete and ends with a point in a fixed closed set. Thus, by Borel determinacy (Martin 1985), the game G_u is determined. The game G_u is obviously harder for Player I than G, so if Player I has a winning strategy in G_u, then he has a winning strategy in G. We must prove that a winning strategy σ_u for Player II in the game G_u yields a winning strategy σ in G for Player II.

To describe a winning strategy σ in the game G, we will let Player II play so that his first move is $A_{z_{\text{ini}}}$, where z_{ini} is the first move of the strategy σ_u, and then in each round n on the side he will produce a point $z(n) \in 2^\omega$ and a play p_v for every function $v : n+1 \to n+1 \cup \{\text{pass}\}$ whose numerical values stay below the identity. The following will be satisfied:

- in the game G, Player II plays a list $\{C_i^n : i \in \omega\}$ so that its union is equal to $A_{z(n)}$;
- p_v is a finite play of the game G_u of $n+1$ rounds played according to the strategy σ_u. In his last move in this play, Player I indicates the point $z(n)$ and the number (or pass) $v(n)$;
- $u \subset v \to p_u \subset p_v$.

This is not difficult to do as follows. Suppose that the play has proceeded to round n. Player I makes his move, a countable list $\{B_i^n : i \in \omega\}$ of Borel I-positive sets. Enumerate all functions v as above by $\langle v_k : k \in l\rangle$ and by induction on $k \in l+1$ construct a decreasing sequence of countable lists $\{B_i^n(k) : i \in \omega\}$ so that $B_i^n(0) = B_i^n$, and the list $\{B_i^n(k+1) : i \in \omega\}$ is the answer of strategy σ after Player I extends the play $p_{v_k \restriction n}$ with the move $\{B_i^n(k) : i \in \omega\}$. In the end, let $z_n \in 2^\omega$ be a point such that $A_{z_n} = \bigcup_i B_i^n(l)$, and let $p_{v_k} = p_{v_k \restriction n}{}^\frown\{B_i^n(k) : i \in \omega\}^\frown\{B_i^n(k+1) : i \in \omega\}^\frown v_k(n)^\frown z(n)$.

In the end of a play of the game G played according to this strategy, Player II has to win. Suppose for contradiction that he lost, and so $C_{\text{ini}} \cap \bigcup_n \bigcup_i C_i^n \notin I$; then there must be a point $y \in \omega^\omega$ such that $\langle\langle z_{\text{ini}}, z(n) : n \in \omega\rangle, y\rangle \in D$. Let $f : \omega \to \omega \cup \{\text{pass}\}$ be some function whose numerical values stay below the identity and their sequence forms the point y. Note that the play $\bigcup_n p_{f\restriction n}$ is a counterplay against the strategy σ_u in the game G_u that Player I won, contradicting the choice of the presumably winning strategy σ_u. □

Proof of Theorem 6.45 Let U be a countable collection of Borel sets and por the porosity function that define the σ-ideal I. Recall the refinement game nomenclature: at round n Player I indicates Borel sets $\{B_i^n : i \in \omega\}$ and Player II responds with $\{C_i^n : i \in \omega\}$. We will describe a winning strategy for Player I. Let Player I play so that, at round $n+1$, the collection $\{B_i^{n+1} : i \in \omega\}$ includes:

(a) some I-positive Borel subset of each set of the form $C_j^n \cap u$ where $j \in \omega$, $u \in U$, and $C_j^n \cap u \notin I$; and also
(b) some I-positive Borel subset of each set of the form $C_j^n \setminus \text{por}(a)$ where $a = \{u \in U : C_j^n \cap u \in I\}$.

Note that the sets in (b) are indeed I-positive. First, the set $C_j^n \setminus \bigcup a$ is I-positive since it is obtained from C_j^n by subtracting countably many I-small sets. Second, $C_j^n \setminus \text{por}(a) \supset C_j^n \setminus (\text{por}(a) \cup \bigcup a)$ and the latter set is obtained from the I-positive set $C_j^n \setminus \bigcup a$ by subtracting the generator $\text{por}(a) \setminus \bigcup a$ of the σ-ideal I.

As the play progresses, Player I will also create an increasing chain of countable elementary submodels M_n of some large structure and he will make sure that the sets B_j^n are all elements of the model M_n and the sets B_j^n belong to the first n many open dense sets of P_I in the models M_m for $m \in n$ under some fixed enumeration.

To show that the previous paragraph indeed constitutes a description of a winning strategy for Player I, let $D = B_{\text{ini}} \cap \bigcap_n \bigcup_i C_i^n$ be the result of a play according to this strategy; we must show that $D \notin I$. To this end, let $\{a_n : n \in \omega\}$ be a countable collection of subsets of U; we will find a point

$x \in D \setminus \bigcup_n(\text{por}(a_n) \setminus \bigcup a_n)$. To find x, by induction on n find numbers $j_n \in \omega$ so that $B^0_{j_0} \supset B^1_{j_1} \supset \cdots$ and $B^n_{j_n} \cap (\text{por}(a_n) \setminus \bigcup a_n) = 0$. This is not difficult to do. At step n, let $b = \{u \in U : C^n_{j_n} \cap u \in I\}$. The induction step divides into two options:

(a) If $a_n \not\subset b$, then find $u \in a_n \setminus b$, and choose j_{n+1} such that $B^{n+1}_{j_{n+1}} \subset C^n_{j_n} \cap u$. This is possible by item (a) in the description of the strategy.
(b) Otherwise, find j_{n+1} so that $B^{n+1}_{j_{n+1}} \subset C^n_{j_n} \setminus \text{por}(b)$. This is again possible by item (b) of the description of the strategy.

In both cases, the induction hypotheses are satisfied, in the latter case because $\text{por}(a_n) \subset \text{por}(b)$ holds, as implied by $a_n \subset b$. In the end, the filter generated by sets $B^n_{j_n}$ for $n \in \omega$ will be P_I-generic over the model $\bigcup_n M_n$; thus, the intersection $\bigcap_n B^n_{j_n}$ is nonempty and contains a single point which is P_I-generic over $\bigcup_n M_n$. The unique point $x \in X$ in that intersection has the desired properties. □

6.2.2 The canonization results

The following theorem in conjunction with the results of the previous section implies Theorem 6.34:

Theorem 6.47 *Let I be a σ-ideal on a Polish space X such that the poset P_I is proper. If Player II has no winning strategy in the refinement game associated with P_I and the quotient poset P_I does not add infinitely equal real, then I has total canonization for the equivalence relations Borel reducible to E_{K_σ} or $=_J$ for some analytic P-ideal J on ω or classifiable by countable structures.*

Proof We will start with a general claim. By a *lower semicontinuous extended-value pseudometric* on a Polish space Y we mean a function $d : Y^2 \to [0, \infty]$ which satifies the triangle inequality, $d(y, y) = 0$ for all $y \in Y$, and for every real $\varepsilon > 0$ the set $\{\langle y, z\rangle \in Y^2 : d(y, z) > \varepsilon\}$ is open in Y^2.

Claim 6.48 *Let d be a lower semicontinuous extended-value pseudometric on a Polish space Y, let $B \subset X$ be Borel I-positive set, and $f : B \to Y$ be a Borel function, and let $c > 0$ be a real number. At least one of the following happens:*

(i) either there is a Borel I-positive set $C \subset B$ such that the d-diameter of $f''C$ is not greater than c;
(ii) or, for every countable collection $\{B_j : j \in \omega\}$ of Borel I-positive subsets of B, there are Borel I-positive sets $\{C_j : j \in \omega\}$ such that $C_j \subset B_j$ for all j, and if $i \neq j \in \omega$ and $x \in C_i$, $y \in C_j$, then $d(f(x), f(y)) > c/2$.

Proof Suppose that (*i*) fails. To prove (*ii*), it is enough to show a weaker version in which i in clause (*ii*) is equal to 0 and then apply the claim repeatedly to the collections $\{C_j : j > k\}$ with increasing $k \in \omega$. For the weaker version, we first claim that for every $j > 0$ there are infinitely many Borel I-positive subsets $\{C_j^k : k \in \omega\}$ of C_j such that for every $k \neq l$ and every $x \in C_j^k$ and every $y \in C_j^l$ we have $d(f(x), f(y)) > c$. To find these subsets, first thin out C_j if necessary so that f-preimages of open sets are either empty or I-positive in C_j. Then, by induction on $k \in \omega$ build points $x_k, y_k \in C_j$ and open sets $O_k, P_k \subset Y$ so that:

I1. $f(x_k) \in O_k$, $f(y_k) \in P_k$, and for every $x' \in O_k$ and $y' \in P_k$ we have $d(x', y') > c$;
I2. O_{k+1}, P_{k+1} are both subsets of P_k.

This is not difficult to do as follows. Suppose that x_k, y_k, O_k, P_k have been constructed. The Borel set $f^{-1}P_k \cap C_j$ is nonempty, containing y_k, and therefore I-positive. The failure of item (*i*) shows that there are points x_{k+1}, y_{k+1} in it so that $d(f(x_{k+1}), f(y_{k+1})) > c$. Use the lower semicontinuity of the pseudometric d find open subsets $O_{k+1}, P_{k+1} \subset P_k$ containing x_{k+1} and y_{k+1} repectively such that for every $x' \in O_{k+1}$ and $y' \in P_{k+1}$ we have $d(x', y') > c$. This completes the induction step. After the induction is performed, the sets $C_j^k = C_j \cap f^{-1}O_k$ for $k \in \omega$ have the desired properties.

Now, if $x \in B_0$ is a P_I-generic point, for every $j > 0$ there can be at most one $k \in \omega$ such that for some $y \in C_j^k$, $d(f(x), f(y)) \leq c/2$. Let $\dot{g}$ be a name for a function from $\omega \setminus \{0\}$ to ω that for each j identifies such k if it exists. Since the poset P_I is assumed not to add an infinitely equal real, we may thin out the set B if necessary to find a ground model function $h \in \omega^\omega$ such that $B \Vdash \check{h} \cap \dot{g} = 0$. Let M be a countable elementary submodel of a large enough structure containing the set B and the point g. The poset P_I is proper, and so the Borel set $C_0 = \{x \in B : x \text{ is } P_I\text{-generic over } M\}$ is I-positive by Fact 2.50. Now, if $x \in C_0$ and $j \in \omega$ then $M[x] \models \forall y \in C_j^{g(j)}\ d(g(x), g(y)) > c/2$ by the forcing theorem. By the coanalytic absoluteness between the model $M[x]$ and V, for every $y \in C_j^{g(j)}$, $d(g(x), g(y)) > c/2$ holds even in V. Therefore, the sets C_0 and $C_j = C_j^{g(j)}$ for $j > 0$ are as required. □

We will now treat the case of equivalence relations Borel reducible to E_{K_σ}. Recall the lower semicontinuous extended-value pseudometric d on ω^ω defined by $d(x, y) = \sup_n |x(n) - y(n)|$; so, E_{K_σ} connects points with finite distance. This is a lower semicontinuous extended-value pseudometric on ω^ω. Let B be a Borel I-positive set and let E be an equivalence relation on the set B, with a Borel reduction $f : B \to \text{dom}(E_{K_\sigma})$ of E to E_{K_σ}. Claim 6.48 gives us two options: either there is a Borel I-positive set $C \subset B$ and a number

$n \in \omega$ such that the d-diameter of C is less than n – in such a case, the set C is a subset of a single E-equivalence class; or clause Claim 6.48(*ii*) is satisfied and so the following description gives a strategy for Player II in the refinement game. He starts with the set $B = B_{\text{ini}}$ and at round n answers with sets C^n_j of diameter $< 2^{-n}$ such that for $i \neq j$ and $x \in C^n_j$, $y \in C^n_i$ it is the case that $d(f(x), f(y)) > n$. Since this is not a winning strategy, there must be a counterplay by Player I in which the result $C \subset B$ is I-positive. This is the desired Borel I-positive set such that $E \restriction C = \text{id}$: if $x \neq y$ are distinct points in the set C, it must be the case that $d(f(x), f(y)) > n$ for every number n and therefore $\neg x\ E\ y$. This concludes the argument in the case of equivalence relations Borel reducible to E_{K_σ}.

To prove the theorem for equivalence relations Borel reducible to $=_J$ for an analytic P-ideal J on ω, first use a theorem of Solecki (1999) to find a lower semicontinuous submeasure ϕ on ω such that $J = \{a \subset \omega : \lim_k \phi(a \setminus k) = 0\}$. For every $n \in \omega$, let d_n be the lower semicontinuous extended-value pseudometric on 2^ω given by $d_n(y, z) = \phi(\{m > n : y(m) \neq z(m)\})$; this is a pseudometric by the subadditivity of ϕ. Let $B \subset X$ be a Borel I-positive set, let E be an equivalence relation on B, and let $f : B \to 2^\omega$ be a Borel function reducing E to $=_J$. There are two distinct cases:

Case 1 For every positive real $\varepsilon > 0$, a natural number $n \in \omega$, and every Borel I-positive set $B' \subset B$ there is a Borel I-positive set $C' \subset B'$ such that the d_n-diameter of $f''C'$ is smaller than ε. In this case, we will find a Borel I-positive set $C \subset B$ and an alternative Polish topology τ on C such that $E \restriction C$ is a closed relation in the topology $\tau \times \tau$. Gao (2009, proposition 5.4.7) shows that every closed equivalence relation is smooth, and then the work below E_{K_σ} above will conclude the canonization in this case.

To find the set C, consider the following strategy of Player II in the refinement game. He starts with $B = B_{\text{ini}}$, and at round n he plays sets C^n_j so that the d_n-radius of $f''C^n_j$ is smaller than 2^{-n}. Since this is not a winning strategy, there must be a counterplay whose result $C \subset B$ is an I-positive set. Consider a Polish topology τ on D which generates the same Borel structure as the original one, and which makes all the sets $C^n_j \cap C$ open. We claim that $E \restriction C$ is a closed relation in the topology $\tau \times \tau$. Suppose that $x, y \in C$ are E-unrelated points, so there is $n \in \omega$ such that $\lim_k d_k(f(x), f(y)) > 2^{-n}$. Let i, j be numbers such that $x \in C^{n+2}_i$, $y \in C^{n+2}_j$. The choice of the sets C^{n+2}_i, C^{n+2}_j together with the triangle inequality shows that $E \cap (C^{n+2}_i \times C^{n+2}_j) = 0$. Since the sets $C^{n+2}_i \cap C$, $C^{n+2}_j \cap C$ are open in the new topology, we produced an open neghborhood of the point $\langle x, y\rangle \notin E$ which is disjoint from E, showing that the complement of $E \restriction D$ is open, and $E \restriction D$ is closed in the new topology as required.

Case 2 If Case 1 fails, then there is a Borel I-positive set $B' \subset B$, a natural number $n \in \omega$, and a positive real $\varepsilon > 0$ such that in every I-positive Borel set $C' \subset B'$ there are points $x, y \in C'$ such that $d_n(f(x), f(y)) > \varepsilon$. Since for every $m \in \omega$, by the σ-additivity of I the set $C' \subset B'$ can be further thinned out so that the finite sequence $f(x) \restriction m$ does not depend on the choice of $x \in C'$, it must be the case that for every $m \in \omega$ there are points $x, y \in C'$ such that $d_m(f(x), f(y)) > \varepsilon$. We will find a Borel I-positive set $C \subset B$ such that $E \restriction B = \mathrm{id}$.

To find the set C, consider the following strategy for Player II in the refinement game. He starts out with $B_{\mathrm{ini}} = B$ and after Player I indicates sets $\{B^n_j : j \in \omega\}$ at round $n \in \omega$, Player II answers with sets $\{C^n_j \subset B^n_j : j \in \omega\}$ of diameter $\leq 2^{-n}$ for some fixed complete metric on the space X such that for every $i \neq j$, every $x \in C^n_i$, and every $y \in C^n_j$ the distance $d_n(f(x), f(y))$ is greater than $\varepsilon/2$. This is possible by Claim 6.48 applied to d_n and the case assumption. The assumption of the theorem shows that this strategy is not winning for Player II and so there is a counterplay in which Player I wins, giving an I-positive outcome $C \subset B$. To see that the set C consists of pairwise E-unrelated points, let $x \neq y \in C$ and $n \in \omega$. Whenever $m > n$ is so large that the distance between x, y is greater than 2^{-m}, then the numbers $i, j \in \omega$ such that $x \in C^m_i$ and $y \in C^m_j$ are distinct, and so $d_n(f(x), f(y)) > \varepsilon/2$. As n was arbitrary, $\neg x \; E \; y$ as desired.

This concludes the argument for the equivalence relations Borel reducible to $=_J$ for an analytic P-ideal J. For the total canonization for equivalence relations classifiable by countable structures, use Corollary 4.10. Note that total canonization for essentially countable equivalence relations follows from the E_{K_σ} case, as essentially countable equivalence relations are Borel reducible to E_{K_σ}. □

Proof of Theorem 6.33 Let I be a σ-ideal on a Polish space X associated with some coanalytic porosity por such that the quotient poset P_I is bounding. The σ-ideal I is $\mathbf{\Pi}^1_1$ on $\mathbf{\Sigma}^1_1$ and the poset P_I preserves Baire category by Fact 6.32(*i*) and it is bounding by the assumption. All such σ-ideals have the rectangular Ramsey property by Fact 2.56.

For the total canonization, let $B \subset X$ be a Borel I-positive set, and let E be an analytic equivalence relation on B. Consider the P_I-extension and consult the trichotomy Theorem 3.5 and its three cases:

Case 1 There is a Borel I-positive set $C \subset B$ such that $E \restriction C = \mathrm{ev}$. In this case, we are happily in one of the clauses of total canonization.

Case 2 The intermediate model case is impossible. By Theorem 6.43, the intermediate model is either given by a single Cohen real (and this contradicts the assumption that P_I is bounding), or it contains all the reals in the extension. Since the P_I extension is given by its generic real, this would mean that the intermediate model must be in fact equal to the whole extension.

Case 3 There is a Borel I-positive set C such that every two I-positive Borel subsets of C are E-connected. We claim that there is an E-class with I-positive intersection with the set C. Suppose for contradiction that this fails. Then, the analytic set $E \subset C \times C$ has I-small vertical sections. Since the σ-ideal I is $\mathbf{\Pi}^1_1$ on $\mathbf{\Sigma}^1_1$, it is possible to use the first reflection theorem to find a Borel superset $E' \supset E$ such that $E' \cap (C \times C)$ has I-small vertical sections. Applying the rectangular Ramsey property of I to E', we see that there are Borel I-positive sets $C_0, C_1 \subset C$ such that $C_0 \times C_1$ is either a subset of or disjoint from E'. Both options lead to a contradiction: the former clashes with the fact that vertical sectons of E' are in I, the latter with the ergodicity of E on C. We conclude that there is an E-class with I-positive intersection with C. This intersection is analytic, and by Fact 6.32 it has a Borel I-positive subset. We are again in one of the clauses of total canonization.

The total canonization is confirmed in each of these cases. For the Silver property of the σ-ideal under the extra assumption that the porosity is Borel, note that I is a game ideal by Fact 6.32(*iii*), use Theorem 5.20 to conclude that I has the selection property and Theorem 5.27 to conclude that it has the Silver property as desired. □

Proof of Theorem 6.34 Let I be a σ-ideal on a Polish space X associated with some porosity por such that the quotient poset P_I does not add an infinitely equal real. The σ-ideal I has total canonization for equivalences Borel reducible to E_{K_σ}, or to $=_J$ for an analytic P-ideal J, or classifiable by countable structures by Theorem 6.47 and Theorem 6.45. If the porosity por is Borel, then the σ-ideal I is a game ideal by Fact 6.32(*iii*), it has the selection property by Theorem 5.20, and the Silver dichotomy for Borel equivalence relations in the given classes follows from Theorem 5.27. □

We conclude this section with some open questions.

Question 6.49 Does the σ-ideal of metrically σ-porous sets on any given compact metric space have the free set property? Does it have a coding function?

Question 6.50 Does the σ-continuity ideal have the rectangular Ramsey property? How about the total canonization for analytic equivalence relations? The free set property? Does it have a rectangular or square coding function?

Question 6.51 Is it true that if Player I has a winning strategy in the refinement game then the σ-ideal I has the selection property? If Player I has a winning strategy and P_I adds no infinitely equal real, does it follow that I has total canonization for analytic equivalence relations?

6.3 Laver forcing

The Laver forcing P is the partial order of *Laver trees*: those trees $T \subset \omega^{<\omega}$ such that there is a finite trunk and all nodes past the trunk split into infinitely many immediate successors. The associated *Laver ideal* I on ω^ω is generated by all sets A_g where $g : \omega^{<\omega} \to \omega$ is a function and $A_g = \{f \in \omega^\omega : \exists^\infty n\ f(n) \in g(f \restriction n)\}$.

Fact 6.52

(i) Every analytic subset of ω^ω is either in I or it contains all branches of a Laver tree, and these two options are mutually exclusive.
(ii) The function $\pi : P \to P_I$ defined by $\pi(T) = [T]$, is a dense embedding of the poset P into P_I.
(iii) The forcing P is proper and adjoins a minimal generic extension.
(iv) I is a game ideal.

Proof (*i*) comes from Brendle *et al.* (1995), see also Zapletal (2008, proposition 4.5.14). (*ii*) follows immediately from (*i*). The minimality of Laver extension is due to Groszek (1987, theorem 7); with some additional work, it can be also derived from Theorem 5.21. The Laver game was in its simplest form presented in Zapletal (2008, section 4.5); see also Example 5.10. □

In the canonization direction, we will prove the following:

Theorem 6.53 *The Laver ideal has total canonization for equivalences classifiable by countable structures. If ω_1 is inaccessible to the reals, then the Laver ideal has the Silver property for Borel equivalences classifiable by countable structures.*

Doucha (2013) extended the conclusions of Theorem 6.53 to equivalence relations that are Borel reducible to $=_J$ for F_σ P-ideals on ω. In the anti-canonization direction, we have:

Theorem 6.54 *E_{K_σ} is in the spectrum of the Laver ideal. The Laver ideal has a rectangular coding function defined on the whole ω^ω.*

Corollary 6.55 *Whenever J is a σ-ideal on a Polish space Y such that the quotient forcing P_J is proper and adds a dominating real, then E_{K_σ} is in the spectrum of the ideal J, and J has a rectangular coding function.*

Proof of Theorem 6.53 The total canonization follows Corollary 4.10 once we verify the (well-known) fact that the Laver forcing is nowhere c.c.c.: given a Laver tree $T \subset \omega^{<\omega}$, for every splitnode $t \in T$ write $b^t \subset \omega$ for the infinite set $\{n \in \omega : t^\frown n \in T\}$, fix an uncountable collection $\{b^t_\alpha : \alpha \in \omega_1\}$ of pairwise almost disjoint infinite subsets of ω, and let $A_\alpha = \{x \in [T]$: for every $n \in \omega$ such that $x \restriction n$ is a splitnode of T, $x(n) \in b^{x\restriction n}_\alpha\}$ for every $\alpha \in \omega_1$. The sets $\{A_\alpha : \alpha \in \omega_1\}$ are easily checked to be I-positive and their pairwise intersections are in the σ-ideal I.

For the Silver property, note that I is not $\mathbf{\Pi}^1_1$ on $\mathbf{\Sigma}^1_1$. Laver forcing adds a dominating real; no ideal whose quotient forcing adds a dominating real cannot be $\mathbf{\Pi}^1_1$ on $\mathbf{\Sigma}^1_1$ (Zapletal 2008, proposition 3.8.15). However, I is a game ideal by Fact 6.52(*iv*) and therefore $\mathbf{\Delta}^1_2$ on $\mathbf{\Sigma}^1_1$ by Theorem 5.13. Thus, the large cardinal assumption yields the Silver property by Theorem 5.20 and Theorem 5.27. □

Proof of Theorem 6.54 To place E_{K_σ} in the spectrum of I, consider the equivalence relation E on the set of increasing functions in ω^ω defined by $x\ E\ y$ if and only if there is a number $n \in \omega$ such that for all m, $x(m) < y(m+n)$ and $x(m) < y(m+n)$. In other words, a finite shift of the graph of x to the right is below the graph of y, and vice versa, a finite shift of the graph of x is below the graph of y.

We claim that $E \leq_B E \restriction B$ for every Borel I-positive set. Such positive set must contain all branches of some Laver tree T; to simplify the notation assume that T has empty trunk. Find a level-preserving injection π from the set of all finite increasing sequences of natural numbers into T so that if the last number on s is less than the last number on t then also the last number on $\pi(s)$ is less than the last number on $\pi(t)$. It is immediate that π extends to a continuous map which embeds E to $E \restriction B$.

The last point is to show that E_{K_σ} is bireducible with E. To embed E_{K_σ} in E, look at the two partitions $\omega = \bigcup_n K_n = \bigcup_n L_n$ into consecutive intervals such that $|K_n| = n+1$, $|L_n| = 2(n+1)$ and for a point $y \in \mathrm{dom}(E_{K_\sigma})$ let $f(y)$ be the function in ω^ω specified by $f(y)(\min(K_n)+i) = \min(L_n)+i+y(n)$ for every n and $i \in n$. It is not difficult to see that f reduces E_{K_σ} to E. To embed E into E_{K_σ}, note that $E = \bigcup_n F_n$ is a countable union of closed sets with compact sections: just write $F_n = \{\langle x, y\rangle$: for every $m \in \omega$, $x(m) < y(m+n)$ and $y(m) < x(m+n)$. Viewing ω^ω as a G_δ subset of a compact metric space Y, consider the closures $\overline{F}_n$ of F_n in $Y \times Y$. Since the vertical sections of F_n are compact, the closures $\overline{F}_n$ have the same sections corresponding to the points in ω^ω as the sets F_n. Thus, $\tilde{E}$, the relation on the space Y defined by $y\ \tilde{E}\ z$ if either $z, y \notin \omega^\omega$ or there is n such that $\langle y, z\rangle \in \overline{F}_n$, is a K_σ equivalence relation equal to E on ω^ω. By a result of Rosendal (2005), it is Borel reducible to E_{K_σ} and the restriction of that Borel reduction reduces E to E_{K_σ} as desired.

To construct the rectangular coding function, it is enough to find a Borel function $f : \omega^\omega \times \omega^\omega \to 2^\omega$ such that the image of any analytic rectangle with I-positive sides has nonempty interior, and then use Proposition 2.67. Define the function f by $f(x, y)(n) = 1 \leftrightarrow x(n) > y(n)$ for $x, y \in \omega^\omega$. To check the requisite properties of f, for any given I-positive analytic sets $B, C \subset \omega^\omega$ find respective Laver trees $T, S \subset \omega^{<\omega}$ with trunk of equal length some $n \in \omega$ such that $[T] \subset B$ and $[S] \subset C$. Then, for every binary sequence $z \in \omega^\omega$, it is easy to find two branches $x \in [T]$, $y \in [S]$ so that $f(x, y) = z$ on all entries past n. So $f''(B \times C)$ contains a nonempty open set. □

Proof of Corollary 6.55 By Propositions 2.69 and 2.70, it is enough to show that there is a Borel J-positive set $B \subset Y$ such that the σ-ideal $J \restriction B$ is Katětov above the Laver ideal. To do this, use the assumptions and Fact 2.51 to find a Borel J-positive set $B \subset Y$ and a Borel function $f : B \to \omega^\omega$ such that for every point $y \in \omega^\omega$, the set $\{x \in B : f(x)$ does not modulo finite dominate $y\}$ is in the σ-ideal I. Let $\pi : \omega^{<\omega} \to \omega$ be a bijection and let $g : B \to \omega^\omega$ be the Borel function defined by $g(x)(n) = f(x)(\pi(x \restriction n))$.

We claim that g is the required Katětov map; in other words, g-preimages of sets in the Laver σ-ideal I are in J. For, if $h : \omega^{<\omega} \to \omega$ is any function and $A_h \subset \omega^\omega$ is the associated generating set for the Laver ideal, $A_h = \{y \in \omega^\omega : \exists^\infty n\ y(n) \in h(y \restriction n)\}$, then the set $B_h = \{x \in B : f(x) \circ \pi$ does not dominate h modulo finite$\}$ is I-small by the properties of the function h, and $g^{-1}A_h \subset B_h$ by the definitions. This concludes the proof of the corollary. □

Question 6.56 Does the Laver ideal have total canonization for equivalence relations Borel reducible to orbit equivalence relations of Polish group actions?

6.4 Fat tree forcings

The purpose of this section is to prove the following theorem, complementing Corollary 4.10:

Theorem 6.57 *There is a nowhere c.c.c. $\mathbf{\Pi}^1_1$ on $\mathbf{\Sigma}^1_1$ σ-ideal I on 2^ω such that the forcing P_I is proper and adds a minimal forcing extension, and E_2 belongs to the spectrum of I. In addition, I consists of null sets and E_2 is nontrivially I-ergodic.*

Question 6.58 Does the theorem remain true with any turbulent orbit equivalence relation in place of E_2?

The σ-ideal in question is obtained so that the quotient poset P_I is one of the *fat tree forcings* of Zapletal (2008, section 4.4.3). Consider the following general definition:

Definition 6.59 Let T_{ini} be a finitely branching tree consisting of finite sequences; for a node $t \in T_{\text{ini}}$ let $a_t = \{i : t^\frown i \in T_{\text{ini}}\}$. Suppose that $\{\phi_t : t \in T_{\text{ini}}\}$ is a collection of submeasures on the respective sets a_t such that $\lim_t \phi_t(a_t) = \infty$. A *fat tree* is a tree $T \subset T_{\text{ini}}$ such that $\lim_{t \in T} \phi_t(b_t) = \infty$, where $b_t = \{i \in a_t : t^\frown i \in T\}$. The *fat tree forcing* P_ϕ (or P if no confusion is possible) is the set of all fat trees ordered by inclusion.

The definition of the fat tree forcing depends on the choice of the submeasures, and so do many of its forcing and canonization properties. We will first record their common forcing features.

Fact 6.60 (Zapletal 2008, theorems 4.4.2 and 4.4.8) *P is $< \omega_1$-proper, bounding, and adds no independent reals.*

The σ-ideals associated with fat tree posets have been identified. For every $k \in \omega$ and every map g on T_{ini} that assigns to each node t a set $g(t) \subset a_t$ such that $\phi_t(g(t)) < k$, write $A_g = \{x \in [T_{\text{ini}}] : \exists^\infty n\ x(n) \in g(x \restriction n)\}$. Write I for the σ-ideal on $[T_{\text{ini}}]$ generated by sets A_g for all numbers k and maps g as above.

Fact 6.61 *For every analytic set $A \subset [T_{\text{ini}}]$, exactly one of the following occurs:*

(i) $A \in I$;
(ii) A contains all branches of some fat tree.

The σ-ideal I is $\mathbf{\Pi}^1_1$ on $\mathbf{\Sigma}^1_1$. It is a game ideal.

Proof The dichotomy is obtained in Zapletal (2008, proposition 4.4.14) and the integer game is a part of the proof. The complexity of the σ-ideal I follows from Zapletal (2008, theorem 3.8.9). $\square$

Corollary 6.62 *The forcing P_I adds a minimal forcing extension and is nowhere c.c.c. The ideal I has total canonization for equivalence relations classifiable by countable structures, and the Silver property for Borel equivalence relations classifiable by countable structures.*

Proof We will first argue for the minimal forcing extension part. The game σ-ideal I has the selection property by Theorem 5.20. Theorem 5.21 then implies that the only nontrivial intermediate extensions of P_I are c.c.c. extensions of the ground model. Such c.c.c. extensions are ruled out though by Corollary 2.63, as P_I adds no independent reals. The same argument shows that P_I is nowhere c.c.c. itself.

Now, the σ-ideal has total canonization for equivalence relations classifiable by countable structures by Corollary 4.10. By the selection property and Theorem 5.27, this imples the Silver property for Borel equivalence relations classifiable by countable structures. □

Further canonization properties of the σ-ideal I depend heavily on the particular choice of the submeasures $\{\phi_t : t \in [T_{\text{ini}}]\}$.

Proof of Theorem 6.57 We will show that there is a choice of submeasures such that E_2 is in the spectrum of I. The underlying space $[T_{\text{ini}}]$ of I will be in natural bijection with 2^ω.

The approach uses a concentration of measure trick. Let $\omega = \bigcup_n J_n$ be a partition of ω into successive finite intervals so long that for every $n \in \omega$, we have $1/(n+1)^2 \geq 2\exp(-2^{-n-1}/8k_{n+1})$, where $k_n = \Sigma_{j\in J_n} 1/(j+1)^2$. Such a partition exists since the sum $\Sigma_{j\in\omega} 1/(j+1)^2$ converges. Consider the set $a_n = 2^{J_n}$ equipped with the normalized counting measure μ_n and the distance $d_n(x, y) = \sum\{\frac{1}{j+1} : x(j) \neq y(j)\}$. The key point behind these initial choices is the following consequence of Corollary 2.66: for every pair $b, c \subset a_n$ of μ_n-mass $> 1/(n+1)^2$ each there are points $x \in b$ and $y \in c$ of d_n-distance $\leq 2^{-n}$. Consider the fat tree forcing P on the initial tree $T_{\text{ini}} = \{t : t$ is a finite sequence such that $\forall n \in \text{dom}(t)\ t(n) \in a_n\}$, and let the submeasure ϕ_n be the normalized counting measure μ_n on a_n multiplied by $(n+1)^2$. We claim that E_2 belongs to the spectrum of the resulting ideal I.

We will first identify a natural isomorphism between the domain $[T_{\text{ini}}]$ of the σ-ideal I and 2^ω, and argue for the second sentence of Theorem 6.57. Let $\pi : [T_{\text{ini}}] \to 2^\omega$ be the map defined by $\pi(x) =$ the concatenation of all sequences on x; it is clear that π is a homeomorphism. Let $E = \pi^{-1} E_2$.

Claim 6.63 *The π-image of the ideal I consists of sets of zero λ-mass, where λ is the usual Borel probability measure on 2^ω.*

Proof It is enough to verify that $\lambda(\pi'' A_g) = 0$ for every generating function g, where $A_g = \{x \in [T] : \exists^\infty n\ x(n) \in g(x \restriction n)\}$ and for some $k \in \omega$, for every $t \in T_{\text{ini}}$, $\phi_{|t|}(g(t)) < k$. For every $n \in \omega$ write $A_n = \{x \in T_{\text{ini}} : x(n) \in g(x \restriction n)\}$. For every sequence $t \in T_{\text{ini}}$ of length n, the normalized counting mass of $g(t)$ inside the set a_n is at most $\frac{k}{(n+1)^2}$, and so $\lambda(A_n) \leq \frac{k}{(n+1)^2}$. Thus, $A_g = \bigcap_{m\in\omega} \bigcup_{n>m} A_n$ is a set of zero λ-mass as required. □

Claim 6.64 *E is nontrivially I-ergodic.*

Proof Let $B, C \subset [T_{\text{ini}}]$ be Borel I-positive sets, and use Fact 6.61 to find fat trees $S, T \in P$ such that $[S] \subset B$, $[T] \subset C$. For every $t \in T$, write $b_t^T = \{u \in a_{|t|} : t^\frown u \in T\}$, and similarly on the S side. Let $t_0 \in T$, $s_0 \in S$ be nodes of equal length, say m, such that for all $t \in T$ extending t_0, we have $\phi_{|t|}(b_t^T) > 1$ and similarly on the S side. To simplify the notation, assume that $n = 0$ and so $s_0 = t_0 = 0$.

To produce a pair of E-related points $x \in [S]$ and $y \in [T]$, by induction on $n \in \omega$ find $t_n \in T$, $s_n \in S$ so that:

I1. t_{n+1} is an immediate successor of t_n, and similarly on the S-side;
I2. writing $u_n, v_n \in 2^{J_n}$ for the binary strings such that $(t_n)^\frown u_n = t_{n+1}$ and $(s_n)^\frown u_n = s_{n+1}$, then $d_n(u_n, v_n) < 2^{-n+1}$.

The inductive construction is enabled by the concentration of measure. Suppose that t_n, s_n have been obtained, and write $c = \{u \in a_n : (t_n)^\frown u \in T\}$ and $d = \{v \in a_n : (s_n)^\frown v \in S\}$. Both of these sets have ϕ_n-mass at least 1, so normalized counting mass inside a_n at least $\frac{1}{(n+1)^2}$, and so by the first paragraph there are $u \in c$ and $v \in d$ such that $d_n(u, v) < 2^{-n+1}$ and the induction step can be performed. In the end, let $x = \bigcup_n t_n$ and $y = \bigcup_n s_n$ and observe that $x \mathrel{E} y$ holds.

To produce a pair of E-unrelated points $x \in [S]$ and $y \in [T]$, consider the tree $\bar{S}$ obtained from S by flipping all bits on all sequences appearing in S. The previous construction applied to $T, \bar{S}$ yields E-related points $y \in [T]$ and $\bar{x} \in [\bar{S}]$. Flip all bits on all sequences appearing in $\bar{x}$, obtaining x. It is not difficult to see that $x \in S$ and $\neg x \mathrel{E} \bar{x}$. Thus, the points $x \in [S]$ and $y \in [T]$ are E-unrelated as desired. □

To prove that E_2 is in the spectrum of I, we will argue that for every fat tree T, $E \restriction [T]$ is bireducible with E_2. Certainly, E is Borel reducible to E_2. Assume for contradiction that for some fat tree $T \in P$, $E_2 \leq_B E \restriction [T]$ fails. Then, by the E_2 dichotomy 2.17, $E_2 \restriction [T]$ is Borel reducible to a countable Borel equivalence relation, and as such it is classifiable by countable structures. By the total canonization for equivalence relations classifiable by countable structures, T can be thinned out if necessary so that $E \restriction [T] = \text{id}$ or $E \restriction [T] = \text{ev}$. This contradicts Claim 6.64, and proves Theorem 6.57. □

6.5 The Lebesgue null ideal

Let λ be the usual Borel probability measure on 2^ω, and let $I = \{B \subset 2^\omega : \lambda(B) = 0\}$. The quotient forcing P_I is the random forcing. It is not difficult to see that the spectrum of the σ-ideal I contains a great number of equivalence relations:

Theorem 6.65 *The spectrum of I contains E_0, E_2, E_{K_σ}, and F_2. It is cofinal in the Borel reducibility ordering of analytic equivalence relations.*

Still, we are able to prove two ergodicity theorem, yielding canonization in certain limited contexts:

Theorem 6.66 *If $E \supset E_2$ is an equivalence relation classifiable by countable structures, then E has an equivalence class of λ-mass* 1.

In other words, for every equivalence relation F classifiable by countable structures, E_2 is F, λ-ergodic.

Let μ be the Borel probability measure on the space $(2^\omega)^\omega$ obtained as the product of countably many copies of λ. We have:

Theorem 6.67 *Let E be an analytic equivalence relation on $X = (2^\omega)^\omega$ such that $F_2 \subset E$. Then:*

(i) either E has an equivalence class of μ-mass 1*;*
(ii) or $F_2 \leq_B E \restriction C$ for every Borel set $C \subset X$ of positive μ-mass.

Restated, for every analytic equivalence relation E, exactly one of the following happens: either $F_2 \leq E$ or else F_2 is E, μ-ergodic.

Proof of Theorem 6.66 Theorem 6.57 and Corollary 6.62 show that there is a σ-ideal $J \subset I$ which has total canonization for equivalence relations classifiable by countable structures, and E_2 is nontrivially ergodic for J. The present theorem follows abstractly from these properties of the σ-ideal J.

Let $E \supset E_2$ be an equivalence relation classifiable by countable structures. Observe that E can have at most one J-positive class. If there were two such classes, then as E_2 is nontrivially J-ergodic, they would have to contain E_2-connected points. Since $E_2 \subset E$, these two points would have to be also E-connected, which is impossible if they come from distinct E-classes. Let B be the complement of the single J-positive equivalence class if such exists, otherwise let $B = X$. We will argue that the set B has zero mass. First of all, it is a coanalytic set and therefore measurable. If B had nonzero mass, it would contain a Borel subset $B' \subset B$ of nonzero mass. The set B' is J-positive and its E-equivalence classes are all in J. Since J has total canonization for equivalence relations classifiable by countable structures, it must be the case that B' contains an J-positive Borel subset B'' on which E is the identity. This is impossible by the J-ergodicity of the equivalence relation $E_2 \subset E$: since the σ-ideal J is nonprincipal, it is possible to divide the set B'' into two disjoint Borel J-positive pieces, these should contain E_2-connected points, which would also be E-connected. This contradiction completes the proof of the theorem. □

For the proof of Theorem 6.67, we will need some preliminary definitions of independent interest. Consider the infinite product μ of ω-many copies of the measure λ; so μ is a Borel probability measure on the space $X = (2^\omega)^\omega$. Let I be the σ-ideal on the space X consisting of sets of μ-mass 0; this is isomorphic to the original definition of the σ-ideal. The poset $P = P_I$ adds a generic point $\dot{x} \in X$. We will also utilize the posets $P^{*n} = P_{I^n}$ for $n \in \omega$, where I_n is the σ-ideal on X^n consisting of the set of μ_n-mass zero, where μ_n is the product measure on X^n. The poset P_n adds an n-tuple of points $\langle \dot{x}_i : i \in n \rangle$. The genericity of the tuple $\langle x_i : i \in n \rangle$ is equivalently restated as $\langle x_i : i \in n \rangle \notin A$ for every μ^n-null set $A \subset X^n$ in the model M.

Definition 6.68 Let M be a transitive model of set theory and $a \subset 2^\omega$ be a countable set. We will say that the set a is *random-symmetric* over M if there is a point $x \in X$ P-generic over M such that $a = \text{rng}(x)$. A finite collection $\{a_i : i \in n\}$ of countable subsets of 2^ω will be called *mutually random-symmetric* over M if there is a tuple $\langle x_i : i \in n \rangle$ P^{*n}-generic over M such that for every $i \in n$, $a_i = \text{rng}(x_i)$.

We will need some elementary manipulations of P-generic and P^{*n}-generic points, as recorded in the following claim. Identifying X with $(2^\omega)^{\mathbb{Z}}$ using any bijection between ω and $\mathbb{Z}$, we define the *shift* $s : X \to X$ by $s(x)(m) = x(m+1)$ for every $m \in \omega$, and similarly for the shift of the finite powers of X^n.

Claim 6.69 *Let M be a transitive model of set theory, let $n \in \omega$ and let $\langle x_i : i \in n \rangle \in X^n$ by a P^{*n}-generic point over M.*

(i) Permuting finitely many points in $\text{rng}(x_i)$ for each $i \in n$ yields another P^{*n}-generic point over M.
(ii) If $\{b_j : j \in m\}$ are pairwise disjoint nonempty subsets of n then there is a P^{*m}-generic point $\langle y_j : j \in m \rangle$ over M such that $\text{rng}(y_j) = \bigcup_{i \in b_j} \text{rng}(x_i)$.
(iii) If $C \in M$ is a Borel μ^n-positive subset of X^n then some shift of $\langle x_i : i \in n \rangle$ is a P^{*n}-generic point over M belonging to C.
(iv) Whenever $l \in \omega$ then $\langle x_i \restriction (l, \omega) : i \in n \rangle$ is P^{*n}-generic over $M[x_i(k) : i \in n, k \in l]$.

Proof To prove (*ii*), choose any bijections $\pi_j : b_j \times \omega \to \omega$ and define $f : X^{\bigcup_j b_j} \to X^m$ by $f(\langle z_k : k \in \bigcup_j b_j \rangle) = \langle y_j : j \in m \rangle$ if $y_j(\pi_j(k, l)) = z_k(l)$ for every $j \in m$, $k \in b_j$ and $l \in \omega$. The map f obviously transfers the product measure $\mu^{\bigcup_j b_j}$ to μ^m and (*ii*) is witnessed by $\langle y_j : j \in m\rangle = f(\langle x_i : i \in \bigcup_j b_j\rangle)$.. To prove (*iii*), observe that the shift preserves the measure μ^n and it is ergodic. Therefore, the shift-saturation of the set C has full μ^n-mass and the point $\langle x_i : i \in n\rangle$ must belong to it. In other words, some shift of $\langle x_i : i \in n\rangle$ belongs to C; (*iii*) follows. (*i*) and (*iv*) are standard and we omit the proofs. □

Corollary 6.70 *Sentences of the P^{*n}-forcing language using only canonical names for elements of the ground model and names for* $\text{rng}(\dot{x}_i)$ *for $i \in n$ are decided by the largest condition in P^{*n}.*

Proof Suppose that ϕ is such a sentence and for contradiction assume that $C \Vdash \phi$ and $D \Vdash \neg\phi$ for some conditions $C, D \in P^{*n}$. Let $\langle x_i : i \in n\rangle \in X^n$ be a P^{*n}-generic point over V in the set D, and use Claim 6.69(*iii*) to shift it to some $\langle y_i : i \in n\rangle \in C$. The models $V[x_i : i \in n]$ and $V[y_i : i \in n]$ are the same and also $\text{rng}(x_i) = \text{rng}(y_i)$ for all $i \in n$. Thus the models have to either both satisfy ϕ or both satisfy $\neg\phi$, contradicting the forcing theorem. □

Proof of Theorem 6.67 Follow the lines of Theorem 6.24. There are four cases; in the first three of them we will end up in the first clause of the theorem, in the last one the second clause prevails. Since $F_2 \subset E$, we will abuse the notation and extend the definition of E to countable subsets of 2^ω.

Case 1 Suppose first $P^{*2} \Vdash \dot{x}_0\ E\ \dot{x}_1$. In such a case, let M be a countable elementary submodel of a large structure and let $B = \{x \in X : x \text{ is } M\text{-generic for } P\}$. This is a Borel co-null subset of X. Any two elements of B must be E-related. To prove this, for points $x_0, x_1 \in B$ pick a point $y \in X$ such that y is P-generic over both models $M[x_0]$ and $M[x_1]$. Both pairs $\langle x_0, y\rangle$ and $\langle x_1, y\rangle$ are P^{*2}-generic over M. The forcing theorem implies that $M[x_0, y] \models x_0\ E\ y$ and $M[x_1, y] \models x_1\ E\ y$. Use the Mostowski absoluteness for wellfounded models to conclude that $x_0\ E\ y\ E\ x_1$, and finally argue that $x_0\ E\ x_1$ by transitivity of E. The first clause of the conclusion of the theorem has just been verified.

Case 2 Suppose instead $P^{*3} \Vdash \text{rng}(\dot{x}_0) \cup \text{rng}(\dot{x}_1)\ E\ \text{rng}(\dot{x}_0) \cup \text{rng}(\dot{x}_2)$. Again, let M be a countable elementary submodel of a large structure and let $B = \{x \in X : x \text{ is } M\text{-generic for } P\}$ and argue that any two elements of B must be E-related. Let $x_0, x_1 \in B$ and pick a point $y \in X$ such that y is P-generic over both models $M[x_0]$ and $M[x_1]$. As in Case 1, we will argue that $x_0\ E\ y\ E\ x_1$ and conclude the argument by the transitivity of E. To see that $x_0\ E\ y$, define points $z_0, z_1, z_2, z_3 \in X$ by $z_0(i) = x_0(2i)$, $z_1(i) = x(2i+1)$, $z_2(i) = y(2i)$ and $z_3(i) = y(2i+1)$, and observe that the tuple $\langle z_0, z_1, z_2, z_3\rangle$ is P^{*4}-generic over the model M. Use the case assumption and the forcing theorem to argue that $\text{rng}(x_0) = \text{rng}(z_0) \cup \text{rng}(z_1)$ is E-related to $\text{rng}(z_0) \cup \text{rng}(z_3)$ and then to $\text{rng}(z_2) \cup \text{rng}(z_3) = \text{rng}(y)$. The first clause of the conclusion of the theorem is satisfied.

Case 3 As another case, suppose that $P^{*2} \Vdash \text{rng}(\dot{x}_0)\ E\ \text{rng}(\dot{x}_0) \cup \text{rng}(\dot{x}_1)$. In this case again, we get a co-null equivalence class similarly to the previous argument.

Case 4 Now assume that all of the previous cases have failed; we will argue that the second clause of the theorem must hold. Let $C \subset X$ be a Borel set of positive mass, and let M be a countable elementary submodel of a large structure containing the set C.

Claim 6.71 *There is a continuous function $h : 2^\omega \to X$ such that for every $n \in \omega$ and every n-tuple $\langle y_i : i \in n \rangle$ of pairwise distinct points in 2^ω, the n-tuple $\langle h(y_i) : i \in n \rangle$ is P^{*n}-generic over M.*

Proof We will start with an independent observation. Let $n \in \omega$ and let $B \subset X^n$ be a set. The *double* of B is the set of all sequences $x \in X^{n \times 2}$ such that for every map $\pi : n \to 2$ the point $y \in X^n$ defined by $y(i) = x(i, \pi(i))$ and denoted by $x \circ \pi$ belongs to B. Note that if $n = 1$, the double is simply $B \times B$. We claim that if $n \in \omega$ and $B \subset X^n$ is a Borel set of positive product mass, then its double has a positive product mass as well. To verify the positivity, use the Lebesgue density theorem to find basic open sets $O_i \subset X$ for $i \in n$ such that the set B_i has relative mass greater than $1 - 2^{-n}$ in the set $\prod_i O_i$ and argue that the double of B has positive mass in the set $\prod_{i \in n, j \in 2} O_{i,j}$ where $O_{i,j} = O_i$ for both $j = 0$ and $j = 1$. Just note that for every map $\pi \in 2^n$ the set $C_\pi = \{x \in X^{n \times 2} : x \circ \pi \in B\}$ has relative mass greater than $1 - 2^{-n}$ in the set $\prod_{i,j} O_{i,j}$, and note that the double of B contains the non-null set $\bigcap_{\pi \in 2^n} C_\pi$.

Now back to the proof of the claim. Let $\langle D_m : m \in \omega \rangle$ enumerate all dense open subsets of all the posets P^{*n} for all $n \in \omega$ that belong to the model M, with infinite repetitions. By induction on $m \in \omega$ build closed sets $C_m \subset X^{2^m}$ in the model M of positive product measure so that:

I1. C_{m+1} is a subset of the double of C_m;

I2. if D_m is an open dense set in P^{*n} for some $n < 2^m$, then for every n-tuple $\langle t_i : i \in n \rangle$ of pairwise distinct elements of 2^m it is the case that the projection of C_m into the coordinates $\langle t_i : i \in n \rangle$ belongs to D_m.

The induction is trivial, at each stage m first passing to the double of C_m, which has positive product mass by the first paragraph of this proof, and then strengthening the sequences $h_m(t)$ for $t \in 2^m$ repeatedly so as to satisfy I2 for all the finitely many choices of the tuples $\langle t_i : i \in n\rangle$. In the end, for every $y \in 2^\omega$ the projections of C_m into the $y \restriction m$-coordinate generate a filter generic over M by I2, and this filter produces a point $h(y) \in X$. The function h is as required. Note that for every $n \in \omega$, every n-tuple $\langle y_i : i \in n\rangle$ of pairwise distinct points in 2^ω, writing m_0 for the least number such that the binary strings $\langle y_i \restriction m_0\rangle$ are pairwise distinct, the projections of the sets C_m into the coordinates $\langle y_i \restriction m : i \in n\rangle$ as $m > m_0$ form a decreasing sequence of compact subsets of X^n by I1, and this sequence generates a filter P^{*n}-generic over M by I2. The tuple $\langle h(y_i) : i \in n\rangle$ is the point in X^n associated with this filter. □

Now, fix a bijection $\pi : \omega \times \omega \to \omega$ and for every $z \in X$ let $k(z) \in X$ be the function defined by $k(z)(\pi(k,l)) = h(z(k))(l)$; k is a Borel function. We will show that for every $z \in X$, $\mathrm{rng}(k(z))$ is a random-symmetric set over the model M. Even more is true:

Claim 6.72 *Let $n \in \omega$ and $\langle z_i : i \in n\rangle$ be points in X with pairwise disjoint ranges. The sets $a_i = \mathrm{rng}(k(z_i))$ for $i \in n$ are mutually random-symmetric over M.*

Proof Let $\langle D_m : m \in \omega\rangle$ enumerate all open dense subsets of P^{*n} in the model M. By induction on $m \in \omega$ build closed sets $p_m \in P^{*n} \cap M$, numbers $l_m \in \omega$ and points $x(i,l) \in 2^\omega$ for $i \in n$ and $l \in l_m$ so that:

I1. $p_0 \geq p_1 \geq \cdots, 0 = l_0 < l_1 < \cdots$;
I2. for every $i \in n$, the points $\{x(i,l) : l \in l_{m+1}\}$ all appear in $\mathrm{rng}(k(z_i))$ and include the m-th point of $\mathrm{rng}(k(z_i))$ under some fixed enumeration;
I3. $p_{m+1} \in D_m$ and the set $p'_{m+1} = \{\langle x_i : i \in n\rangle \in X^n : \langle\langle x(i,l) : l \in l_{m+1}\rangle^\frown x_i : i \in n\rangle \in p_{m+1}\}$ has positive μ^n-mass.

Once the induction is performed, let g be the filter on P generated by the conditions $\{p_m : m \in \omega\}$, and for every $i \in n$ let $x_i \in X$ be the point defined by $x_i(l) = x(i,l)$ for every $l \in \omega$. Item I3 shows that g is a filter generic over M and $\langle x_i : i \in n\rangle$ is the generic tuple associated with the filter g, and I2 shows that for every $i \in n$, $\mathrm{rng}(x_i) = \mathrm{rng}(k(z_i))$ holds. This will complete the proof of the claim.

The induction starts with letting p_0 to be the largest condition in P^{*n} and $l_0 = 0$. Suppose that p_m and $x(i,l)$ for $i \in n, l \in l_m$ have been obtained. Let $\{b_i : i \in n\}$ be finite subsets of 2^ω such that $b_i \subset \mathrm{rng}(z_i)$ and the set $\bigcup\{\mathrm{rng}(h(y)) : y \in b_i\}$ includes all the points $x(i,l)$ for $l \in l_m$ as well as the m-th element of $\mathrm{rng}(k(z_i))$ under some fixed enumeration. Now, the tuple

$\langle h(y) : y \in \bigcup_i b_i \rangle$ are mutually generic over M by the choice of the function h. By Claim 6.69(*ii*), there is an n-tuple $\langle x_i^m : i \in n \rangle$ P^{*n}-generic over M such that $\mathrm{rng}(x_i^m) = \bigcup_{y \in b_i} \mathrm{rng}(h(y))$. Permuting the range of x_i^m as in Claim 6.69(*i*) if necessary, we can obtain $x_i^m(l) = x(i, l)$ for all $l \in l_m$. By I3 and Claim 6.69(*iii*) applied in $V[x(i, l) : i \in n, l \in l_m]$, permuting the rest of the sequences x_i^m, we can obtain $\langle x_i^m : i \in n \rangle \in p_m$. The filter $g \subset P^{*n}$ generic over M associated with the tuple $\langle x_i^m : i \in \omega \rangle$ contains the condition p_m, and by genericity also contains a condition $p_{m+1} \leq p_m$ in the open dense set D_m. Choose a large enough number $l_{m+1} \in \omega$ and let $x(i, l) = x_i^m(l)$ for all $i \in n$ and $l \in l_{m+1}$. This completes the induction step. □

Theorem 2.29 shows that the set $C' = \{y \in X : \exists x \in C$ P-generic over M such that $x\ F^2\ y\}$ is Borel and there is a Borel function $g : C' \to C$ which for every $y \in C'$ produces a P-generic point over M such that $x\ F_2\ y$. Claim 6.72 for $n = 1$ shows that $\mathrm{rng}(k) \subset C'$. We claim that the function $f = g \circ k$ is a Borel reduction from F_2 to $E \restriction C$. It is clear that the function f is an F_2-homomorphism since k is, and so F_2-equivalent points are sent to E-equivalent points. We must prove that if $z_0, z_1 \in X$ are not F_2-equivalent then their f-images are not E-equivalent. There are several cases depending on the inclusion comparison of $\mathrm{rng}(z_0)$ and $\mathrm{rng}(z_1)$; we will handle the case when the sets $\mathrm{rng}(z_0) \cap \mathrm{rng}(z_1)$, $\mathrm{rng}(z_0) \setminus \mathrm{rng}(z_1)$, $\mathrm{rng}(z_1) \setminus \mathrm{rng}(z_0)$ are all nonempty. In such a case, let $z_0', z_1', z_2' \in X$ enumerate these three respective pairwise disjoint sets and use Claim 6.72 to find points $\langle x_i : i \in 3 \rangle$ P^{*3}-generic over M such that $\mathrm{rng}(x_i)$ enumerates $\mathrm{rng}(k(z_i'))$ for every $i \in 3$. By the failure of Case 2 and Corollary 6.70 we see that $\neg x_0 \cup x_2\ E\ x_1 \cup x_2$, in other words $\neg f(z_0)\ E\ f(z_1)$. The other cases are dealt with similarly. □

Proof of Theorem 6.65 To show that E_0 is in the spectrum of I, we will argue that E_0 maintains its complexity on each Borel I-positive set. As in Section 7.1, consider the E_0 σ-ideal J of Borel subsets of 2^ω on which E_0 is smooth. By Fact 7.2, it is σ-generated by partial Borel E_0-selectors; these are Lebesgue null by the Steinhaus theorem (Bartoszynski and Judah 1995, theorem 3.2.10), and so $J \subset I$. Thus, if $B \subset 2^\omega$ is a set of positive λ-mass, then $B \notin J$ and then Claim 7.4 yields a continuous reduction of E_0 to $E_0 \restriction B$.

To show that E_2 is in the spectrum of I, we will show that E_2 maintains its complexity on each I-positive set. This follows from the treatment of the ideal J associated with E_2 of Definition 7.41 in Section 7.3. As Claim 7.47 shows, $J \subset I$. Thus, if $B \subset 2^\omega$ is a Borel set of positive λ-mass, then $B \notin J$, and Theorem 7.43 yields a continuous reduction of E_2 to $E_2 \restriction B$.

To show that the spectrum of I is cofinal in the Borel equivalence relations under the Borel reducibility order, consider the space $X = (2^\omega)^\omega$ with the product Borel probability measure μ instead, yielding an isomorphic copy of the σ-ideal I. Consider the ideal J associated with the product of Sacks forcing, as defined in Section 9.3. Then $J \subset I$ by Proposition 9.31, in other words the identity function is a Katětov map from I to J. Then, the spectrum of J is included in the spectrum of I by Proposition 2.70. We conclude that the spectrum of I cofinal among the Borel equivalence relations and contains E_{K_σ} by Theorem 9.29.

To place F_2 in the spectrum of I, argue that F_2 maintains its complexity on every Borel set $B \subset X$ of positive μ-mass. This follows immediately from Theorem 6.67 applied to $E = F_2$, since the equivalence classes of F_2 are μ-null. □

7

Benchmark equivalence relations

7.1 The E_0 ideal

Let E_0 be the equivalence on 2^ω defined by $x \; E_0 \; y$ if $x \Delta y$ is a finite set. This is a basic example of a nonsmooth equivalence relation, as evidenced by the Glimm–Effros dichotomy. The purpose of this section is to prove a canonization theorem for it. Let I be the collection of all Borel sets $B \subset 2^\omega$ such that $E_0 \restriction B$ is smooth.

Theorem 7.1 *Let $B \subset 2^\omega$ be a Borel set for which E_0 is not smooth. Let E be an analytic equivalence relation on B. Then there is a Borel set $C \subset B$ on which E_0 is still not smooth, and $E \restriction C$ is equal to either* id, *E_0, or* ev. *Restated,*

$$\text{analytic} \to_I \{\text{id}, \text{ev}, E_0\}.$$

The σ-ideal I has been studied in Zapletal (2008, section 4.7.1), Zapletal (2004, section 2.3), and Kanovei (2008, section 10.9), which have established its basic properties.

Fact 7.2

(i) I is a σ-ideal, every analytic subset of 2^ω is either in the ideal I, or it contains a Borel I-positive subset.
(ii) I is σ-generated by Borel partial E_0-selectors.
(iii) The σ-ideal I on 2^ω is $\mathbf{\Pi}^1_1$ on $\mathbf{\Sigma}^1_1$.
(iv) The forcing P_I is proper, bounding, and preserves Baire category.

The main theorem of this section will be superceded in Section 9.2, where we prove a multidimensional version of it; however, the present proof in a single dimension is much simpler and so we keep it.

7.1.1 The reduced product

The feature that sets apart the case of the E_0 equivalence from the more complex E_1 and E_2 equivalences studied in later sections is the simple form of a reduced product of P_I. We will start with a combinatorial representation of the poset P_I.

Definition 7.3 The partial order P consists of pairs $p = \langle t_p, \vec{c}_p \rangle$ where t_p is a finite binary string (the *trunk* of p), $\vec{c}_p$ is an infinite sequence, and each entry of $\vec{c}_p$ is a pair of finite binary strings of the same length that differ at the first entry.

For a finite sequence $\langle c_i : i \in n \rangle$ of pairs of finite binary strings, let $[c_i : i \in n]$ be the set of all binary strings obtained as all concatenations $s_0{}^\frown s_1{}^\frown \cdots {}^\frown s_{n-1}$ where $s_i \in c_i$ for every $n \in \omega$.

The order on P is defined by $q \leq p$ if there is a strictly increasing sequence $\langle i_n : n \in \omega \rangle$ of natural numbers such that $t_q = t_p{}^\frown s$ for some $s \in [\vec{c}_p(i) : i < i_0]$ and $\vec{c}_q(n)$ consists of strings in $[\vec{c}_p(i) : i_n \leq i < i_{n+1}]$.

If $p \in P$ then $[p] \subset 2^\omega$ is the compact set of all sequences obtained as concatenation of t_p with some binary sequences $\langle s_i : i \in \omega \rangle$ where for every $i \in \omega$, $s(i) \in c_p(i)$.

It is clear that the poset P adds a point $\dot{x}_{\text{gen}} \in 2^\omega$, the unique point in the set $\bigcap\{[p] : p$ is in the generic filter$\}$. Conditions in the poset P are in natural bijection with E_0 trees of Zapletal (2004, theorem 2.3.29), and so the following holds:

Fact 7.4 (Zapletal 2004, theorem 2.3.29) *For every analytic set $A \subset 2^\omega$, exactly one of the following happens:*

(i) $A \in I$;
(ii) there is a condition $p \in P$ such that $[p] \subset A$.

The map $p \mapsto [p]$ is an isomorphism of the poset P with a dense subset of P_I.

Now, we will describe a combinatorial form of a reduced product of the poset P.

Definition 7.5 $P \times_{E_0} P$ is the collection of all pairs $\langle p, q \rangle \in P \times P$ whose trunks have equal length and for every $i \in \omega$, $c_p(i) = c_q(i)$. $P \times_{E_0} P$ is equipped with the coordinatewise ordering. If $\langle p, q \rangle \in P \times_{E_0} P$ then $[p, q]$ is defined as the set $([p] \times [q]) \setminus E_0$.

It is not difficult to see that if $G \subset P \times_{E_0} P$ is a filter generic over V then both $G_{\text{l}} = \{p \in P : \exists q\ \langle p, q \rangle \in G\}$ and $G_{\text{r}} = \{q \in P : \exists p\ \langle p, q \rangle \in$

$G\}$ are P-generic filters over V. We will denote their associated points with $\dot{x}_{\mathrm{lgen}}, \dot{x}_{\mathrm{rgen}} \in 2^\omega$. Note that $\dot{x}_{\mathrm{lgen}}, \dot{x}_{\mathrm{rgen}}$ are forced to be E_0-inequivalent.

The poset $P \times_{E_0} P$ is proper. For the canonization result, we will need a stronger statement than that:

Proposition 7.6 *Whenever M is a countable elementary submodel of a large enough structure, and $\langle p, q\rangle \in (P \times_{E_0} P) \cap M$ is a condition, then there is an M-master condition $\langle p', q'\rangle \leq \langle p, q\rangle$ with the same trunks such that the set $[p', q']$ consists of pairs $P \times_{E_0} P$-generic over the model M.*

Proof We will need some notation. If $p \in P$ is a condition, then a *splitnode* of p, is the concatenation of the trunk t_p with some binary sequence in $[c_p(i) : i \in n]$ for some $n \in \omega$; such a splitnode is said to be at n-th splitting level of p. If t is a splitnode of p at splitting level n, then $p \restriction t$ is the condition $\langle t, c_p(i) : i \geq n\rangle$; so $p \restriction t \leq p$. Two splitnodes s, t at the same splitting level $n+1$ will be called *forked* if the sequences in $c_p(n)$ which are subsequences of s and t respectively are distinct.

To get the condition $\langle p', q'\rangle$, by induction on $n \in \omega$ build a decreasing sequence of conditions $\langle p_n, q_n\rangle \in (P \times_{E_0} P) \cap M$ so that:

I1. $t_{p_n} = t_p, t_{q_n} = t_q, \vec{c}_{p_n} \restriction n = \vec{c}_{p_{n+1}} \restriction n$;
I2. whenever u, v are two forked sequences at $n+1$-st splitting level of p_n, q_n respectively then the condition $\langle p_n \restriction u, q_n \restriction v\rangle$ is in all the first n open dense subsets of $P \times_{E_0} P$ in the model M under some fixed enumeration.

The induction is elementary. In the end, consider conditions $p'q'$ given by $t_{p'} = t_p$ and $\vec{c}_{p'}(n) = \vec{c}_{p_{n+1}}(n)$ and similarly on the q side. I1 shows that the pair $\langle p', q'\rangle$ is indeed a condition in $P \times_{E_0} P$. I2 shows that all elements of $[p', q']$ are M-generic. Clearly, $\langle p', q'\rangle \in P \times_{E_0} P$ is an M-master condition with the desired properties. □

7.1.2 The canonization result

For the proof of Theorem 7.1, suppose that $B \subset X$ is a Borel I-positive set and E is an analytic equivalence relation on it. Let $p_0 \in P$ be a condition such that $[p_0] \subset B$. The treatment divides into two cases:

Case 1 Suppose that there is a condition $\langle p, q\rangle \leq \langle p_0, p_0\rangle$ in the reduced product which forces $\dot{x}_{\mathrm{lgen}}\ E\ \dot{x}_{\mathrm{rgen}}$. Let M be a countable elementary submodel of a large structure containing $\langle p, q\rangle$. Use Proposition 7.6 to find a condition $\langle p'q'\rangle \leq \langle p, q\rangle$ such that $[p', q']$ consists of pairs reduced product generic over M. We claim that $C = [p'] \subset B$ is an I-positive subset of B consisting of pairwise E-related points. Indeed, if $x_0, x_1 \in C$ then there is a point $x_2 \in [q']$ which is not E_0-related to either x_0 or x_1, by the forcing theorem it is the case that $x_0\ E\ x_2$ and $x_1\ E\ x_2$, and by transitivity $x_0\ E\ x_2$ as desired. So in this case we are done.

Case 2 If Case 1 fails, then $\langle p_0, p_0 \rangle$ forces in the reduced product that $\dot{x}_{\text{lgen}}\ E\ \dot{x}_{\text{rgen}}$ fails. In this case, we will first thin out p_0 to p_1 so that $E \restriction [p_1] \subset E_0$, and $E \restriction [p_1]$ is a Borel equivalence relation. Let M be a countable elementary submodel of a large structure containing p_0. Use Proposition 7.6 to find a condition $p_1 \leq p_0$ such that $[p_1, p_1]$ consists of pairs reduced product generic over the model M. The forcing theorem applied in the model M to $P \times_{E_0} P$ then implies that non-E_0-related points of $[p_1]$ are also not E-related. Therefore, ($\star$) $E \restriction [p_1] \subset E_0$. By the forcing theorem applied to P, if $x \in [p_1]$ and $t \in 2^{<\omega}$ then $x\ E\ (x \text{ rew } t)$ if and only if there is a condition $q \in P \cap M$ such that $x \in [q]$ and $q \Vdash \dot{x}_{\text{gen}}\ E\ (\dot{x}_{\text{gen}} \text{ rew } t)$. Since there are only countably many conditions in $P \cap M$, this means that $E \restriction [p_1]$ is a Borel relation. An alternative approach would use Theorem 4.27, thinning down p_1 to secure the Borelness of $E_1 \restriction [p_1]$.

Case 2a If the equivalence relation $E \restriction [p_1]$ is smooth then (as the equivalence classes of E are countable) $[p_1] = \bigcup_n C_n$ where C_n are all Borel partial E-selectors. One of these sets, $C = C_n$, must be I-positive. We conclude that $E \restriction C = \text{id}$ and we are done.

Case 2b If Case 2a fails then the Glimm–Effros dichotomy (Fact 2.16) yields a continuous one-to-one reduction $\pi : 2^\omega \to [p_1]$ of E_0 to E. Look at the equivalence relation $F = \pi^{-1} E_0$. This is a Borel equivalence relation with countable classes and $E_0 \subset F$. Applying Case 1 and Case 2 up to ($\star$) of the present proof to F, we see that Case 1 is impossible due to the countable classes of F, therefore Case 2 must prevail and there is a Borel I-positive set $C' \subset 2^\omega$ such that $F \restriction C' \subset E_0$, and therefore $F \restriction C' = E_0$. The image $C = \pi'' C'$ is a Borel set on which $E = E_0$, and E_0 is still nonsmooth on C since π reduces the nonsmooth equivalence relation $E_0 \restriction C'$ to $E_0 \restriction C$. We conclude that $E = E_0$ on the I-positive Borel set $C \subset B$ and we are done again!

7.2 The E_1 ideal

E_1 is the equivalence relation on the Polish space $X = (2^\omega)^\omega$ defined by $x\ E_1\ y$ if for all but finitely many $n \in \omega$, $x(n) = y(n)$. This is the canonical example of a Borel equivalence relation that is not Borel reducible to an orbit equivalence relation of a Polish group action; its behavior is governed by the Kechris–Louveau dichotomy (Fact 2.18). Let I be the collection of Borel sets $B \subset X$ for which $E_1 \restriction B$ is not bireducible with E_1. It turns out that this is a σ-ideal I of Borel sets, and we prove several canonization results for it:

Theorem 7.7 *Let $B \subset X$ be a Borel set such that $E_1 \restriction B$ is bireducible with E_1. Whenever E is a hypersmooth equivalence relation on B then there is a compact set $C \subset B$ such that $E_1 \restriction C$ is still bireducible with E_1, and $E \restriction C$ is equal to either* id, ev, *or E_1. In other words,*

$$\text{hypersmooth} \to_I \{\text{id}, \text{ev}, E_1\}.$$

The conjunction of this result with the canonization theorems proved in Section 9.3 yields total canonization for other classes of equivalence relations:

Theorem 7.8 *Let $B \subset X$ be a Borel set such that $E_1 \restriction B$ is bireducible with E_1. Whenever E is an equivalence relation on B classifiable by countable structures or Borel reducible to $=_J$ for an analytic P-ideal J on ω, then there is a compact set $C \subset B$ such that $E_1 \restriction C$ is still bireducible with E_1, and $E \restriction C$ is equal either to* id *or to* ev.

An additional trick yields the best canonization result for subequivalence relations of E_1:

Theorem 7.9 *Let $B \subset X$ be a Borel set such that $E_1 \restriction B$ is bireducible with E_1. Whenever $E \subset E_1$ is an analytic equivalence relation then there is a compact set $C \subset B$ such that $E_1 \restriction C$ is still bireducible with E_1, and $E \restriction C$ is equal either to* id *or E_1. In other words,*

$$\subset E_1 \to_I \{\text{id}, E_1\}$$

We will also provide a simple set of generators for I and show that every analytic I-positive set has an I-positive compact subset of a fairly canonical kind:

Theorem 7.10 *Whenever $A \subset X$ is an analytic set then:*

(i) either $A \in I$, and this happens exactly in case there is $z \in 2^\omega$ such that for every $x \in A$, for all but finitely many $n \in \omega$ it is the case that $x(n) \in \Delta^1_1(x \restriction (n, \omega), z)$;
(ii) or $A \notin I$, and this happens exactly if there is a (compact) s-cube $C \subset A$ for some infinite set $s \subset \omega$ as in Definition 7.13.

As a corollary, we obtain a new proof of the Kechris–Louveau dichotomy (Fact 2.18).

As for the open questions, it is clear from the definitions that E_1 belongs to the spectrum of the σ-ideal I. We do not see any other features of the spectrum:

Question 7.11 Is there any other analytic equivalence relation in the spectrum of the σ-ideal I besides E_1?

From the forcing point of view, the σ-ideal I is somewhat awkward as the quotient poset P_I is not proper (Zapletal 2004, lemma 2.3.33; Kanovei 2008, theorem 11.7.4). At the same time, there is a closely related σ-ideal whose quotient poset is proper and previously independently studied. It is the iteration of Sacks forcing along the reverse ordering on ω, $\cdots \prec 2 \prec 1 \prec 0$, at which a number of authors arrived from different angles and at different times (Groszek 1994; Kanovei 1999; Kellner and Shelah 2011). The associated ideal was computed in Zapletal (2008, section 5.4). The connection between the E_1 equivalence and reverse iterations should not be surprising in view of Theorem 4.23. Here, we show that there is a natural collection of σ-ideals between the E_1 ideal and the reverse iteration ideal that allow uniform treatment.

Regarding the notation in this section, for a natural number $n \in \omega$ we set $(n, \omega) = \{m \in \omega : n < m\}$ and $[n, \omega) = \{m \in \omega : n \leq m\}$.

7.2.1 The σ-ideals and their dichotomies

Definition 7.12 Let $S \subset [\omega]^{\aleph_0}$ be a nonempty analytic set. The σ-ideal I_S on $X = (2^\omega)^\omega$ is σ-generated by the sets A_z where z ranges over all elements of 2^ω and $A_z = \{x \in X : \forall s \in S\, \exists n \in s\; x(n) \in \Delta^1_1(z, x \restriction (n, \omega)\}$.

Equivalently, a set $A \subset X$ is I_S positive if for every $z \in 2^\omega$ there is $x \in A$ such that the set $\{n : x(n) \notin \Delta^1_1(z, (n, \omega))\}$ has a subset in S. Note that the generators of the σ-ideal I_S are coanalytic. This is in contradistinction to other σ-ideals considered in this book, where the generators are typically Borel or analytic, but it will cause only very minor annoyance. It will turn out that I contains the same analytic sets as $I_{[\omega]^{\aleph_0}}$, while the reverse Sacks iteration ideal is equal to $I_{\{\omega\}}$. Before we can arrive at this conclusion though, we must show that I_S-positive analytic sets come in a specific form. That is the subject of the following definition and theorem.

Definition 7.13 Let $s \subset \omega$ be an infinite set, with an increasing enumeration $\pi : \omega \to s$. An *s-keeping homeomorphism* is a continuous injective map $f : X \to X$ such that for every $x_0, x_1 \in X$ and every $i \in \omega$:

(a) if $x_0(i) \neq x_1(i)$ then $f(x_0)(\pi(j)) \neq f(x_1)(\pi(j))$ for all $j \leq i$;
(b) if $x_0(j) = x_1(j)$ for all $j > i$ then $f(x_0)(n) = f(x_1)(n)$ for all $n > \pi(i)$.

An *s-cube* is the range of an s-keeping homeomorphism. If $S \subset \mathcal{P}(\omega)$ is a collection of infinite sets, an *S-cube* is a s-cube for some $s \in S$. A *reverse cube* is an ω-cube.

Note that if f is an s-keeping homeomorphism for an infinite $s \subset \omega$ then f is a one-to-one continuous reduction of E_1 to $E_1 \restriction \mathrm{rng}(f)$ by the two demands (a) and (b). Kanovei (1999) used a definition of a projection-keeping homeomorphism that differs in an insignificant technical detail from our definition of an ω-keeping homeomorphism.

Theorem 7.14 *Let $S \subset [\omega]^{\aleph_0}$ be a nonempty analytic set. The collection I_S is a $\mathbf{\Pi}^1_1$ on $\mathbf{\Sigma}^1_1$ σ-ideal. For every analytic set $A \subset X \times \omega^\omega$:*

(i) either $p(A) \in I_S$, and this happens if and only if there is $z \in 2^\omega$ such that for every $x \in p(A)$ the set $\{n : x(n) \notin \Delta^1_1(z, x \restriction (n, \omega))\}$ has no subset in S;
(ii) or $p(A) \notin I_S$, and this happens if and only if there is an S-cube $C \subset p(A)$ and a continuous function $g : C \to \omega^\omega$ such that $g \subset A$.

The quotient poset P_{I_S} is typically not proper. The statement (*ii*) above may be viewed as the remnants of the bounding property.

Proof The difficult part is inscribing an S-cube into an I_S-positive analytic set. After that, a couple of routine applications of the Kuratowski–Ulam theorem complete the proof.

So let $S \subset [\omega]^{\aleph_0}$ be an analytic set and $A \subset X \times \omega^\omega$ be an analytic set. In order to promote the similarity with the canonization arguments in the next section, we will divide the treatment into two cases, even though here one of the cases is nearly trivial. The cases correspond to the two clauses of the theorem.

Case 1 There is a parameter $z \in 2^\omega$, a set $s \in S$ and a point $\langle x, y\rangle \in A$ such that S, A are both $\Sigma^1_1(z)$ and for every $n \in s$, $x(n) \notin \Delta^1_1(x \restriction (n, \omega), z)$. In this case, we will arrive in the second clause of the theorem: we will inscribe an S-cube C into $p(A)$ and find a continuous function $g : C \to \omega^\omega$ such that $g \subset A$. For simplicity assume $z = 0$.

First, some notation. Let $a \subset \omega \times \omega$ be a finite set. Write $\max(a) = \max\{m : \exists k\ \langle m, k\rangle \in a\}$, $a \restriction (n_0, n_1] = \{\langle m, k\rangle \in a : n_0 < m \leq n_1\}$, and for functions $\sigma \neq \tau \in 2^a$ let $d(\tau, \sigma) = \max\{m : \exists k\ \tau(m, k) \neq \sigma(m, k)\}$. For an infinite set $s \subset \omega$ and $j \in \omega$, the symbol $s(j)$ stands for the j-th element of s in its increasing enumeration.

Let $a \subset \omega \times \omega$ be a finite set. A *precondition* with domain a is a tuple $p = \langle S(p), A(p, \tau) : \tau \in 2^a \rangle$ such that for all $\tau \in 2^a$, $A(p, \tau) \subset A$ is a nonempty Σ^1_1 set and $S(p) \subset S$ is a nonempty Σ^1_1 set as well. Write $a = \mathrm{dom}(p)$. Preconditions are ordered by $q \leq p$ if $\mathrm{dom}(p) \subseteq \mathrm{dom}(q)$, $S(q) \subset S(p)$, and for every $\tau \in 2^{\mathrm{dom}(q)}$, $A(q, \tau) \subset A(p, \tau \restriction \mathrm{dom}(p))$. A *witness* for p is a tuple $w = \langle s(w), x(w, \tau), y(w, \tau) : \tau \in 2^a \rangle$ such that:

W1. $s(w) \in S(p)$, $\langle x(w, \tau), y(w, \tau) \rangle \in A(p, \tau)$;
W2. for every $\tau \in 2^a$ and every $j \in \omega$, $x(w, \tau)(s(w)(j)) \notin \Delta^1_1(x(w, \tau) \restriction (s(w)(j), \omega))$;

and for all $\tau \neq \sigma \in 2^a$,

W3. for all $j \leq d(\sigma, \tau)$, $x(w, \sigma)(s(w)(j)) \neq x(w, \tau)(s(w)(j))$;
W4. for all $j > s(w)(d(\sigma, \tau))$, $x(w, \sigma)(j)) = x(w, \tau)(j)$.

Note that the set of all witnesses for a given precondition is Σ^1_1 by Fact 2.8(*i*). A *stem* of a witness w is the pair consisting of the common value of $x(w, \tau) \restriction (s(w)(\max(a)), \infty)$ for all $\tau \in 2^a$, and of the sequence of the first $\max(a)$ elements of $s(w)$. A *condition* is a precondition with a witness.

Claim 7.15 *For every condition p with domain a and every open dense subset D of the Gandy–Harrington forcing there is a condition $q \leq p$ with the same domain such that $S(q) \in D$ and $A(q, \tau) \in D$ for all $\tau \in 2^a$.*

Proof Enumerate the functions in 2^a by $\langle \tau_i : i \in j \rangle$ and by induction on $i \in j$ construct Σ^1_1 sets $A(q, \tau_i) \subset A(p, \tau_i)$ in the set D so that there is still a witness for $p_i = \langle S(p), A(q, \tau_k) : k \in i, A(p, \tau_k) : k > i \rangle$. To see how the induction step is performed, consider the set B_i of all witnesses for p_i; this is a nonempty Σ^1_1 set. The projection C_i of B_i into τ_i-th coordinate is again a nonempty Σ^1_1 set with $C_i \subset A(p, \tau_i)$; let $A(q, \tau_i)$ be any of its nonempty Σ^1_1 subsets in D. This completes the induction step. After the induction on i is performed, thin out $S(p)$ to some $S(q) \in D$ in a similar way, obtaining the condition q. □

Claim 7.16 *For every condition p with domain a, every witness w for p, and every countable set $b \subset 2^\omega$ there is a witness v for p such that* $\mathrm{stem}(v) = \mathrm{stem}(w)$ *and for every $\tau \in 2^a$ and for every $j \leq \max(a)$, we have $x(v, \tau(s(v)(j)) \notin b$.*

Proof Let $B = \{\langle x^v_\tau(s(v)(j)) : j \leq \max(a), \tau \in 2^a \rangle : v$ is a witness for p with $\mathrm{stem}(v) = \mathrm{stem}(w)\}$. This is a nonempty $\Sigma^1_1(\mathrm{stem}(w))$ set consisting of tuples whose entries are not $\Delta^1_1(\text{stem}(w))$ by item W2. By Proposition 2.10, there is a tuple in the set B neither of whose entries belongs to the set b. Let v be the witness corresponding to that tuple. □

Write $\omega \times \omega$ as an increasing union $\bigcup_n a_n$ of finite sets such that $a_0 = 0$ and $|a_{n+1}| = |a_n| + 1$. Let M be a countable elementary submodel of a large structure and let $\langle D_n : n \in \omega \rangle$ enumerate all the open dense subsets of the Gandy–Harrington forcing in the model M. By induction on $n \in \omega$ build conditions $p_n \in M$ so that:

I1. $\text{dom}(p_n) = a_n$, $p_{n+1} \leq p_n$;
I2. for all $\tau \in 2^{a_n}$ we have $A(p_n, \tau) \in D_n$;
I3. for all $\tau \in 2^{a_n}$ and all $j \in n$, the pair $\langle x(j) \restriction n, y(j) \restriction n \rangle$ is the same for every choice of $\langle x, y \rangle \in A(p_n, \tau)$;
I4. for all $\sigma \neq \tau \in 2^{a_n}$, for all $j \leq d(\sigma, \tau)$ and all $\langle x_\sigma, y_\sigma \rangle \in A(p_n, \sigma)$ and $\langle x_\tau, y_\tau \rangle \in A(p_n, \tau)$ and all $s \in S(p_n)$, we have $x_\sigma(s(j)) \neq x_\tau(s(j))$.

Once this induction succeeds, then for every $x \in X$ let $G_x = \{A(p_n, x \restriction a_n) : n \in \omega\}$. Item I2 shows that the upward closure of G_x is a Gandy–Harrington generic filter over M and so $\bigcap G_x$ contains a singleton $\langle g(x), h(x) \rangle \in A$. Similarly, let $s \subset \omega$ be the range of the unique element in $\bigcap \{S(p_n) : n \in \omega\}$. Item I3 shows that functions $g, h, h \circ g^{-1}$ are continuous, items I4 and W4 show that g is an s-keeping homeomorphism from X to X. This will complete the proof of Case 1.

The case assumption gives us the basis of the induction, a witness for the condition $p_0 = \langle S, A \rangle$. To perform the induction step, suppose that p_n has been found. We will first show that $p'_n = \langle S(p_n), A(p_n, \tau \restriction a_n) : \tau \in 2^{a_{n+1}} \rangle$ is a condition with domain a_{n+1}. For that, we must produce a witness w' for p'_n, obtained by *splitting* a fixed witness w for p_n. Let $\langle m, l \rangle \in \omega \times \omega$ be the unique pair in $a_{n+1} \setminus a_n$. Formally, the treatment divides into three similar cases, $m < \max(a_n)$, $m = \max(a_n)$, and $m > \max(a_n)$; for definiteness, we will handle the hardest first case. Write $a^0_n = a_n \restriction [0, m]$ and $a^1_n = a_n \restriction (m, \max(a_n))$

For every $\tau \in 2^{a^1_n}$ write p_τ for the condition with $\text{dom}(p_\tau) = a^0_n$ given by $S(p_\tau) = S(p)$ and $A(p_\tau, \sigma) = A(p, \tau \cup \sigma)$. Note that the tuple $w_\tau = \langle s(w), x(w, \tau \cup \sigma), y(w, \tau \cup \sigma) : \sigma \in 2^{a^0_n} \rangle$ is a witness for p_τ. Use Claim 7.16 repeatedly at each $\tau \in 2^{a^1_n}$ to find witnesses $v(\tau, 0)$ and $v(\tau, 1)$ with the same stem as w_τ, such that the sets $\{x(v(\tau, i), \sigma)(s(j)) : j \leq m\}$ are pairwise disjoint for $\tau \in 2^{a^1_n}$ and $i \in 2$. Now *glue* these newly found witnesses together: let $w' = \langle s(w), x(v(\sigma \restriction a^1_n, \sigma(m, l))(\sigma \restriction a^0_n) : \sigma \in 2^{a_{n+1}} \rangle$. A review of definitions shows that w' is a witness for p'_n.

Finally, we will thin down p'_n to p_{n+1} so that items I2, I3, and I4 are satisfied. Let w' be any witness for p'_n. Choosing k sufficiently large and replacing $A(p'_n, \tau)$ with $\{\langle x, y\rangle \in A(p'_n, \tau) : \forall j \leq \max(a_n)\ x(j) \restriction k = x(w', \tau)(j) \restriction k$ as well as $y \restriction k = y(w', \tau) \restriction k\}$, items I3, I4 will be satisfied for the resulting condition p''_n. Finally, use Claim 7.15 in the model M to thin down p''_n to p_{n+1} so that for every $\tau \in 2^{a_{n+1}}$, $A(p_{n+1}, \tau) \in D_n$. Then p_{n+1} satisfies I2 as well, concluding the induction step and the proof in Case 1.

Case 2 If Case 1 fails, then let $z \in 2^\omega$ be a parameter such that S, A are both $\Sigma^1_1(z)$ and note that then $p(A)$ is a subset of a generator of the σ-ideal I_S, namely the set $\{x \in X : \forall s \in S\ \exists n \in s\ x(n) \in \Delta^1_1(x \restriction (n, \omega), z)\}$. We conclude that $p(A) \in I_S$ and arrive at the first clause of the theorem.

The rest of the argument for Theorem 7.14 is now easy. Suppose that $A \subset X \times \omega^\omega$ is an analytic set. If $p(A) \in I_S$, then (*i*) is just a restatement of the definition of I_S. If $p(A) \notin I_S$, then the above argument provides an S-cube C and a continuous function $g : C \to \omega^\omega$ such that $g \subset A$. If, on the other hand, $p(A)$ contains an S-cube C, a range of an s-keeping homeomorphism f, then C and so $p(A)$ must be I_S-positive. To verify the positivity, note that whenever $z \in 2^\omega$ is a parameter and $n \in s$, then the set $B_{n,z} = \{x \in X : f(x) \in \Delta^1_1(f(x) \restriction (n, \omega), z)\}$ is meager in X by the Kuratowski–Ulam theorem – it has countable sections in the ni-th coordinate where i is such that n is i-th element of the set s. This means that the f-preimages of the generators of the σ-ideal I_S are meager (the generator associated with z has f-preimage which is a subset of the union $\bigcup_n B_{n,z}$) and so $C = \operatorname{rng}(f) \notin I_S$.

For the complexity estimate, suppose that $A \subset 2^\omega \times X$ is an analytic set, and choose a parameter $z \in 2^\omega$ such that A, S are both $\Sigma^1_1(z)$. Case 1 above then shows that for every $y \in 2^\omega$, the set A_y is I_S-positive if and only if there is $x \in X$ and $s \in S$ such that $\langle y, x\rangle \in A$ and for every $n \in s$, $x(n) \notin \Delta^1_1(x \restriction (n, \omega), z, y)$. This is a $\Sigma^1_1(z)$ condition by Fact 2.8(*i*). □

There are two interesting special cases that we will consider here. If $S = \{\omega\}$ then the quotient poset P_{I_S} has a dense subset naturally isomorphic to the *reverse iteration of Sacks forcing* (Groszek 1994; Kanovei 1999), which is defined to consist of reverse cubes ordered by inclusion. A more interesting case occurs when $S = [\omega]^{\aleph_0}$; it corresponds to the σ-ideal I and the Kechris–Louveau dichotomy (Fact 2.18).

Corollary 7.17 *Let $S = [\omega]^{\aleph_0}$. The ideals I_S and I contain the same analytic sets. In other words, for every analytic set $A \subset X$, $A \notin I_S$ if and only if $E_1 \restriction A$ is not Borel reducible to E_0.*

Proof Suppose that $A \notin I_S$; we will conclude that E_1 is Borel reducible to $E_1 \restriction A$. Use Theorem 7.14 to find an infinite set $s \subset \omega$ and an s-keeping homeomorphism $f : X \to A$ whose compact range C is a subset of A. Clearly, f is a continuous injective reduction of E_1 to $E_1 \restriction C$ as required.

Suppose that $A \in I_S$; we must conclude that $E_1 \restriction B$ is Borel reducible to E_0 for some Borel superset $B \supset A$. For simplicity assume that A is Σ^1_1. Use the first reflection theorem 2.2 to find a Borel set $B \supset A$ such that for every $x \in B$ the set $\{n : x(n) \notin \Delta^1_1(x \restriction (n, \omega))\}$ is finite. Use the Kreisel selection (Kanovei 2008, theorem 2.4.5) to find a Borel function $h : B \to \omega$ such that this finite set is always bounded by $h(x)$. Let $i : B \to X$ be the Borel function assigning to each $x \in B$ the sequence $i(x)$ given by $i(x)(m) = 0$ if $m < h(x)$ and $i(x)(m) = x(m)$ if $m \geq h(x)$. Note that the E_1-equivalence classes on $\mathrm{rng}(i)$ are countable. Use the first reflection theorem to produce a Borel superset $C \supset \mathrm{rng}(i)$ on which the E_1 classes are countable. Now, $E_1 \restriction C$ is a hypersmooth equivalence relation with countable classes, and by a result of Dougherty, Jackson, and Kechris (Kanovei 2008, theorem 8.1.1(iii)), such an equivalence is Borel reducible to E_0 via some Borel function $j : C \to 2^\omega$. The composition $j \circ i : B \to 2^\omega$ reduces $E_1 \restriction B$ to 2^ω as desired. □

Corollary 7.18 (Kechris–Louveau) *If E is a hypersmooth equivalence relation on a Polish space Y then either $E \leq_B E_0$ or $E_1 \leq_B E$.*

Proof Fix Borel maps f_n for $n \in \omega$, $f_n : Y \to 2^\omega$ such that for every $y_0, y_1 \in Y$, $y_0 \; E \; y_1$ iff the set $\{n : f_n(y_0) = f_n(y_1)\} \subset \omega$ is co-finite. Let $g : Y \to X$ be the map defined by $g(y) = \langle f_n(y) : n \in \omega \rangle$, and look at the set $A = \mathrm{rng}(g)$. The set $A \subset X$ is analytic; we discern the cases where $A \in I$ or not.

If $A \notin I$ then by Theorem 7.14(*ii*) there is an s-cube $C \subset A$ for some infinite set $s \subset \omega$, a continuous map $g : C \to Y$ such that $f \circ g = \mathrm{id}$, and an s-keeping homeomorphism $h : X \to C$ whose range C is. Note that $g \circ h : X \to Y$ is a continuous injective reduction of E_1 to $E \restriction A$.

If $A \in I$ then it is covered by a Borel set $B \in I$. By Corollary 7.17, $E_1 \restriction B$ is Borel reducible to E_0 via a map $h : B \to 2^\omega$. It is immediate that $h \circ f$ is a reduction of E to E_0. □

Corollary 7.19 *Let $S = \{\omega\}$. The poset P_{I_S} has a dense subset naturally isomorphic to the Sacks iteration along ω with the inverted ordering.*

Proof Up to an insignificant technical detail, the iteration isolated in Kanovei (1999) consists of ω-cubes ordered by inclusion. Theorem 7.14 precisely states that this is a dense subset of the ordering P_{I_S}. □

7.2.2 Canonization results

On the road to Theorem 7.7, we will prove the following somewhat stronger statement. Let $X = (2^\omega)^\omega$ and let J be the σ-ideal of analytic sets which contain no reverse cube. Let id_n is the smooth equivalence relation connecting points $x, y \in X$ if for all $m \geq n$, $x(m) = y(m)$ – so that $E_1 = \bigcup_n \mathrm{id}_n$.

Theorem 7.20 *Let $B \subset X = (2^\omega)^\omega$ be a reverse cube, and let E be a hypersmooth equivalence relation on B. Then there is a reverse cube $C \subset B$ such that $E \restriction C$ equals to* id *or to* ev *or to E_1, or to id_n for some $n \in \omega$. Restated,*

$$\text{hypersmooth} \to_J \{E_1, \mathrm{ev}, \mathrm{id}_n : n \in \omega\}.$$

Theorem 7.7 is a formal consequence of Theorem 7.20.

The arguments below follow the same pattern. There is a split into cases, in one of the cases we use a Gandy–Harrington forcing just like in the previous section, and in the other a fusion argument appears. Thus, we need to make preliminary fusion definitions.

Definition 7.21 Let $B, C \subset X$ be reverse cubes and $n \in \omega$. We write $C \leq_n B$ if $C \subset B$ and C is id_n-invariant inside B. A *fusion sequence* is a sequence $\langle C_j : j \in \omega\rangle$ of reverse cubes such that there are increasing numbers $\langle n_j : j \in \omega\rangle$ such that $C_{j+1} \leq_{n_j} C_j$.

Proposition 7.22 *Suppose that $\langle C_j : j \in \omega\rangle$ is a fusion sequence of reverse cubes. Then $\bigcap_j C_j \notin J$.*

Proof By compactness, $C = \bigcap_j C_j$ is a nonempty compact set. To show that $C \notin J$, we will show that the generating sets of the ideal J are relatively meager in C. Consider a typical generator: let $z \in 2^\omega$ be an arbitrary parameter, let $n \in \omega$ and consider the coanalytic set $B = \{x \in C : x(n) \in \Delta^1_1(x \restriction (n, \omega), z)\}$. Let $D \subset 2^\omega \times (2^\omega)^{(n,\omega)}$ be the nonempty compact projection of the set C into $(2^\omega)^{[n,\omega)}$. By a *horizontal section* of the set D we mean a set of the form $\{y \in 2^\omega : \langle y, x\rangle \in D\}$ for some $x \in (2^\omega)^{(n,\omega)}$. The horizontal sections of the set D are all perfect: if nonempty, every such horizontal section is equal to a corresponding section in the cube C_n since $C \leq_n C_n$, and the section in the cube C_n is perfect by the definition of an ω-keeping homeomorphism. The horizontal sections corresponding to the set B are countable, and therefore meager in the horizontal sections of the set D. By the Kuratowski–Ulam theorem, the projection of B into $(2^\omega)^{[n,\omega)}$ is meager in D and so B is meager in C as desired. □

As a useful warm-up, we will reprove the canonization of smooth equivalence relations achieved in Kanovei (1999).

Theorem 7.23 *For every reverse cube $A \subset X$ and a smooth equivalence E on A there is a reverse cube $C \subset A$ such that either $E \restriction C = \mathrm{id}_n$ for some $n \in \omega$ or $E \restriction C = \mathrm{ev}$.*

Proof Let $A \subset X$ be a J-positive Borel set and E a smooth equivalence relation. Let $f : A \to 2^\omega$ be the Borel reduction of E to the identity. By Theorem 7.14, passing to an J-positive subset if necessary, we may assume that $f \restriction A$ is continuous.

Case 1 Suppose that there exists a parameter $z \in 2^\omega$ such that $f \in \Pi^0_1(z)$ and there is a set $B \in \Sigma^1_1(z)$ with $B \subset A$, a point $x \in B$ and a number $m \in \omega$ such that for every $n \in \omega$, $x(n) \notin \Delta^1_1(x \restriction (n, \omega))$, and $f(x) \notin \Delta^1_1(x \restriction [m, \omega), z)$. Choose the objects so that the number m is minimal possible. In this case, we will find a reverse cube $C \subset B$ such that $E \restriction C = \mathrm{id}_{m-1}$. For simplicity assume that $z = 0$ and $B = A$.

Recall the notation from Case 1 of Theorem 7.14. Let $a \subset \omega \times \omega$ be a finite set. A *precondition* with domain a is a tuple $p = \langle A(p, \tau) : \tau \in 2^a \rangle$ such that for all $\tau \in 2^a$, $A(p, \tau) \subset A$ is a nonempty Σ^1_1 set. Write $a = \mathrm{dom}(p)$. Preconditions are ordered by $q \leq p$ if $\mathrm{dom}(p) \subseteq \mathrm{dom}(q)$ and for every $\tau \in 2^{\mathrm{dom}(q)}$, $A(q, \tau) \subset A(p, \tau \restriction \mathrm{dom}(p))$. A *witness* for p is a tuple $w = \langle x(w, \tau) : \tau \in 2^a \rangle$ such that:

W1. $x(w, \tau) \in A(p, \tau)$,
W2. for every $\tau \in 2^a$, for every $j \in \omega$ we have $x(w, \tau)(j) \notin \Delta^1_1(x(w, \tau) \restriction (j, \omega))$ and $f(x(w, \tau)) \notin \Delta^1_1(x(w, \tau)) \restriction (m, \omega))$,

and for all $\tau \neq \sigma \in 2^a$,

W3. for all $j \leq d(\tau, \sigma)$, $x(w, \sigma)(j)) \neq x(w, \tau)(j)$,
W4. for all $j > d(\tau, \sigma)$, $x(w, \sigma)(j)) = x(w, \tau)(j)$,
W5. if $d(\tau, \sigma) \geq m$ then $f(x(w, \tau)) \neq f(x(w, \sigma))$.

Note that the set of all witnesses for a given precondition is Σ^1_1 by Fact 2.8. A *stem* of a witness is the common value of $x(w, \tau) \restriction (s(\max(a), \infty)$ for all $\tau \in 2^a$. A *condition* is a precondition has a witness.

Claim 7.24 *For every condition p with domain a and every open dense subset D of the Gandy–Harrington forcing there is a condition $q \leq p$ with the same domain such that $A(q, \tau) \in D$ for every $\tau \in 2^a$.*

Proof Enumerate the functions in 2^a by $\langle \tau_i : i \in j \rangle$ and by induction on $i \in j$ construct Σ^1_1 sets $A(q, \tau_i) \subset A(p, \tau_i)$ in the set D so that there is still a witness for $p_i = \langle A(q, \tau_k) : k \in i, A(p, \tau_k) : k > i \rangle$. To see how the induction step is performed, consider the set B_i of all witnesses for p_i; this is a nonempty Σ^1_1 set. The projection C_i of B_i into the τ_i-th coordinate is again a nonempty Σ^1_1 set with $C_i \subset A(p, \tau_i)$; let $A(q, \tau_i)$ be any of its nonempty Σ^1_1 subsets in D. This completes the induction step. □

Claim 7.25 *For every condition p with domain a with $\max(a) \leq m$, every witness w for p, and every countable set $b \subset 2^\omega$ there is a witness v for p such that $\mathrm{stem}(v) = \mathrm{stem}(w)$ and for every $\tau \in 2^a$ and for every $j \leq \max(a)$, we have $x(v, \tau)(j) \notin b$.*

Proof Let $B = \{\langle x^v_\tau(j) : j \leq \max(a), \tau \in 2^a\rangle : v$ is a witness for p with $\mathrm{stem}(v) = \mathrm{stem}(w)\}$. This is a nonempty $\Sigma^1_1(\mathrm{stem}(w))$ set consisting of tuples whose entries are not $\Delta^1_1(\mathrm{stem}(w))$ by W2. By Proposition 2.10, there is a tuple in the set B, neither of whose entries belongs to the set b. Let v be the witness corresponding to that tuple. □

Claim 7.26 *For every condition p with domain a with $\max(a) \geq m$, every witness w for p, and every countable set $b \subset 2^\omega$ there is a witness v for p such that $\mathrm{stem}(v) = \mathrm{stem}(w)$ and for every $\tau \in 2^a$ and for every $j \leq \max(a)$, $x(v, \tau)(j) \notin b$ and $f(x(v, \tau)) \notin b$.*

Proof Let $B = \{\langle x(v, \tau)(j), f(x(v, \tau)) : j \leq \max(a), \tau \in 2^a\rangle : v$ is a witness for p with $\mathrm{stem}(v) = \mathrm{stem}(w)\}$. This is a nonempty $\Sigma^1_1(\mathrm{stem}(w))$ set consisting of tuples whose entries are not $\Delta^1_1(\mathrm{stem}(w))$ by W2. By Proposition 2.10, there is a tuple in the set B neither of whose entries belongs to the set b. Let v be the witness corresponding to that tuple. □

Write $\omega \times \omega$ as an increasing union $\bigcup_n a_n$ of finite sets such that $a_0 = 0$ and $|a_{n+1}| = |a_n| + 1$. Let M be a countable elementary submodel of a large structure and let $\langle D_n : n \in \omega\rangle$ enumerate all the open dense subsets of the Gandy–Harrington forcing in the model M. By induction on $n \in \omega$ build conditions $p_n \in M$ so that:

I1. $\mathrm{dom}(p_n) = a_n$, $p_{n+1} \leq p_n$;
I2. for every $\tau \in 2^{a_n}$ we have $A(p_n, \tau) \in D_n$;
I3. for all $\tau \in 2^{a_n}$ and all $j \in n$, the string $x(j) \restriction n$ is the same for every choice of $x \in A(p_n, \tau)$;
I4. for all $\sigma \neq \tau \in 2^{a_n}$, for all $j \leq d(\tau, \sigma)$ and all $x_\sigma \in A(p_n, \sigma)$ and $x_\tau \in A(p_n, \tau)$, we have $x_\sigma(j) \neq x_\tau(j)$. If $d(\tau, \sigma) \geq m$ then also $f_k(x_\tau) \neq f(x_\sigma)$.

Once this induction succeeds, then for every $x \in X$ let $G_x = \{A(p_n, x \restriction a_n) : n \in \omega\}$. Item I2 shows that the upward closure of G_x is a Gandy–Harrington generic filter over M and so $\bigcap G_x$ contains a singleton $g(x) \in A$.

Item I4 shows that the function g is continuous, item I4 shows that it is an ω-keeping homeomorphism from X to X. We will obtain the cube C by massaging the cube $\mathrm{rng}(g)$.

Towards the induction, the case assumption gives us the basis of the induction, a witness for the condition $p_0 = \langle A \rangle$. To perform the induction step, suppose that p_n has been found. We will first show that $p'_n = \langle A(p_n, \tau \restriction a_n) : \tau \in 2^{a_{n+1}} \rangle$ is a condition with domain a_{n+1}. For that, we must produce a witness w' for p'_n, obtained by *splitting* a fixed witness w for p_n. Let $\langle m', l \rangle \in \omega \times \omega$ be the unique pair in $a_{n+1} \setminus a_n$. The treatment divides into several cases according to the ordering of the numbers m, m and $\max(a_n)$; we will treat the hardest case where $m < m < \max(a_n)$. Write $a_n^0 = a_n \restriction [0, m']$, $a_n^1 = a_n \restriction (m', m]$, $a_n^2 = (m, \max(a_n)]$.

First, for every $\tau \in 2^{a_n^2}$ write p_τ for the condition with $\mathrm{dom}(p_\tau) = a_n \restriction [0, m]$ given by $A(p_\tau, \sigma) = A(p_n, \tau \cup \sigma)$. Note that the tuple $w_\tau = \langle x(w, \tau \cup \sigma) : \sigma \in 2^{a_n^0 \cup a_n^1} \rangle$ is a witness for p_τ. We will show that p'_τ defined by $\mathrm{dom}(p'_\tau) = a_{n+1} \restriction [0, m]$ and $A(p'_\tau, \sigma) = A(p_\tau, \sigma \restriction a_n)$ is a condition with some witness w'_τ such that $\mathrm{stem}(w'_\tau) = \mathrm{stem}(w_\tau)$. Towards this goal, fix τ. For every $\sigma \in a_n^1$, let p_σ be the condition with domain a_n^0 given by $A(p_{\sigma\chi}) = A(p_n, \tau \cup \sigma \cup \chi)$. Note that the tuple $w_\sigma = \langle x(w, \tau \cup \sigma \cup \chi) : \chi \in 2^{a_n^0} \rangle$ is a witness for p_σ. Use Claim 7.25 repeatedly to find witnesses $v(\sigma, 0)$, $v(\sigma, 1)$ for p_σ with the same step as w_σ such that the sets $\{x(v(\sigma, i), \tau \cup \sigma \cup \chi)(j) : i \in 2, \chi \in 2^{a_n^0}, j \leq m\}$ are pairwise disjoint for $\sigma \in 2^{a_n^1}$ and $i \in 2$. Then, *glue* these witnesses together to get the desired witness for the condition p'_τ: write w'_τ for the tuple defined by $x(w'_\tau, \eta) = x(v(\eta \restriction a_n^1, \eta(m', l)), \eta \restriction a_n^0)$, and check that this is indeed the desired witness.

Now, use Claim 7.26 repeatedly at each $\tau \in 2^{a_n^2}$ to find witnesses $v(\tau)$ for p'_τ such that $\mathrm{stem}(v_\tau) = \mathrm{stem}(w_\tau)$ and the sets $b_\tau = \{x(v(\tau), \sigma)(j), f(x(v, \sigma)) : j \leq m, \sigma \in 2^{a_n^0 \cup a_n^1}\}$ are pairwise disjoint for $\tau \in 2^{a_n^2}$. Now *glue* these newly found witnesses together to get a witness for p'_n: let w' be the tuple defined by $x(w', \sigma \cup \tau) = x(v_\tau, \sigma)$. A review of definitions shows that w' is a witness for p'_n.

Finally, we will thin down p'_n to p_{n+1} so that items I2, I3, and I4 are satisfied. Let w' be any witness for p'_n. Choosing h sufficiently large and replacing $A(p'_n, \tau)$ with $\{x \in A(p'_n, \tau) : \forall j \leq \max(a_n)\ x(j) \restriction h = x(w', \tau)(j) \restriction h$ and $f(x) \restriction h = f(x(w', \tau) \restriction h\}$, items I3, I4 will be satisfied. Then w' is still a witness for p'_n and the condition p'_n satisfies I3 and I4 at $n + 1$. Finally, use Claim 7.24 in the model M to thin down p'_n so that for every $\tau \in 2^{a_{n+1}}$, $A(p'_n, \tau) \in D_n$. The resulting condition $p_{n+1} = p'_n$ will satisfy I2 as well, concluding the induction step.

Let $C = \mathrm{rng}(g)$; this is a reverse cube such that for all $x_0, x_1 \in C$ with $x_0 \restriction [m, \omega) \neq x_1 \restriction [m, \omega)$ it is the case that $f(x_0) \neq f(x_1)$, and by the minimality choice of m, for every $x \in C$, $f(x) \in \Delta^1_1(x \restriction [m, \omega))$. Now choose any reasonable effective enumeration of $\Delta^1_1(\cdot)$ singletons by natural numbers such as in Fact 2.8(*iv*). For every natural number e, the set $B_e = \{x \in C : e$ codes a $\Delta^1_1(x \restriction [m, \omega))$-singleton and that singleton is equal to $f(x)\}$ is coanalytic, and $C \subset \bigcup_e B_e$. By the reduction property of $\mathbf{\Pi}^1_1$ (Kechris 1994, theorem 35.1(ii)), the sets B_e can be thinned out to Borel sets whose union still covers C. By Theorem 7.14(*ii*) and the σ-additivity of the σ-ideal I_S we can then further thin out C so that it is wholly included in one of these sets. After that, $E \restriction C = \mathrm{id}_m$ as required.

Case 2 If Case 1 fails, we will construct a fusion sequence $C_0 \geq_0 C_1 \geq_1 C_2 \geq_2 \cdots$ so that for every $j \in \omega$, for every $x \in C_j$, the functional value $f(x)$ depends only on $x \restriction (j, \omega)$. In the end, the intersection $\bigcap_j C_j$ is I-positive by Proposition 7.22, and so it contains a reverse cube C; note that f is constant on E_1 classes of C. Since the equivalence classes of E_1 are dense in C and the function f is continuous, this means that $f \restriction C$ is constant and $E \restriction C = \mathrm{ev}$ as desired.

The construction of C_j proceeds by induction. Suppose that C_j has been constructed. Find $z \in 2^\omega$ such that $C_j \in \Sigma^1_1(z)$. The case assumption shows that whenever $x \in C_j$ is such that for every $n \in \omega$, $x(n) \notin \Delta^1_1(x \restriction (n, \omega))$, then $f(x) \in \Delta^1_1(x \restriction [j+1, \omega), z)$. Just as in Case 1, one can form the coanalytic sets $B_e = \{x \in C_j : e$ codes a $\Delta^1_1(x \restriction [j+1, \omega), z)$-singleton and that singleton is equal to $f(x)\}$. The union $\bigcup_e B_e$ covers the analytic I-positive set $C'_j = \{x \in C_j : \forall n\ x(n) \notin \Delta^1_1(x \restriction (n, \omega))\}$. By the reduction property of $\mathbf{\Pi}^1_1$, the sets B_e can be refined to Borel sets covering C'_j, one of these Borel sets has to contain a reverse cube C'_{j+1} by Theorem 7.14, and then for $x \in C'_{j+1}$, $f(x)$ depends only on $x \restriction [j+1, \omega)$. Let $C_{j+1} = \{x \in C_j : x \restriction [j, \omega) = y \restriction [j, \omega)$ for some $y \in C'_{j+1}\}$; this is a reverse cube such that $C_{j+1} \leq_j C_j$. Since $f(x)$ depends only on $x \restriction [j, \omega)$ for $x \in C_j$, it follows that $f(x)$ depends only on $x \restriction [j+1, \omega)$ for $x \in C_{j+1}$. This completes the induction step and the proof. □

Proof of Theorem 7.20 Let $A \subset X$ be a reverse cube. Let $E = \bigcup F_n$ for some inclusion increasing sequence of smooth equivalence relations, and let $f_n : A \to 2^\omega$ be Borel functions reducing each F_n to the identity.

Case 1 There is a parameter $z \in 2^\omega$ such that the functions f_n are all $\Pi^0_1(z)$, a set $B \in \Sigma^1_1(z)$ with $B \subset A$, a number $m \in \omega$, and a point $x \in B$ such that for all $n \in \omega$, $x(n) \notin \Delta^1_1(x \restriction (n, \omega), z)$ and for all $k \in \omega$, $f_k(x) \notin \Delta^1_1(x \restriction (m, \omega), z)$. We will assume that m is minimal possible, we will make the harmless assumption that $B = A$ and that thinning out if necessary, that $\forall x \in A\ \forall n\ x(n) \notin \Delta^1_1(x \restriction (n, \omega), z)$ holds. Here we extract a reverse cube $C \subset A$ such that on it $E \restriction C \subseteq \mathrm{id}_m$, and then thin it out further so that on it either $E \restriction C = \mathrm{id}_m$ or $E \restriction C = \mathrm{id}_{m-1}$. Assume for simplicity that the parameter z is in fact 0.

The construction of a reverse cube $C \subset A$ such that on it $E \restriction C \subseteq \mathrm{id}_m$ follows almost literally the proof of Case 1 of Theorem 7.23 with several minor changes. Let $a \subset \omega \times \omega$ be a finite set. A *precondition* with domain a is a tuple $p = \langle A(p, \tau) : \tau \in 2^a \rangle$ such that for all $\tau \in 2^a$, $A(p, \tau) \subset A$ is a nonempty Σ^1_1 set. Write $\mathrm{dom}(p) = a$. Preconditions are ordered by $q \leq p$ if $\mathrm{dom}(p) \subseteq \mathrm{dom}(q)$ and for every $\tau \in 2^{\mathrm{dom}(q)}$, $A(q, \tau) \subset A(p, \tau \restriction \mathrm{dom}(p))$.

A *witness* for p is a tuple $w = \langle x(w, \tau) : \tau \in 2^a \rangle$ such that:

W1. $x(w, \tau) \in A(p, \tau)$,
W2. for every $\tau \in 2^a$, for every $j \in \omega$ we have $x(w, \tau)(j) \notin \Delta^1_1(x(w, \tau) \restriction (j, \omega))$ and for every $n \in \omega$, we have $f_n(x(w, \tau)) \notin \Delta^1_1(x(w, \tau)) \restriction (m, \omega))$,

and for all $\tau \neq \sigma \in 2^a$,

W3. for all $j \leq d(\sigma, \tau)$, $x(w, \sigma)(j)) \neq x(w, \tau)(j)$,
W4. for all $j > d(\sigma, \tau)$, $x(w, \sigma)(j)) = x(w, \tau)(j)$,
W5. if $d(\tau, \sigma) \geq m$ then for every $n \in \omega$ we have $f_n(x(w, \tau)) \neq f_n(x(w, \sigma))$.

Note that the set of all witnesses for a given precondition is Σ^1_1 by Fact 2.8(*i*). A *stem* of a witness is the common value of $x(w, \tau) \restriction (s(\max(a), \infty)$ for all $\tau \in 2^a$. A *condition* is a precondition which has a witness. The following claims have proofs identical with their counterparts in Theorem 7.23.

Claim 7.27 *For every condition p with domain a and every open dense subset D of the Gandy–Harrington forcing there is a condition $q \leq p$ with the same domain such that $A(q, \tau) \in D$ for every $\tau \in 2^a$.*

Claim 7.28 *For every condition p with domain a with $\max(a) \leq m$, every witness w for p, and every countable set $b \subset 2^\omega$ there is a witness v for p such that* $\mathrm{stem}(v) = \mathrm{stem}(w)$ *and for every $\tau \in 2^a$ and for every $j \leq \max(a)$, we have $x(v, \tau)(j) \notin b$.*

Claim 7.29 *For every condition p with domain a with $\max(a) \geq m$, every witness w for p, and every countable set $b \subset 2^\omega$ there is a witness v for p*

such that $\mathrm{stem}(v) = \mathrm{stem}(w)$ *and for every* $\tau \in 2^a$ *and for every* $j \leq \max(a)$, $x(v, \tau)(j) \notin b$ *and for every* $n \in \omega$ $f_n(x(v, \tau)) \notin b$.

Write $\omega \times \omega$ as an increasing union $\bigcup_n a_n$ of finite sets such that $a_0 = 0$ and $|a_{n+1}| = |a_n| + 1$. Let M be a countable elementary submodel of a large structure and let $\langle D_n : n \in \omega \rangle$ enumerate all the open dense subsets of the Gandy–Harrington forcing in the model M. By induction on $n \in \omega$ build conditions $p_n \in M$ so that:

I1. $\mathrm{dom}(p_n) = a_n$, $p_{n+1} \leq p_n$;
I2. $\forall \tau \in 2^{a_n}$ $A(p_n, \tau) \in D_n$;
I3. for all $\tau \in 2^{a_n}$ and all $j \in n$, the string $x(j) \restriction n$ is the same for every choice of $x \in A(p_n, \tau)$;
I4. for all $\sigma \neq \tau \in 2^{a_n}$, for all $j \leq d(\tau, \sigma)$ and all $x_\sigma \in A(p_n, \sigma)$ and $x_\tau \in A(p_n, \tau)$, we have $x_\sigma(j) \neq x_\tau(j)$. If $d(\tau, \sigma) \geq m$ then also $f_k(x_\tau) \neq f_k(x_\sigma)$ for all $k \in n$.

Once this induction succeeds, then for every $x \in X$ let $G_x = \{A(p_n, x \restriction a_n) : n \in \omega\}$. Item I2 shows that the upward closure of G_x is a Gandy–Harrington generic filter over M and so $\bigcap G_x$ contains a singleton $g(x) \in A$. Item I4 shows that the function g is continuous, item I4 shows that its is an ω-keeping homeomorphism from X to X such that $E \restriction \mathrm{rng}(g) \subset \mathrm{id}_m$.

The case assumption gives us the basis of the induction, a witness for the condition $p_0 = \langle A \rangle$. To perform the induction step, suppose that p_n has been found. We will first show that $p'_n = \langle A(p_n, \tau \restriction a_n) : \tau \in 2^{a_{n+1}} \rangle$ is a condition with domain a_{n+1}. For that, we must produce a witness w' for p'_n, obtained by *splitting* a fixed witness w for p_n. Let $\langle m', l \rangle \in \omega \times \omega$ be the unique pair in $a_{n+1} \setminus a_n$. The treatment divides into several cases according to the ordering of the numbers m', m and $\max(a_n)$; we will treat the hardest case where $m' < m < \max(a_n)$. Write $a_n^0 = a_n \restriction [0, m']$, $a_n^1 = a_n \restriction (m', m]$, $a_n^2 = (m, \max(a_n)]$.

First, for every $\tau \in 2^{a_n^2}$ write p_τ for the condition with $\mathrm{dom}(p_\tau) = a_n \restriction [0, m]$ given by $A(p_\tau, \sigma) = A(p_n, \tau \cup \sigma)$. Note that the tuple $w_\tau = \langle x(w, \tau \cup \sigma) : \sigma \in 2^{a_n^0 \cup a_n^1} \rangle$ is a witness for p_τ. We will show that p'_τ defined by $\mathrm{dom}(p'_\tau) = a_{n+1} \restriction [0, m]$ and $A(p'_\tau, \sigma) = A(p_\tau, \sigma \restriction a_n)$ is a condition with some witness w'_τ such that $\mathrm{stem}(w'_\tau) = \mathrm{stem}(w_\tau)$. Towards this goal, fix τ. For every $\sigma \in a_n^1$, let p_σ be the condition with domain a_n^0 given by $A(p_{\sigma\chi}) = A(p_n, \tau \cup \sigma \cup \chi)$. Note that the tuple $w_\sigma = \langle x(w, \tau \cup \sigma \cup \chi) : \chi \in 2^{a_n^0} \rangle$ is a witness for p_σ. Use Claim 7.28 repeatedly to find witnesses $v(\sigma, 0)$, $v(\sigma, 1)$ for p_σ with the same step as w_σ such that the sets

$\{x(v(\sigma, i), \tau \cup \sigma \cup \chi)(j) : i \in 2, \chi \in 2^{a_n^0}, j \leq m'\}$ are pairwise disjoint for $\sigma \in 2^{a_n^1}$ and $i \in 2$. Then, *glue* these witnesses together to get the desired witness for the condition p'_τ: write w'_τ for the tuple defined by $x(w'_\tau, \eta) = x(v(\eta \restriction a_n^1, \eta(m', l)), \eta \restriction a_n^0)$ for $\eta \in 2^{a_{n+1} \restriction [0,m]}$, and check that this is indeed the desired witness.

Now, use Claim 7.29 repeatedly at each $\tau \in 2^{a_n^2}$ to find witnesses $v(\tau)$ for p'_τ such that $\mathrm{stem}(v_\tau) = \mathrm{stem}(w_\tau)$ and the sets $b_\tau = \{x(v(\tau), \sigma)(j), f_k(x(v, \sigma)) : j \leq m, k \in \omega, \sigma \in 2^{a_n^0 \cup a_n^1}\}$ are pairwise disjoint for $\tau \in 2^{a_n^2}$. Now *glue* these newly found witnesses together to get a witness for p'_n: let w' be the tuple defined by $x(w', \sigma \cup \tau) = x(v_\tau, \sigma)$. A review of definitions shows that w' is a witness for p'_n.

Finally, we will thin down p'_n to p_{n+1} so that items I2, I3, and I4 are satisfied. Let w' be any witness for p'_n. Choosing h sufficiently large and replacing $A(p'_n, \tau)$ with $\{x \in A(p'_n, \tau) : \forall j \leq \max(a_n)\ x(j) \restriction h = x(w', \tau)(j) \restriction h$ and $f_k(x) \restriction h = f_k(x(w', \tau)) \restriction h$ for all $k \in n\}$, items I3 and I4 will be satisfied. Then w' is still a witness for p'_n and the condition p'_n satisfies I3 and I4 at $n + 1$. Finally, use Claim 7.27 in the model M to thin down p'_n so that for every $\tau \in 2^{a_{n+1}}$, $A(p'_n, \tau) \in D_n$. The resulting condition $p_{n+1} = p'_n$ will also satisfy I2 concluding the induction step.

This completes the construction of a reverse cube C such that $E \restriction C \subseteq \mathrm{id}_m$. By the minimal choice of the number m, for every $x \in C$ there is a number $n \in \omega$ such that $f_n(x) \in \Delta^1_1(x \restriction [m-1, \omega))$. The sets $B_{n,e} = \{x \in C : e$ is a code for a $\Delta^1_1(x \restriction [m-1, \omega))$ singleton which is equal to $f_n(x)\}$ are coanalytic and their union covers C. By the reduction property of $\mathbf{\Pi}^1_1$, these sets can be refined to Borel sets whose union still covers C. Theorem 7.14 then shows that we may thin out C into a smaller reverse cube which is a subset of one of these Borel sets. It follows that $\mathrm{id}_{m-1} \subseteq E \restriction C$. We will now thin out C further to fully canonize $E \restriction C$.

Write C_m for the projection of C to $(2^\omega)^{[m,\omega)}$ and similarly for C_{m-1}; these are reverse cubes in the appropriate sense. Note that for all $n > n_0$ and $x \in C$, the functional value $f_n(x)$ depends on $x \restriction [m-1, \omega)$ only, and write f_n for the function from C_{m-1} to 2^ω. Use Theorem 7.14(*ii*) to thin out C so that the functions f_n for $n \in \omega$ are all continuous on C_{m-1}.

Consider the sets $D_0 = \{\langle x, P\rangle : x \in C_m, P \subset 2^\omega$ is a nonempty perfect set, $\{x\} \times P \subset C_{m-1}$, and for some $n \geq n_0$, f_n is constant on $\{x\} \times P\}$ and $D_1 = \{\langle x, P\rangle : x \in C_m, P \subset 2^\omega$ is a nonempty perfect set, $\{x\} \times P \subset C_{m-1}$, and for every $n \geq n_0$, f_n is one-to-one on $\{x\} \times P\}$. Since the functions f_n are continuous, both of these sets are analytic in $(2^\omega)^{[m,\omega)} \times K(2^\omega)$. We will show that the union of projections of D_0 and D_1 covers the whole cube C_m. For every $x \in C_m$, write $Q_x = \{y \in C_{m-1} : y \restriction [m, \omega) = x\}$; this is a perfect set. Either there is $n \geq n_0$ such that the function f_n is not countable-to-one on Q_x; then, the function is constant on a perfect set and so x belongs to the projection of D_0. Or, all the functions $f_n : n \geq n_0$ are countable-to-one on Q_x, and then there is a perfect set on which all of these functions are injective, for example by Mycielski's theorem (Mycielski 1964). Thus, by Theorem 7.14 we can thin out C_m into a reverse subcube if necessary and find a continuous function $g : C_m \to K(2^\omega)$ such that (a) $g \subset D_0$ or (b) $g \subset D_1$. In either case, it is possible to thin out C further so that for every $x \in C$, $x(m-1) \in g(x \restriction [m, \omega))$. In case (a), we get $E \restriction C = \mathrm{id}_m$, in case (b) we get $E \restriction C = \mathrm{id}_{m-1}$ as desired.

Case 2 There is a reverse cube $C_0 \subset A$, a parameter $z \in 2^\omega$ and a number n such that for every $x \in C_0$ and every $m \in \omega$, we have $f_n(x) \in \Delta^1_1(x \restriction (m, \omega), z)$. In this case, use Theorem 7.23 to thin down C_0 to a reverse cube $C \subset C_0$ so that the equivalence relation $F_n \restriction C$ is equal to id_m for some $m \in \omega$ or to ev. We will show that only the latter case is possible, meaning that $E \restriction C = \mathrm{ev}$.

Suppose for contradiction that $F_n \restriction C = \mathrm{id}_m$ for some $m \in \omega$. Since C is a reverse cube, there is an uncountable set $Q \subset C$ such that the elements of Q are pairwise id_m-inequivalent and id_{m+1}-equivalent. Write y for the common value of $x \restriction (m, \omega)$ for $x \in Q$, and observe that the uncountably many distinct values $f(x)$ for $x \in Q$ must fit into the countable set $\Delta^1_1(y, z)$. This is a contradiction.

Case 3 Suppose now that both Cases 1 and 2 fail. We will build a fusion sequence whose limit, a reverse cube $C \subset X$, will have $E \restriction C = E_1$. By induction on $j \in \omega$ build numbers $m_j, n_j \in \omega$ and cubes C_j so that $m_0 < m_1 < m_2 < \cdots$, $C_j \geq_{m_j} C_{j+1}$, and $f_{n_j} \restriction C_j = \mathrm{id}_{m_j}$. If this succeeds, let $C = \bigcap_j C_j$. This is a reverse cube; the assumptions immediately imply that $n_0 < n_1 < \cdots$ and $E \restriction C = E_1$. To perform the induction, suppose C_j, m_j, n_j have been constructed. Let $z \in 2^\omega$ be such that $C_j \in \Sigma^1_1(z)$. Since Case 1 fails, for every $x \in C_j$ such that $\forall i\ x(i) \notin \Delta^1_1(x \restriction (i, \omega), z)$ there must be n such that $f_n(x) \in \Delta^1_1(x \restriction (m_j, \omega), z)$. The sets $B_n = \{x \in C_j : f_n(x) \in \Delta^1_1(x \restriction (m_j, \omega), z)\}$ are coanalytic for each $n \in \omega$ and their union covers the analytic I-positive set $C'_j = \{x \in C_j : \forall i\ x(i) \notin \Delta^1_1(x \restriction (i, \omega), z)\}$. By the reduction property of $\mathbf{\Pi}^1_1$, the sets B_n can be refined to Borel sets whose union still covers C'_j, and then Theorem 7.14 yields a reverse cube $C'_j \subset C_j$ which is a subset of one of the sets B_n, say for $n = n_{j+1}$. Find a new parameter $z' \in 2^\omega$ coding z and such that C'_j is $\Sigma^1_1(z')$. Since Case 2 fails, there must be $x \in C_j$ such that for every $i \in \omega$, $x(i) \notin \Delta^1_1(x \restriction (i, \omega), z')$ and for some $m > m_j$, $f_{n_{j+1}}(x) \notin \Delta^1_1(x \restriction (m, \omega), z')$. Let $m_{j+1} > m_j$ be the least number m for which such a point x exists. Case 1 of Theorem 7.23 shows that there is a subcube $C'_{j+1} \subset C'_j$ such that $E_{n_{j+1}} \restriction C'_{j+1} = \mathrm{id}_{m_j}$. Let $C_{j+1} = \{x \in C_j : x \restriction (m_j, \omega) = y \restriction (m_j, \omega)$ for some $y \in C'_{j+1}\}$; this is a reverse cube and $C_{j+1} \leq_{m_j} C_j$. Since $E_{n_j} \subset E_{n_{j+1}}$ and $E_{n_j} \restriction C_j = \mathrm{id}_{m_j}$, it is still the case that $E_{n_{j+1}} \restriction C_{j+1} = \mathrm{id}_{m_j}$. This concludes the induction step and the proof. □

Proof of Theorem 7.7 Let $B \subset X$ be an analytic set such that $E_1 \restriction B$ is not essentially countable, and let E be a hypersmooth equivalence on B. Using Theorem 7.14(*ii*), thinning out the set B if necessary we may assume that B is an s-cube for some infinite set $s \subset \omega$, with an s-keeping homeomorphism $f : X \to B$. Let $F = f^{-1}E$ be the (still hypersmooth) pullback of E, and use Theorem 7.20 to find a reverse cube $D \subset X$ such that $F \restriction D$ is either equal to E_1 or to ev or to id_n for some $n \in \omega$. In the first two cases, let $C = f''D$. This is a Borel J-positive set since $E_1 \leq_B E_1 \restriction D \leq_B E_1 \restriction C$, the latter reduction witnessed by $f \restriction D$. Clearly, $E \restriction C$ is equal to either E_1 or to ev. In the last case, write D_n for the projection of D to $(2^\omega)^{[n,\omega)}$; this is a reverse cube in a suitable sense. Use the Borel uniformization for Borel sets with compact sections to find a Borel injection $g : D_n \to D$ such that $g(x) = x \restriction [n, \omega)$, and then note that $\mathrm{rng}(g)$ is a Borel J-positive set consisting of pairwise F-inequivalent elements. Let $C = f''\mathrm{rng}(g)$. This is a Borel J-positive set since $E_1 \leq_B E_1 \restriction \mathrm{rng}(g) \leq_B E_1 \restriction C$, the latter reduction witnessed by $f \restriction \mathrm{rng}(g)$. Also, $E \restriction C = \mathrm{id}$. □

Proof of Theorem 7.8 The beginning of the proof is the same as above. Let $B \subset X$ be an analytic set such that $E_1 \restriction B$ is not essentially countable, and let E be an equivalence relation on B classifiable by countable structures, or Borel reducible to $=_K$ for an analytic P-ideal K on ω. Using Theorem 7.14(*ii*), thinning out the set B if necessary we may assume that B is an s-cube for some infinite set $s \subset \omega$, with an s-keeping homeomorphism $f : X \to B$. Let $F = f^{-1}E$ be the pullback of E. Use Theorem 9.27 or 9.26 to find a product $D' \subset X$ of nonempty perfect sets such that $F \restriction D'$ is smooth. Note that D' is a reverse cube, so we can use Theorem 7.20 to find a reverse cube $D \subset X$ with a ω-keeping homeomorphism $g : X \to D$ such that $F \restriction D$ is either equal to E_1 or to ev or to id_n for some $n \in \omega$. The E_1 option is certainly impossible: the composition $g \circ f$ would be a Borel reduction of E_1 to E, and so E_1 would be either classifiable by countable structures or Borel reducible to $=_K$ for an analytic P-ideal K on ω. Since both equivalences classifiable by countable structures and $=_K$ are reducible to orbit equivalences (Solecki 1999), this would contradict the fact that E_1 is not reducible to an orbit equivalence. As in the proof of Theorem 7.7, the ev option provides a

Borel set $C \subset B$ on which E_1 is still bireducible with E_1 and $E \restriction C = \text{ev}$, and the id_n option provides a Borel set $C \subset B$ on which E_1 is still bireducible with E_1 and $E \restriction C = \text{id}$. □

Sketch of proof of Theorem 7.9 We will treat only the case when $\text{dom}(E) = X$. The following reduced product definition is key.

Definition 7.30 Let $n \in \omega$ and $z \in 2^\omega$. The poset Q_n^z consists of pairs $\langle A, B\rangle$ such that both $A, B \subset X$ are Σ^1_1 sets such that there is a *witness*: a pair of points $x \in A$ and $y \in B$ such that:

U1. $\forall m \in \omega\, x(m) \notin \Delta^1_1(z, x \restriction (m, \omega))$ and $y(m) \notin \Delta^1_1(z, y \restriction (m, \omega))$;
U2. $x \restriction (n, \omega) = y \restriction (n, \omega)$.

The ordering is the coordinatewise inclusion.

It is not difficult to see that the poset Q_n^z is countable and adds two points $\dot{x}_{\text{gen}}$ and $\dot{y}_{\text{gen}}$ which are in the intersections of the left and right coordinates in the conditions in the generic filter respectively, and $\dot{x}_{\text{gen}} \restriction (n, \omega) = \dot{y}_{\text{gen}} \restriction (n, \omega)$.

Lemma 7.31 *For every reverse cube $C \subset X$ and for every countable elementary submodel M of a large structure containing B there is $z \in 2^\omega$ and a reverse cube $C' \subset C$ such that for every pair $x, y \in C'$ and every $n \in \omega$ such that $x \restriction (n, \omega) = y \restriction (n, \omega)$ and $x(n) \neq y(n)$, the pair x, y is Q_n^z-generic over the model M.*

Proof Let z be such that B is $\Pi^0_1(z)$; for simplicity assume that $z = 0$. The fusion argument for the construction of C follows almost literally the argument for Theorem 7.14, with the s-coordinate eliminated. The only significant change is that Claim 7.15 must be replaced with:

Claim 7.32 *Whenever p is a condition, $a = \text{dom}(p)$, $n \in \max(a)$ and $D \subset Q_n$ an open dense set then there is a condition $q \leq p$ with the same domain such that for every pair $\sigma, \tau \in 2^a$, if $d(\sigma, \tau) = n$ then $\langle A(q, \sigma), A(q, \tau)\rangle \in D$.*

To prove the claim, choose $\sigma, \tau \in 2^a$ such that $d(\sigma, \tau) = n$. We will produce a condition $q \leq p$ with $\text{dom}(q) = a$ such that $\langle A(q, \sigma), A(q, \tau)\rangle \in D$. The full statement of the claim follows from a repetition of this single step.

Let $A = \{x \in A(p, \sigma) : \text{there is a witness } w \text{ for } p \text{ such that } w(\sigma) = x\}$ and let $B = \{y \in A(p, \tau) : \text{there is a witness } w \text{ for } p \text{ such that } w(\tau) = y\}$. This is a pair of Σ^1_1 sets in Q_n as witnessed by any pair $\langle x, y\rangle$ coming from a single witness for p. Find a pair of sets $\langle A', B'\rangle$ in Q_n below $\langle A, B\rangle$ in D. We will show that the precondition $q \leq p$ obtained by replacing $A(p, \sigma)$ with A' and $A(p, \tau)$ with B' is a condition; this will conclude the argument. We must produce a witness for q.

Let $x, y \in X$ be a pair of points witnessing $\langle A', B' \rangle \in Q_n$. Use the definitions of the sets A and B to find witnesses v, w for p such that $v(\sigma) = x$ and $w(\tau) = y$. Use (in essence) Claim 7.16 to adjust the witness w so that $w(\tau) \in B'$, $w(\tau) \restriction (n, \omega) = x \restriction (n, \omega)$ and for every $\eta \in 2^a$ such that $\tau \restriction [n, \max(a)] \subset \eta$ and every $m \in n$, the point $w(\eta)(m)$ is not in the countable set $b = \{v(\chi)(k) : \chi \in 2^a \text{ and } k \in n\}$. Now *glue* these witnesses together: consider the function u given by $u(\eta) = w(\eta)$ if $\tau \restriction [n, \max(a)] \not\subset \eta$, and $u(\eta) = v(\eta)$ if $\tau \restriction [n, \max(a)] \subset \eta$. It is not difficult to check that u is a witness for q as desired. □

Lemma 7.33 *For every reverse cube C and a countable collection of analytic sets $\{A_n : n \in \omega\}$ there is a reverse cube $C' \subset C$ such that for every $n \in \omega$, A_n is relatively clopen in C'.*

Proof Since analytic sets have the Baire property, there is a comeager subset of C in which all the set A_n are relatively clopen. The comeager set is J-positive, since the generators of the σ-ideal J are meager in C. Inscribe a reverse cube C' into the comeager set using Theorem 7.14. This completes the proof. □

Now let $E \subset E_1$ be an analytic equivalence relation on X. We will produce a reverse cube $C \subset X$ on which for every number $n \in \omega$ the equivalence relation $\text{id}_n \cap E$ is smooth. Then, $E \restriction C = \bigcup_n (\text{id}_n \cap E)$ is a hypersmooth equivalence relation. Theorem 7.7 then shows that the cube C can be thinned out further so that $E \restriction C$ attains one of the required forms.

To find the reverse cube C, let M be a countable elementary submodel of a large structure. Use Lemma 7.31 to find C such that for every $n \in \omega$ and every $x, y \in C$ such that $x \restriction (n, \omega) = y \restriction (n, \omega)$ and $x(n) \neq y(n)$, the pair x, y is Q_n-generic over the model M. Use Lemma 7.33 to thin out the cube C further so that all Σ^1_1 sets are relatively open in it. Fix a number $n \in \omega$ and argue that $E \cap \text{id}_n$ is a G_δ relation on X; in fact, it is a finite combination of open and closed sets. Namely, distinct points $x, y \in C$ are in the relation $E \cap \text{id}_n$ if for some $m \leq n$, $x(k) = y(k)$ for all $k > m$, $x(m) \neq y(m)$, and there is a condition $\langle A, B \rangle \in Q_m$ such that $x \in A$, $y \in B$ and $\langle A, B \rangle \Vdash_{Q_m} \dot{x}_{\text{gen}}\ E\ \dot{y}_{\text{gen}}$. Since the sets A, B are relatively open in C, the equivalence relation $E \cap \text{id}_n$ is G_δ. All G_δ relations are smooth by Gao (2009, theorem 6.4.4) and so $E \cap \text{id}_n$ is a smooth equivalence relation as required. □

7.2.3 Borel functions on pairs

The methods of this section make it possible to identify coding functions for the σ-ideals discussed above. This makes it very unlikely that any reduced product forcing argument could be used to produce further canonization results.

Theorem 7.34 *Let J be the σ-ideal on X consisting of Borel sets with no reverse subcube. Then J has a square coding function defined on X^2.*

A rectangular coding function for the σ-ideal I does not exist. To exclude it, note that the σ-ideal J is $\mathbf{\Pi}^1_1$ on $\mathbf{\Sigma}^1_1$ and its quotient poset is proper, bounding, and category preserving, as shown in Zapletal (2008, theorems 5.4.11 and 5.4.12). Therefore, the σ-ideal has the rectangular Ramsey property by Fact 2.56, which excludes the existence of a rectangular coding function.

Proof Let $f(x, y)(m) = x(m)(k)$ where $k \in \omega$ is the least number such that $x(m)(k) \neq y(m)(k)$ if such number exists; if such a k does not exist, then let $f(x, y)(m) = \text{trash}$. We will show that this function works.

Let $C \subset X$ be a reverse cube and let $z \in 2^\omega$ be an arbitrary point. We have to produce points $x, y \in C$ such that $f(x, y) = z$. Assume for simplicity that both C, z are Δ^1_1. A *condition* is a triple $p = \langle A(p), B(p), n(p)\rangle$ where $A(p), B(p) \subset C$ are nonempty Σ^1_1 sets, $n(p) \in \omega$, such that there is a *witness* $w = \langle x(w), y(w)\rangle$, a pair such that:

W1. $x(w) \in A(p)$, $y(p) \in B(p)$;
W2. for every $n \in \omega$ we have $x(w)(n) \notin \Delta^1_1(x(w) \restriction (n, \omega))$ and $y(w)(n) \notin \Delta^1_1(y(w) \restriction (n, \omega))$;
W3. for every $m < n(p)$, $x(w)(m) \neq y(w)(m)$ and $x(m)(k) = z(m)$ where k is the least number such that $x(w)(m)(k) \neq y(w)(m)(k)$;
W4. for every $m \geq n(p)$, $x(w)(m) = y(w)(m)$.

Let M be a countable elementary submodel of a large structure and let $\langle D_n : n \in \omega\rangle$ enumerate all dense open subsets of Gandy–Harrington forcing in the model M. By induction on $n \in \omega$ find conditions $p_n \in M$ so that:

I1. $p_{n+1} \leq p_n$, $n(p_n) = n$;
I2. $A(p_{n+1}), B(p_{n+1}) \in D_n$;
I3. for every $x \in A(p_n)$ and every $y \in B(p_n)$ and every $m < n$, $f(x, y)(m) = z(m)$.

Once the induction is completed, the sets $\{A(p(n)) : n \in \omega\}$ and $\{B(p(n)) : n \in \omega\}$ generate Gandy–Harrington generic filters over the model M, so their intersections will contain two respective points $x, y \in C$. Item I3 shows that $f(x, y) = z$ as desired.

The induction starts with $p_0 = \langle C, C, 0\rangle$. To construct p_{n+1}, find a witness $w \in M$ for p_n and let $B = \{u \in 2^\omega : \exists v\ v$ is a witness for p_n such that $x(w) \restriction (n+1, \omega) = x(v) \restriction (n+1, \omega)$ and $u = x(v)(n)\}$. This is a $\Sigma^1_1(x(w) \restriction (n+1, \omega))$ set by Fact 2.8(*ii*), and it contains an element $x(w)(n)$ which is not $\Delta^1_1(x(w) \restriction (n+1, \omega))$ by item W2. Thus, it contains uncountably many elements, of which we use two distinct ones, $u_0 \neq u_1$, indexed so that writing k for the least number such that $u_0(k) \neq u_1(k)$, $u_0(k) = z(n)$ holds. We claim that $p'_n = \langle \{x \in A(p_n) : x(n) \restriction k+1 = u_0 \restriction k+1\}, \{x \in B(p_n) : x(n) \restriction k+1 = u_1 \restriction k+1\}, n+1\rangle$ is a condition. To produce a witness w for it, let v_0, v_1 be two witnesses for p_n such that $x(v_0)(n) = u_0$ and $x(v_1)(n) = u_1$ and let $w = \langle x(v_0), y(v_1)\rangle$; the pair w is a witness for p'_n. The final task in the induction step is to thin out p'_n to p_{n+1} so that item I2 is satisfied. First let $A = \{x \in A(p'_n) : x = v(x)$ for some witness v for $p'_n\}$; this is a nonempty Σ^1_1 set. Thin it down to some $A' \in D$ and observe that $p''_n = \langle A', B(p'_n), n+1\rangle$ is still a condition. Perform the same task on the B side; the resulting condition p_{n+1} is as required. This completes the induction step and the proof. □

Corollary 7.35 *Let L be a σ-ideal on a Polish space Y such that the quotient forcing P_L is proper. If E_1 is in the spectrum of the σ-ideal L then L has a square coding function.*

Proof We will show that the assumptions imply that there is a Borel L-positive set $B \subset Y$ such that $L \restriction B \geq_K J$, where J is the σ ideal on X generated by Borel sets containing no reverse cube. The corollary then follows from Theorem 7.34 and Proposition 2.69.

By Theorem 4.23, if E_1 is in the spectrum then there is a condition $B \in P_L$ and names $\langle \dot{y}_n : n \in \omega\rangle$ for elements of the Cantor space such that $B \Vdash \forall n\ y_n \notin V[\langle y_m : m > n\rangle]$. Using the properness of the quotient poset, thinning out the condition B if necessary it is possible to find Borel functions $g_n : B \to 2^\omega$ representing the names $\dot{y}_n$ for every $n \in \omega$, and combine these functions to $g : B \to (2^\omega)^\omega$. We claim that the function g is the required Katětov map, i. e., preimages of L-small sets are K-small. To this end, suppose that $C \subset B$ is a L-positive Borel set and argue that the analytic set $g''C \subset (2^\omega)^\omega$ is J-positive. Whenever $z \in 2^\omega$ is a parameter, M is a countable elementary submodel of a large enough structure containing z, K, C, and $x \in C$ is a P_L-generic point over M, it is the case for every $n \in \omega$ that $g_n(x) \notin M(\langle g_m(x) : m > n\rangle)$ by the forcing theorem. Therefore, $g_n(x) \notin \Delta^1_1(z, g_m(x) : m > n)$ and so $g(x)$ does not belong to the generator set of the σ-ideal J associated with the parameter z. □

7.3 The E_2 ideal

E_2 is the equivalence relation on $X = 2^\omega$ defined by $x E_2 y$ if $\Sigma\{\frac{1}{n+1} : x(n) \neq y(n)\} < \infty$. It is Borel bireducible with the ℓ_1 cosets inside the Banach space ℓ_∞. It is a basic example of a turbulent equivalence relation, and therefore it is not classifiable by countable structures (Kanovei 2008, theorem 13.9.1). It turns out that the collection I of Borel sets $B \subset 2^\omega$ such that $E_2 \restriction B$ is not bireducible with E_2 is a σ-ideal. We will prove several canonization results for it.

The first canonization result concerns, among others, the ℓ_p equivalence relations, those Borel reducible to the ℓ_p cosets inside $\mathbb{R}^\omega$, for all $p \geq 1$. If $p > 1$, the coset equivalence relations are strictly above E_2 in the Borel reducibility order (Kanovei 2008, theorem 6.5.1). It is easy to see that every ℓ_p equivalence relation is Borel reducible to $=_J$ for some F_σ P-ideal J.

Theorem 7.36 *Suppose that $B \subset 2^\omega$ is a Borel set such that $E_2 \restriction B$ is bireducible with E_2. Suppose that E is an equivalence relation on B Borel reducible to $=_J$ for an F_σ P-ideal J on 2^ω. Then there is a compact set $C \subset B$ such that $E_2 \restriction C$ is still bireducible with E_2, and $E \restriction C$ is equal to either* id, *E_2, or* ev. *In particular,*

$$\ell_p \text{ equivalences} \to_I \{\text{id}, E_2, \text{ev}\}.$$

Theorem 7.37 *Suppose that $B \subset 2^\omega$ is a Borel set such that $E_2 \restriction B$ is bireducible with E_2. Suppose that E is an equivalence relation on B classifiable by countable structures. Then there is a compact set $C \subset B$ such that $E_2 \restriction C$ is still bireducible with E_2, and $E \restriction C$ is equal to either* id *or* ev. *Restated, I has total canonization for equivalence relations classifiable by countable structures, or*

$$\text{classifiable by countable structures} \to_I \{\text{id}, \text{ev}\}.$$

Basic properties of the σ-ideal I were investigated earlier.

Fact 7.38 (Kanovei 2008, section 15.2) *The collection I is a $\mathbf{\Pi}^1_1$ on $\mathbf{\Sigma}^1_1$ σ-ideal, and every analytic I-positive set has a Borel I-positive subset. A Borel set $B \subset X$ is in the ideal I if and only if the equivalence $E_2 \restriction B$ is essentially countable if and only if E_2 is not reducible to $E_2 \restriction B$.*

The quotient poset P_I is fairly well behaved from the forcing point of view, and its properties were surveyed in Kanovei (2008) and Zapletal (2008). Here, we will provide a more thorough treatment:

Theorem 7.39 *The forcing P_I is proper, bounding, preserves Baire category and outer Lebesgue measure.*

As for the related open questions, it is clear from the definitions that E_2 is in the spectrum of the σ-ideal I. We do not see any other features of the spectrum:

Question 7.40 Is there any other equivalence relation beyond E_2 in the spectrum of the σ-ideal I?

For the notation in this section, we will write d for the usual metric on 2^ω: if $x\ E_2\ y$ then let $d(x, y) = \sum\{\frac{1}{n+1} : x(n) \neq y(n)\}$, otherwise let $d(x, y) = \infty$. Thus, d is a metric on each equivalence class.

7.3.1 The ideal

We will first identify the generators of the σ-ideal I:

Definition 7.41 (Kanovei 2008, definition 15.2.2) Let $\varepsilon > 0$ be a real number. An *ε-walk* is a finite sequence $\langle x_i : i \leq n\rangle$ of points in 2^ω such that $d(x_i, x_{i+1}) < \varepsilon$ for all $i < n$. A set $A \subset 2^\omega$ is *grainy* if there is a real number $\varepsilon > 0$ such that A contains no ε-walk whose endpoints are at a d-distance at least 1 from each other.

Fact 7.42 (Kanovei 2008, theorem 15.2.1) *The σ-ideal σ-generated by the analytic grainy sets is equal to I.*

Here, we will provide a part of that argument that will serve as a useful warm-up for the later canonization results:

Theorem 7.43 *Let $A \subset 2^\omega \times \omega^\omega$ be an analytic set. Then exactly one of the following happens:*

(i) either $p(A) \in I$;
(ii) or there is a one-to-one continuous E_2-preserving function $f : 2^\omega \to p(A)$ and a continuous function $e : \mathrm{rng}(f) \to \omega^\omega$ such that $e \subset A$.

In particular, every I-positive analytic set contains a compact I-positive subset.

Proof Let $z \in 2^\omega$ be a parameter such that $A \in \Sigma^1_1(z)$. To simplify the notation, assume that $z = 0$. Let N be the union of all Σ^1_1 grainy sets. The set N is Π^1_1 by Fact 2.8(v).

Claim 7.44 *Let $A' \subset 2^\omega \setminus N$ be a nonempty Σ^1_1 set. Then for every nonempty open subset O of the real number interval $(0, 1)$ there are points $x, y \in A'$ such that $x\ E_2\ y$ and $d(x, y) \in O$.*

Proof Pick an open interval $K \subset O$. Let $\varepsilon > 0$ be a positive real smaller than the length of K. Since the set A' is not grainy, there is an ε-walk consisting of elements of A' such that its endpoints are at least d-distance 1 away from each other. Let x be the first point on that walk, and let y be the first point on the walk whose distance from x is greater than the leftmost point of the interval K. The points x, y are as desired. □

Let $B = p(A) \setminus N$; this is a Σ^1_1 set. If $B = 0$ then we are in the first case of the theorem; if $B \neq 0$ then we must verify the conclusion of the second case. Let M be a countable elementary submodel of a large enough structure, and enumerate all the open dense subsets of the Gandy–Harrington forcing on the space $2^\omega \times \omega^\omega$ by $\{D_i : i \in \omega\}$. By induction on $k \in \omega$ build numbers $n_k \in \omega$ and functions $g_k : 2^k \to 2^{n_k}$, $h_k : 2^k \to \omega^k$ and sets A_t for $t \in 2^k$ so that:

I1. $0 = n_0 < n_1 < \cdots$ and $A_0 = B$;
I2. $s \subset t$ implies $g(s) \subset g(t)$, $h(s) \subset h(t)$ and $A_s \supset A_t$;
I3. for every $s, t \in 2^k$, the following two statements hold:

$$\sum\left\{\tfrac{1}{n+1} : n_k \leq n < n_{k+1}, g(s^\frown 0)(n) \neq g(t^\frown 0)(n)\right\} < 2^{-k},$$

$$\left|\tfrac{1}{k+1} - \sum\left\{\tfrac{1}{n+1} : n_k \leq n < n_{k+1}, g(s^\frown 0)(n) \neq g(t^\frown 1)(n)\right\}\right| < 2^{-k-1};$$

I4. $A_s \subset [g(s)] \times [h(s)]$ and $A_s \in D_{|s|-1}$;
I5. there is a *witness* for the sequence $\langle A_t : t \in 2^k\rangle$. A witness is a sequence $\langle x_t, y_t : t \in 2^k\rangle$ such that for each $t \in 2^k$, $\langle x_t, y_t\rangle \in A_t$ holds, and

$$\sum\left\{\frac{1}{n+1} : n \geq n_k \text{ and } \exists t, s \in 2^k\ x_t(n) \neq x_s(n)\right\} < 2^{-k-1}.$$

Once the induction is performed, for every $y \in 2^\omega$ let $g(y) = \bigcup_k g(y \restriction k)$ and $h(y) = \bigcup_k h(y \restriction k)$. For every $y \in 2^\omega$, the sets $\{A_{y \restriction k} : k \in \omega\}$ generate a Gandy–Harrington generic filter over the model M and the pair $\langle g(y), h(y)\rangle$ is its associated generic point by I4. It follows then from I1 that $\langle g(y), h(y)\rangle \in A$. It follows from I3 that the map $g : 2^\omega \to p(A)$ is a one-to-one continuous E_2-preserving map, and the map $e = h \circ g^{-1}$ is the required continuous map defined on $\operatorname{rng}(g)$ whose graph is a subset of A. This will complete the proof of the theorem.

To perform the induction step, suppose that $g \restriction 2^k$, $h \restriction 2^k$ and $\langle A_t : t \in 2^k\rangle$ have been constructed. Write $\vec{0}$ for the sequence in 2^k with only 0 entries. Let $W \subset (2^\omega \times \omega^\omega)^{2^k}$ be the Σ^1_1 set of all witnesses for the sequence $\langle A_t : t \in 2^k\rangle$. Its projection to the 2^ω coordinate of the $\vec{0}$-th coordinate is a nonempty Σ^1_1 subset of B and therefore contains points x^0, x^1 such that $|d(x^0, x^1) - \frac{1}{k+1}| < 2^{-k-1}$ by Claim 7.44. Let $\langle x^0_t, y^0_t : t \in 2^k\rangle$ and $\langle x^1_t, y^0_t : t \in 2^k\rangle$ be the witnesses in W such that $x^0 = x^0_t$ and $x^1 = x^1_t$. Define the sequence $\langle x_t, y_t : t \in 2^{k+1}\rangle$ by setting $x_t = x^{t(k)}_{t\restriction k}$ and $y_t = y^{t(k)}_{t\restriction k}$ for every $t \in 2^{k+1}$. Find $n_{k+1} > n_k$ such that

$$\sum\left\{\frac{1}{n+1} : n \geq n_{k+1} \text{ and } \exists s, t \in 2^{k+1}\ x_s(n) \neq x_t(n)\right\} < 2^{-k-1}.$$

Such a number exists as all the points $\{x_t : t \in 2^{k+1}\}$ are E_2-related. For every sequence $t \in 2^{k+1}$ let $g(t) = x_t \restriction n_{k+1}$ and $h(t) = y_t \restriction k+1$. For $t \in 2^{k+1}$ let $A'_t = \{\langle x, y\rangle \in A_t \restriction k : x \restriction n_{k+1} = g(t) \text{ and } y \restriction k+1 = h(t)\}$. The triangle inequality for the metric d shows that I3 is satisfied at k and the sequence $\langle x_t, y_t : t \in 2^{k+1}\rangle$ is a witness for $\langle A'_t : t \in 2^{k+1}\rangle$ in the sense of I5. As the last concern in the induction step, we must make sure that I4 is satisfied at $k+1$. Let $W'_{k+1} \subset (2^\omega \times \omega^\omega)^{2^{k+1}}$ be the nonempty Σ^1_1 set of witnesses for $\langle A'_t : t \in 2^{k+1}\rangle$. Thin it out repeatedly to a smaller nonempty Σ^1_1 set W_{k+1} whose Σ^1_1 projections into the various coordinates associated with $t \in 2^{k+1}$ belong to the open dense set D_k. In the end, for every $t \in 2^{k+1}$, let $A_t \subset A'_t$ be the nonempty Σ^1_1 projection of the set W_{k+1} into the coordinate t. This concludes the induction step. □

Proof of Theorem 7.39 First, we deal with properness. It is possible to give a fusion-style argument, but we will instead provide an abstract and more flexible argument. It makes use of the following claim of independent interest:

Claim 7.45 *Let $B \subset 2^\omega$ be a Borel set, and $f : B \to 2^\omega$ be a Borel function whose graph is a subset of E_2. The following holds:*

(i) $B \in I$ if and only if $f''B \in I$;
(ii) if $B \notin I$ then there is a Borel I-positive set $C \subset B$ on which f is injective;
(iii) if $B \notin I$ then $B \Vdash \dot{f}(\dot{x}_{\text{gen}})$ is a P_I-generic real over V.

Proof For the first item, for the left-to-right direction assume that $f''B \in I$. Then the analytic set $f''B$ is covered by the union D of countably many grainy sets, and $E_2 \restriction D$ is essentially countable. But then, the function f Borel reduces $E_2 \restriction B$ to an essentially countable equivalence relation and this means that $B \in I$. For the right-to-left direction, assume now that $f''B \notin I$. Use Theorem 7.43 to thin the set $f''B$ to a Borel I-positive set A on which there is a Borel function $g : A \to B$ such that $f \circ g = \text{id}$. Then, apply the left-to-right direction to the function g to conclude that $\text{rng}(g) \subset B$ is I-positive. This completes the proof of the first item.

For the second item, as in the previous paragraph find a Borel I-positive set $A \subset f''B$ on which there is a Borel function $g : A \to B$ such that $f \circ g = \text{id}$.

$\text{rng}(g) \notin I$, and we may use Theorem 7.43 to find a Borel I-positive subset $C \subset \text{rng}(g)$. It is immediate that $f \restriction C$ is injective.

For the last item, use the second item to thin down B if necessary so that the function f becomes injective. Then, the first item shows that the map $\hat{f} : P_I \restriction B \to P_I \restriction f''B$ is an isomorphism of partial orders, where $\hat{f}(C) = f''C$. Note that one-to-one Borel images of Borel sets are again Borel. □

Now, suppose that M is a countable elementary submodel of a large structure, and $B \in P_I \cap M$ is a Borel I-positive set in M. We must show that the set $C = \{x \in B : x \text{ is } P_I\text{-generic over } M\}$ is I-positive. Suppose for contradiction that it is not, and it is covered by the union of countably many Borel grainy sets $\{D_n : n \in \omega\}$. The grainy property of each set D_n is witnessed by some rational number $\varepsilon_n > 0$. Consider the countable notion of forcing $P_I \cap M$. It introduces a filter on P_I generic over M, with its associated M-generic point $\dot{x}_{\text{gen}}$. We will show that in the forcing $P_I \cap M$, the condition B forces $\dot{x}_{\text{gen}} \notin \bigcup_n D_n$. Once this is proved, just let N be a larger countable elementary submodel of a large structure containing M and the sets D_n for $n \in \omega$, and let $g \subset P_I \cap M$ be a filter generic over N. Then, the point $\dot{x}_{\text{gen}}/g$ is a point in the set B, P_I-generic over M, and by the forcing theorem applied in the model N to the poset $P_I \cap M$, it does not belong to $\bigcup_n D_n$. This contradicts the assumption that $C \subset \bigcup_n D_n$.

To show that in the forcing $P_I \cap M$, $B \Vdash \dot{x}_{\text{gen}} \notin \bigcup_n D_n$, assume for contradiction that some condition $B_0 \in P_I \cap M$ below B forces the opposite, and strengthen it if necessary to find a definite number n such that $B_0 \Vdash \dot{x}_{\text{gen}} \in \dot{D}_n$. Now work in the model M for a while. Consider the set $B_1 = \{x \in B_0 :$ there is an ε_n-walk in the set B_0 starting with x, whose endpoint is in d-distance at least 1 from $x\}$. We claim that the analytic set B_1 must be I-positive. If $B_1 \in I$ then it is covered by a Borel I-small set A, and then $B_0 \subseteq A \cup (B_0 \setminus A)$. The Borel set $B_0 \setminus A$ is grainy as witnessed by the real number ε_n, and $A \in I$. This of course contradicts the assumption that $B_1 \notin I$. Now, applying the σ-additivity of the σ-ideal I and Theorem 7.43, we can find a Borel I-positive set $B_2 \subset B_1$, a number $m \in \omega$, and Borel functions f_i for $i \in m$ with domain B_2 such that for every $x \in B_2$, the sequence $\langle f_i(x) : i \in m \rangle$ is an ε_n-walk in the set B_0 starting with x and ending with a point of distance at least 1 from x. Note that the functions f_i for $i \in m$ are all subsets of the equivalence relation E_2, and so by Claim 7.45 we can thin the set B_2 to an I-positive Borel set B_3 on which they are injective.

Look at the posets $P_I \cap M \restriction B_3$ (the part of $P_I \cap M$ below B_3) and $P_I \cap M \restriction f_i''B_3$ for $i \in m$. Since the previous construction took place in the model M, the functions $\hat{f}_i : (P_I \cap M) \restriction B_3 \to (P_I \cap M) \restriction f_i''B_3$ given by $\hat{f}_i(A) = f_i''A$ are isomorphisms of partial orders, for each $i \in m$. It follows that if $g \subset P_I \cap M$ is a generic filter with the associated point x_{gen}, so are the filters $g_i = \hat{f}_i'' g$ with the associated generic points $x_i = f_i(x_{\mathrm{gen}})$. The points $\langle x_i : i \in m \rangle$ form an ε_m-walk starting form x_0, and the endpoint of that walk is in distance at least 1 from x_0. For every $i \in m$, the condition $f_i'' B_3 \in g_i$ is below the condition B_0, and so by the forcing theorem it follows that $x_i \in D_n$. This, however, contradicts the assumption that the set D_n is grainy and so does not contain any such walk.

The bounding property of the poset P_I is obtained as a conjunction of Theorem 7.43 – which provides both density of compact sets in P_I and the continuous reading of names – and the bounding criterion (Fact 2.52). For the preservation of Baire category, it is enough to show that on the collection of analytic sets, the σ-ideal I is the intersection of meager ideals for some set of Polish topologies on 2^ω yielding the same Borel structure. Then, a reference to the Kuratowski–Ulam theorem and/or Zapletal (2008, corollary 3.5.8) will finish the argument. Let T be the set of all Polish topologies t giving the same Borel structure such that all sets in I are t-meager. Let $A \subset 2^\omega$ be an analytic I-positive set; we must prove that there is a topology in T which makes A nonmeager. This uses a simple claim of independent interest:

Claim 7.46 *In the usual topology on 2^ω, all sets in I are meager.*

Proof Argue by repeatedly applying the Pettis theorem (Kechris 1994, theorem 9.9) to the compact Cantor group $\langle 2^\omega, + \rangle$. Let $B \subset 2^\omega$ be a Borel nonmeager set; we must show that B is not grainy; in other words, given $\varepsilon > 0$, we must find an ε-walk in B whose endpoints have distance at least 1. Fix $\varepsilon > 0$ and by induction on $i \in \omega$ build Borel nonmeager sets $B_i \subset 2^\omega$ and finite sets $a_i \subset \omega$ so that:

I1. $B = B_0 \supset B_1 \supset \cdots$;

I2. the sets a_i are pairwise disjoint and for every $i \in \omega$ it is the case that $\varepsilon/2 < \sum_{n \in a_i} \frac{1}{n+1} < \varepsilon$;

I3. $B_{i+1} \Delta \chi_{a_i} \subset B_i$ where χ_{a_i} is the characteristic function of a_i.

To perform the induction step, suppose that the set B_{i+1} and a_i do not exist to satisfy the requirements in I1, I2, and I3. Write z for the collection of those sets $b \subset \omega$ such that I2 is satisfied with $b = a_i$. Thus, the set $C_b = \{x \in B_i : x \Delta \chi_b \in B_i\}$ is meager for every $b \in z$, and the set $B_i' = B_i \setminus \bigcup_{b \in z} C_b$ is Borel and nonmeager. Now use the Pettis theorem to argue that $B_i' \Delta B_i'$ contains an open neighborhood of 0, and so contains the characteristic function of some set b satisfying the second item. This is of course a contradiction.

In the end, choose $j \in \omega$ such that $\frac{j\varepsilon}{2} > 1$ and a point $x \in B_{j+1}$. The points $x, x\Delta\chi_{a_j}, x\Delta\chi_{a_j}\Delta\chi_{a_{j-1}} \cdots$ form the required ε-walk whose endnodes have distance greater than 1. □

Now, if $A \subset 2^\omega$ is an analytic I-positive set and $\pi : 2^\omega \to A$ is a continuous one-to-one reduction of E_2 to $E_2 \restriction A$ as in Theorem 7.43, then the transported topology will make A comeager and all sets in the ideal I meager.

For the preservation of outer Lebesgue measure, it is enough to argue that the ideal I is polar in the sense of Zapletal (2008, section 3.6.1); i. e., on the collection of all analytic sets, it is an intersection of null ideals for some set of Borel probability measures. In other words, we need to show that every Borel I-positive set $B \subset 2^\omega$ carries a Borel probability measure that vanishes on the ideal I. Just as in the treatment of Baire category, it is enough to show:

Claim 7.47 *For the usual Haar measure μ on 2^ω, the ideal I consists of μ-null sets.*

The proof of Claim 7.47 is the same as for Claim 7.46, using the Steinhaus theorem (Bartoszynski and Judah 1995, theorem 3.2.10) instead of the Pettis theorem. □

7.3.2 The canonization results

We will now tackle the main results of this section.

Proof of Theorem 7.36 Use a theorem of Solecki (1999) to find a lower semicontinuous submeasure μ on ω such that $J = \{a \subset \omega : \mu(a) < \infty\} = \{a \subset \omega : \lim_n \mu(a \setminus n) = 0\}$. Let $B \subset 2^\omega$ be a Borel I-positive set, E a Borel equivalence relation on it, and let $f : B \to 2^\omega$ be a Borel function reducing E to $=_J$. Let $z \in 2^\omega$ be a parameter such that B, f, μ are $\Delta^1_1(z)$ sets; to simplify the notation, assume that $z = 0$. Remove all Σ^1_1 grainy sets from B to form a set $B' \subset B$, which will still be Σ^1_1 by Fact 2.8(*v*). There are two cases:

Case 1 For every nonempty Σ^1_1 set $A \subset B'$ and every nonempty open set $O \subset (0, 1)$ there are points $x, y \in A$ such that $d(x, y) \in O$ and $x \ E \ y$. Here, all the three canonization outcomes are still possible, and we will have a split into three subcases. First, find a suitable continuous reduction of E_2 to $E_2 \restriction B'$. Namely, there will be numbers $n_k, m_k \in \omega$ and finite sets $b_k \subset \omega$ for $k \in \omega$ and functions $g, h : 2^{<\omega} \to 2^{<\omega}$ so that:

I1. $0 = n_0 < n_1 < n_1 < \cdots$, $0 = m_0 < m_1 < m_2 < \cdots$ and $b_k \subset [m_k, m_{k+1})$;

I2. for every sequence $s \in 2^k$, we have $g(s) \in 2^{n_k}$ and $h(s) \in 2^{m_k}$. For finite binary sequences $s \subset t$, it is the case that $g(s) \subset g(t)$ and $h(s) \subset h(t)$;
I3. for every $y \in 2^\omega$, the point $g(y) = \bigcup_k g(y \restriction k) \in 2^\omega$ is an element of B', and the point $\bigcup_k h(y \restriction k)$ is its f-image.

Moreover, whenever $s, t \in 2^k$ are finite binary sequences of the same length, we will have:

I4. $\sum\{\frac{1}{n+1} : n_k \leq n < n_{k+1}, g(s^\frown 0)(n) \neq g(t^\frown 0)(n)\} < 2^{-k}$;
I5. $\mu(\{n : m_k \leq n < m_{k+1}, h(s^\frown 0)(n) \neq h(t^\frown 0)(n)\}) < 2^{-k}$;
I6. $\left|\frac{1}{k+1} - \sum\{\frac{1}{n+1} : n_k \leq n < n_{k+1}, g(s^\frown 0)(n) \neq g(t^\frown 1)(n)\}\right| < 2^{-k}$;
I7. $\mu(b_k \Delta \{n : m_k \leq n < m_{k+1}, h(s^\frown 0)(n) \neq h(t^\frown 1)(n)\}) < 2^{-k}$.

Assume for the moment that the construction has been performed. There is a split into three subcases:

Case 1a There is an infinite set $a \subset \omega$ such that the sum $\sum\{\frac{1}{k+1} : k \in a\}$ diverges, while $\bigcup_{k \in a} b_k \in J$. In this case, let $D = \{y \in 2^\omega : \forall k \notin a\ y(k) = 0\}$. This is a compact subset of 2^ω for which $E_2 \restriction D$ is still bireducible with E_2 by Kanovei (2008, lemma 6.2.4), as the set a is not summable. Let $C = g''D$. This is an I-positive compact subset of B' and every two elements $x_0, x_1 \in C$ are E-connected: the set $\{n \in \omega : f(x)(n) \neq f(y)(n)\} \setminus \bigcup_{k \in a} b_k$ has μ-mass at most 1 since it is a subset of the union of the sets in items I5 and I7. Thus, $E \restriction C = \text{ev}$ and we are done in this case.

Case 1b There is an infinite set $a \subset \omega$ such that $\bigcup_{k \in a} b_k \notin J$, while the sum $\sum\{\frac{1}{k+1} : k \in a\}$ converges. Here, for every $n \notin a$ find a set $c_n \subset a$ such that $\bigcup_{k \in c_n} b_k \notin J$, so that the sets $\{c_n : n \notin a\}$ form a partition of the set a. Let $D = \{y \in 2^\omega : \forall n \notin a \forall m \in c_n\ y(n) = y(m)\}$. This is a compact subset of 2^ω and $E_2 \restriction D$ is still bireducible with E_2: every $y \in 2^\omega$ has an E_2-equivalent in D obtained by adjusting the values of y on the summable set a. Let $C = g''D$. This is an I-positive compact subset of B' and every two distinct elements of it are E-unrelated: if $y_0, y_1 \in D$ are distinct binary sequences then they differ on some entry $n \notin a$, and then the set $\{m \in \omega : f(g(y_0))(m) \neq f(g(y_1))(m)\}$ includes the J-positive set $\bigcup_{k \in c_n} b_k$ up to a μ-mass 1 set by items I5 and I7. Thus, $E \restriction C = \text{id}$ and we are done in this case.

Case 1c If both Cases 1a and 1b fail, then we are happily in the situation that for every set $a \subset \omega$, the sum $\sum\{\frac{1}{k+1} : k \in a\}$ converges if and only if $\bigcup_{k \in a} b_k \in J$. Let $C = g''2^\omega$. This is an I-positive compact subset of B'. The case assumption shows that $E \restriction C = E_2 \restriction C$ and we are done.

Now for the construction of the functions g, h. Let M be a countable elementary submodel of a large structure, and let $\{D_n : n \in \omega\}$ be an enumeration of all open dense subsets of the Gandy–Harrington forcing that appear in the model M. By induction on $k \in \omega$ construct the numbers $n_k, m_k \in \omega$, the functions $g, h \restriction 2^k$, and the sets b_k as in I1–I7 (obviously with the exception of I3), and in addition at stage k, build nonempty Σ^1_1 sets $A_s \subset B'$ for $s \in 2^k$ so that:

I8. $A_s \in D_{|s|-1}$, and if $s \subset t$ then $A_t \subset A_s$;
I9. $A_s \subset [g(s)]$ and also $f''A_s \subset [h(s)]$;
I10. there is a *witness* for $\langle A_t : t \in 2^k\rangle$, which is a sequence $\langle x_t : t \in 2^k\rangle$ such that for every $t \in 2^k$ we have $x_t \in A_t$, and moreover,

$$\sum\left\{\tfrac{1}{n+1} : n \geq n_k \text{ and } \exists s, t \in 2^k\ x_s(n) \neq x_t(n)\right\} < 2^{-k},$$
$$\mu\left(\left\{m : m \geq m_k \text{ and } \exists s, t \in 2^k\ f(x_s)(m) \neq f(x_t)(m)\right\}\right) < 2^{-k}.$$

Once this induction is performed, item I3 will follow: for every $y \in 2^\omega$ it is the case that the sets $A_{y\restriction k}$ form a Gandy–Harrington generic filter over M by I8, so their intersection is a singleton $g(y)$ which satisfies I3 by I9. To perform the induction step at $k \in \omega$, write $\vec{0}$ for the sequence consisting of k many zeroes, look at the set $W_k \subset (2^\omega)^{2^k}$ of all witnesses for the sequence $\langle A_t : t \in 2^k\rangle$ and project it to the $\vec{0}$-th coordinate, obtaining a nonempty Σ^1_1 set $A' \subset A_{\vec{0}}$. The case assumption shows that there are two points $x^0, x^1 \in A'$ such that $|d(x^0, x^1) - \frac{1}{k+1}| \leq 2^{-k-1}$ and $x^0\ E\ x^1$. Let $\langle x^0_t : t \in 2^k\rangle$ be a witness starting with $x^0_0 = x^0$, let $\langle x^1_t : t \in 2^k\rangle$ be a witness starting with $x^1_0 = x^1$, and define a sequence $\langle x_t : t \in 2^{k+1}\rangle$ by setting $x_t = x^{t(k)}_{t\restriction k}$ for every $t \in 2^{k+1}$. Find a number $n_{k+1} > n_k$ such that

$$\sum\left\{\frac{1}{n+1} : n \geq n_{k+1} \text{ and } \exists t, s \in 2^{k+1}\ x_t(n) \neq x_s(n)\right\} < 2^{-k-1},$$

and define the function $g : 2^{k+1} \to 2^{n_{k+1}}$ by setting $g(t) = x_t \restriction n_{k+1}$. Find a number $m_{k+1} > m_k$ such that

$$\mu(\{m : m \geq m_{k+1} \text{ and } \exists s, t \in 2^{k+1}\ f(x_s)(m) \neq f(x_t)(m)\}) < 2^{-k-1}.$$

Such a number exists by the choice of the submeasure μ and the fact that the points on the sequence $\langle x_t : t \in 2^{k+1}\rangle$ are all E-related. For each sequence $t \in 2^{k+1}$ let $h(t) = f(x_t) \restriction m_{k+1}$, and let $b_k = \{m : m_k \leq m < m_{k+1}$ and $f(x_{\vec{0}^\frown 0})(m) \neq f(x_{\vec{0}^\frown 1})(m)\}$. Items I4–I7 are satisfied by the subadditivity of the submeasure μ. For every sequence $t \in 2^{k+1}$ let $A'_t = \{x \in A_{t\restriction k} : g(t) \subset x$ and $h(t) \subset f(x)\}$; these are nonempty Σ^1_1 sets and the Σ^1_1 set W'_{k+1} of witnesses for them is nonempty, containing the sequence $\langle x_t : t \in 2^{k+1}\rangle$. Thin out the set W'_{k+1} repeatedly to get a nonempty Σ^1_1 set $W_{k+1} \subset W'_{k+1}$ whose projections into all coordinates belong to the open dense set D_k, and let A_t be the projection of W_k into the t-th coordinate, for every sequence $t \in 2^{k+1}$. This completes the induction step and the proof of Case 1.

Case 2 There is a nonempty Σ^1_1 set $A \subset B'$ and an open subset O of the interval $(0, 1)$ such that for every two points $x, y \in A$ with $d(x, y) \in O$ it is the case that $x\ E\ y$ fails. In this case, we will find an I-positive compact set $C \subset A$ such that $E \restriction C = \text{id}$. First, note an important consequence of the case assumption:

Claim 7.48

(i) *For every nonempty Σ^1_1 set $A' \subset A$ and every positive real $\varepsilon > 0$ there are points $x, y \in A'$ such that $d(x, y) < \varepsilon$ and $x\ E\ y$ fails.*

(ii) *Suppose that $\varepsilon > 0$, $k \in \omega$, and $\langle A_i : i \in j\rangle$ are nonempty Σ^1_1 subsets of A. Then, for every sequence $\langle x_i : i \in j\rangle$ of pairwise E_2-equivalent points such that $x_i \in A_i$ there is a sequence $\langle x'_i : i \in j\rangle$ such that $x'_i \in A_i$, $|d(x_{i_0}, x_{i_1}) - d(x'_{i_0}, x'_{i_1})| < \varepsilon$ and $\mu(\{n : f(x'_{i_0})(n) \neq f(x'_{i_1})(n)\}) > k$ holds for all $i_0, i_1 \in j$.*

Proof For the first item, make ε smaller if necessary so that the open set O from the case assumption contains an interval K of length at least ε. Since the set A' is not grainy, there is a finite sequence $\langle x_i : i \in n\rangle$ of points in A' such that the d-distance between its successive points is at most ε and the d-distance between the endpoints of the sequence is at least 1. Let $i \in n$ be the smallest number such that $d(x_0, x_i)$ is greater than the leftmost point of the interval K. Since the length of K is greater than ε, it follows that $d(x_0, x_i) \in K$ and therefore $x_0\ E\ x_i$ fails by the case assumption. The transitivity of the relation E shows that for some successive points x, y on the ε-walk indexed between 0 and i, $x\ E\ y$ must fail. This completes the proof of the first item.

For the second item, as the first approximation we will show how to get points $\langle x'_i : i \in j\rangle$ such that the following holds:

($\star$) $\forall i \in j\ x'_i \in A_i$, $\forall i_0, i_1 \in j\ |d(x_{i_0}, x_{i_1}) - d(x'_{i_0}, x'_{i_1})| < \varepsilon$ and $\mu(\{n : f(x'_0)(n) \neq f(x'_1)(n)\}) > k$.

Once ($\star$) is obtained, the full statement is proved in the following way. Let $z \subset [j]^2$ be an inclusion maximal set such that there is a sequence $\langle x''_i : i \in j\rangle$ such that $x''_i \in A_i$, $|d(x_{i_0}, x_{i_1}) - d(x''_{i_0}, x''_{i_1})| < \varepsilon$ holds for all $\{i_0, i_1\} \in [j]^2$ and $\mu(\{n : f(x''_{i_0})(n) \neq f(x''_{i_1})(n)\}) > k$ holds for all pairs $\{i_0, i_1\} \in z$. We will show that $z = [j]^2$ must hold. If this is not the case, we can reindex the sets so that $\{0, 1\} \notin z$, find $m \in \omega$ so large that for all pairs $\{i_0, i_1\} \in z$ we have $\mu(\{n \in$

$m : f(x''_{i_0})(n) \neq f(x''_{i_1})(n)\}) > k$, and find $\delta > 0$ so small that $|d(x_{i_0}, x_{i_1}) - d(x''_{i_0}, x''_{i_1})| < \varepsilon - \delta$ for all $i_0, i_1 \in j$. Then, apply ($\star$) with A_i replaced with $\{x \in A_i : f(x) \restriction m = f(x''_i) \restriction m\}$, x_i replaced with x''_i and ε replaced with δ to find a sequence contradicting the maximal choice of the set z.

To prove ($\star$), for each pair $i_0, i_1 \in j$ let $O_{i_0 i_1} \subset (0, 1)$ be a rational interval of diameter less than $\varepsilon/4$ including the number $d(x_{i_0}, x_{i_1})$. Let $W \subset (2^\omega)^j$ be the nonempty Σ^1_1 set of all tuples $\langle y_i : i \in j\rangle$ such that for every $i \in j$,we have $y_i \in A_i$ and for every $i_0, i_1 \in j$, we have $d(y_{i_0}, y_{i_1}) \in O_{i_0 i_1}$. Let A' be the nonempty Σ^1_1 projection of W into the first coordinate. The first item shows that the set A' contains points x, y such that $d(x, y) < \varepsilon/4$ and $\mu(\{n : f(x)(n) \neq f(y)(n)\}) > 2k$. Let $\langle x''_i : i \in j\rangle$ be an element of the set W such that $x''_0 = x$. The subadditivity of the submeasure μ shows that either $\mu(\{n : f(y)(n) \neq f(x''_1)(n)\}) > k$ or $\mu(\{n : f(x'_0)(n) \neq f(x'_1)(n)\}) > k$ holds. If the sequence $\langle x'_i : i \in j\rangle$ satisfies ($\star$) then we are done. Otherwise, replace its first entry with y. This does not change the d-distances between the points on the sequence by more than $\varepsilon/4$ and so they still remain within the required tolerance, and the distance between the first two entries of this improved sequence is at least k and ($\star$) is satisfied again. $\square$

We now find a suitable continuous reduction of E_2 to $E_2 \restriction A$. Namely, there will be numbers $n_k, m_k \in \omega$ for $k \in \omega$ and functions $g, h : 2^{<\omega} \to 2^{<\omega}$ so that:

I1. $0 = n_0 < n_1 < n_1 < \cdots, 0 = m_0 < m_1 < m_2 < \cdots$;

I2. for every sequence $s \in 2^k$, we have $g(s) \in 2^{n_k}$ and $h(s) \in 2^{m_k}$. For finite binary sequences $s \subset t$, it is the case that $g(s) \subset g(t)$ and $h(s) \subset h(t)$;

I3. for every $y \in 2^\omega$, the point $g(y) = \bigcup_k g(y \restriction k) \in 2^\omega$ is an element of A, and the point $\bigcup_k h(y \restriction k)$ is its f-image.

Moreover, whenever $s, t \in 2^k$ are finite binary sequences of the same length, we will have:

I4. $\sum\{\frac{1}{n+1} : n_k \leq n < n_{k+1}, g(s^\frown 0)(n) \neq g(t^\frown 0)(n)\} < 2^{-k}$;

I5. $\left|\frac{1}{k+1} - \sum\{\frac{1}{n+1} : n_k \leq n < n_{k+1}, g(s^\frown 0)(n) \neq g(t^\frown 1)(n)\}\right| < 2^{-k}$;

I6. if $s \neq t$, then $\mu(\{n < m_{k+1} : h(s)(n) \neq h(t)(n)\}) > k$.

Once the construction has been performed, let $C = g''2^\omega$. Items I4 and I5 show that g is a reduction of E_2 to $E_2 \restriction C$. This is a compact I-positive subset of A, and the last item shows that any two distinct points of it are E-unrelated. Thus, $E \restriction C = \mathrm{id}$ and this will complete the proof of the theorem.

Now for the construction of the functions g, h. Let M be a countable elementary submodel of a large structure, and let $\{D_n : n \in \omega\}$ be an enumeration of all open dense subsets of the Gandy–Harrington forcing that appear in the model M. By induction on $k \in \omega$ construct the numbers $n_k, m_k \in \omega$, the functions $g, h \restriction 2^k$ as in I1–I6 (obviously with the exception of I3), and in addition at stage k, build nonempty Σ^1_1 sets $A_s \subset A$ for $s \in 2^k$ so that:

I7. $A_s \in D_{|s|-1}, s \subset t \to A_t \subset A_s$;
I8. $A_s \subset [g(s)]$ and also $f''A_s \subset [h(s)]$;
I9. there is a *witness* for $\langle A_t : t \in 2^k \rangle$, which is a sequence $\langle x_t : t \in 2^k \rangle$ such that for every $t \in 2^k$ we have $x_t \in A_t$, and moreover

$$\sum \left\{ \frac{1}{n+1} : n \geq n_k \text{ and } \exists s, t \in 2^k \; x_s(n) \neq x_t(n) \right\} < 2^{-k}.$$

Once this induction is performed, item I3 will follow: for every $y \in 2^\omega$ it is the case that the sets $A_{y \restriction k}$ form a Gandy–Harrington generic filter over M by I7, so their intersection is a singleton $g(y)$ which satisfies I3 by I8. To perform the induction step at $k \in \omega$, suppose that the numbers n_k, m_k, the functions $g \restriction 2^k$ and $h \restriction 2^k$, and the set A_t for $t \in 2^k$ have been obtained. Write $\vec{0}$ for the sequence consisting of k many zeroes, look at the set $W_k \subset (2^\omega)^{2^k}$ of all witnesses for $\langle A_t : t \in 2^k \rangle$ and project it to the $\vec{0}$-th coordinate, obtaining a nonempty Σ^1_1 set $A'_{\vec{0}} \subset A_{\vec{0}}$. Claim 7.44 shows that there are two points $x^0, x^1 \in A'_{\vec{0}}$ such that $|d(x^0, x^1) - \frac{1}{k+1}| \leq 2^{-k}$. Let $\langle x^0_t : t \in 2^k \rangle$ be a witness starting with $x^0_{\vec{0}} = x^0$, let $\langle x^1_t : t \in 2^k \rangle$ be a witness starting with $x^1_{\vec{0}} = x^1$. Define the sequence $\langle x_t : t \in 2^{k+1} \rangle$ by putting $x_t = x^{t(k)}_{t \restriction k}$ for every $t \in 2^{k+1}$. Find a number $n_{k+1} > n_k$ such that

$$\sum \left\{ \frac{1}{n+1} : n \geq n_{k+1} \text{ and } \exists s, t \in 2^{k+1} \; x_t(n) \neq x_s(n) \right\} < 2^{-k-1}.$$

Define $g : 2^{k+1} \to 2^{n_{k+1}}$ by putting $g(t) = x_t \restriction n_{k+1}$. Thus, the sequence $\langle x_t : t \in 2^{k+1} \rangle$ defined by $x_t = x^{t(k)}_{t \restriction k}$ is a witness in the sense of item I9 for the collection $\langle A'_t : t \in 2^{k+1} \rangle$ of nonempty Σ^1_1 sets defined by $A'_t = A_{t \restriction k} \cap [g(t)]$. Now, invoke Claim 7.48(*ii*) to find an improved witness $\langle x'_t : t \in 2^{k+1} \rangle$ for the collection $\langle A'_t : t \in 2^{k+1} \rangle$ such that for every $s \neq t \in 2^{k+1}$ it is the case that $\mu(\{m \in \omega : f(x'_s)(m) \neq f(x'_t)(m)\}) \geq k + 2$. Find a number $m_{k+1} > m_k$ such that

$$\forall s \neq t \in 2^{k+1} \;\; \mu(\{m < m_{k+1} : f(x'_s)(m) \neq f(x'_t)(m)\}) \geq k + 1$$

holds. For each sequence $t \in 2^{k+1}$ let $h(t) = f(x'_t) \restriction m_{k+1}$ and $A''_t = \{x \in A'_t : h(t) \subset f(x)\}$; these are nonempty Σ^1_1 sets and the Σ^1_1 set W'_{k+1} of witnesses for them as in I9 is nonempty, containing the sequence $\langle x'_t : t \in 2^{k+1} \rangle$. Thin out the set W'_{k+1} repeatedly to get a nonempty Σ^1_1 set $W_{k+1} \subset W'_{k+1}$

whose projections into all coordinates belong to the open dense set D_k, and let A_t be the projection of W_{k+1} into the t-th coordinate, for every sequence $t \in 2^{k+1}$. This completes the induction step and the proof of Case 2. □

Proof of Theorem 7.37 Note that equivalences classifiable by countable structures may not be reducible to $=_J$ for F_σ-ideals as the example of F_2 shows, and so Theorem 7.36 cannot be used directly. Suppose that E is an equivalence relation on some Borel I-positive set $B \subset 2^\omega$, classifiable by countable structures. Theorem 7.36 shows that the poset P_I adds a minimal real degree, since the middle E_2 option is impossible for canonization of smooth equivalence relations. Thus, Theorem 4.9 shows that we can thin B to a smaller Borel I-positive set on which E is essentially countable. The poset P_I preserves Baire category by Theorem 7.39, and thus Theorem 4.7 shows that we can thin B to a smaller Borel I-positive set on which E is reducible to E_0. Another application of Theorem 7.36 yields the set $C \subset B$ on which $E \restriction C$ is either id or ev: the middle E_2 option is impossible as E_2 is not Borel reducible to E_0! □

It is quite obvious from the definitions that E_2 belongs to the spectrum of the σ-ideal I. The last theorem of this section quantifies the position of E_2 precisely.

Theorem 7.49 *E_2 is in the spectrum of I. Moreover, $P_I^{E_2}$ is a regular $\aleph_0$-distributive subposet of P_I, and it generates the model $V[\dot{x}_{\text{gen}}]_{E_2}$.*

Proof The first sentence follows from the definition of the σ-ideal I. For the regularity of $P_I^{E_2}$ in P_I, first argue that E_2-saturations of analytic I-small sets are I-small: if $A \subset 2^\omega$ is an analytic set with I-positive saturation $[A]_{E_2}$, Theorem 7.43 shows that there is a compact I-positive set $C \subset [A]_{E_2}$ and a continuous function $f : C \to A$ such that the graph of f is a subset of E_2, and then Claim 7.45 shows that $A \notin I$. Now, Corollary 3.20 implies that $P_I^{E_2}$ is regular in P_I. For the $\aleph_0$-distributivity, first verify that E_2 is I-dense and then apply Proposition 3.23. To see the I-density, if $A \subset 2^\omega$ is an I-positive set, then Theorem 7.43 provides a continuous injection f from 2^ω to a compact subset $C \subset A$. Since all equivalence classes of E_2 are dense in 2^ω, their f-images will be relatively dense in C as desired.

To show that the poset P_I^E generates the $V[\dot{x}_{\text{gen}}]_{E_2}$ model, first verify the I-homogeneity of the equivalence relation E_2 and then apply Theorem 3.25. For the I-homogeneity, just observe that if $B, C \subset 2^\omega$ are analytic I-positive sets with the same E_2-saturation, then one can thin B if necessary to get a function $f : B \to C$ whose graph is a subset of E_2. Then, Claim 7.45 shows that the function f transports I-positive sets to I-positive sets, as required in the definition of I-homogeneity. □

7.3.3 Borel functions on pairs

Theorem 7.50 *There is a square coding function for the σ-ideal I.*

There is no rectangular coding function for I: the σ-ideal I is $\mathbf{\Pi}^1_1$ on $\mathbf{\Sigma}^1_1$, the quotient poset P_I is proper, bounding, and preserves Baire category, and so it has the rectangular Ramsey property by Fact 2.56.

Proof The target space for the coding function will be $\mathbb{R}^+$. Let $f(x, y) = \sum\{1/(n+1) : x(n) \neq y(n)\}$ if the indicated sum is finite, and let $f(x, y) =$trash otherwise. We claim that this function works.

For suppose that $B \subset 2^\omega$ is a Borel I-positive set. The fusion argument in Theorem 7.43 yields numbers $\{n_i : i \in \omega\}$ and a function $g : 2^{<\omega} \to 2^{<\omega}$ such that:

(a) $0 = n_0 < n_1 < n_2 < \cdots$, for every $i \in \omega$ and every $t \in 2^i$ we have $g(t) \in 2^{n_i}$, and $t \subset s \in 2^{<\omega}$ implies $g(t) \subset g(s)$;
(b) for every $i \in \omega$ and every $s, t \in 2^{i+1}$, write $\varepsilon(s, t) = \sum\{\frac{1}{n+1} : n_i \leq n < n_{i+1}, g(s) \neq g(t)\}$. Then, it is the case that $|\varepsilon(s, t) - \frac{1}{i}| \leq 2^{-i-2}$ whenever the last entries on s, t differ, and $\varepsilon(s, t) \leq 2^{-i-2}$ if the last entries of s, t are the same;
(c) for every $x \in 2^\omega$, $g(x) = \bigcup_i g(x \restriction i) \in B$.

Now suppose that $r \in \mathbb{R}$ is a positive real number. By induction on $m \in \omega$ build numbers $i_m \in \omega$ and binary sequences $s_m, t_m \in 2^{i_m}$ so that:

I1. $s_m \subset s_{m+1}, t_m \subset t_{m+1}$;
I2. $2^{-i_{m-1}} < r - d(g(s_m), g(t_m)) < 2^{-i_m}$, except for 0 where $d(g(s_0), g(t_0)) < r - 2^{-i_0}$.

In the end, let $x = \lim_m g(s_m)$ and $y = \lim_m g(t_m)$; it immediately follows from I2 that $d(x, y) = r$. The induction starts with choosing i_0 so that $r > 2^{-i_0}$ and letting $s_0 = t_0$ be any binary sequence of length i_0. Suppose that s_m, t_m have been successfully found. Find infinite binary sequences z_0, z_1 extending s_m such that $d(z_0, z_1) = r - 2^{-i_m-1} - d(g(s_m), g(t_m))$. Let z_2 be an infinite binary sequence extending t_m such that for every $j > i_m$, $z_2(m) = z_1(m)$. It follows from (b) that $r - 2^{-i_m} < d(g(z_0), g(z_2)) < r$. Find i_{m+1} so large that

$$r - 2^{-i_m} < d(g(z_0 \restriction i_{m+1}), g(z_2 \restriction i_{m+1})) < r - 2^{-i_{m+1}},$$

let $s_{m+1} = z_0 \restriction i_{m+1}$, $t_{m+1} = z_2 \restriction i_{m+1}$, and continue to the next step of the induction! □

Corollary 7.51 *If E_2 is in the spectrum of a σ-ideal J then there is a square coding function for J.*

Proof Let X be the Polish space on which J lives. Let $B \subset X$ be a Borel J-positive set such that there is an equivalence relation E on B Borel reducible via some function $g : B \to 2^\omega$ to E_2, such that for every J-positive Borel set $C \subset B$, $E_2 \leq_B E \restriction C$. It is clear from the definitions that g-preimages of I-small sets are J-small; in other words, $J \geq_K I$ and so the corollary follows from Proposition 2.69. □

8

Ramsey-type ideals

8.1 Halpern–Läuchli cubes

The Halpern–Läuchli forcing P (Rosłanowski 2006) consists of those trees $T \subset 2^{<\omega}$ such that there is an infinite set $a_T \subset \omega$ such that $t \in T$ is a splitnode if and only if $|t| \in a_T$. The ordering is that of reverse inclusion. The terminology alludes to the Halpern–Läuchli theorem, which for partitions of $2^{<\omega}$ provides homogeneous sets that are exactly sets of splitnodes of trees in P. The poset P is quite similar to the Sacks or Silver forcing notions, and the standard fusion arguments show that it is proper and has the continuous reading of names. Consequently, the computation of the associated ideal can be found in Zapletal (2008, proposition 2.1.6). Let $X = 2^\omega$, and let I be the σ-ideal generated by those Borel sets A such that for no tree $T \in P$, $[T] \subset A$. Then I is a $\mathbf{\Pi}^1_1$ on $\mathbf{\Sigma}^1_1$ σ-ideal, every positive analytic set contains a positive Borel subset, and the map $T \mapsto [T]$ is a dense embedding from P to P_I.

This forcing serves as a good example refuting several natural conjectures; its features are summed up in the main theorem of this section.

Theorem 8.1

(i) The σ-ideal I has the free set property but not the mutual generics property.

(ii) I has the total canonization but not the Silver property.

(iii) The P_I extension has a nontrivial intermediate σ-closed extension.

Regarding the last item, note that the intermediate extension cannot be obtained as a model of the form $V[\dot{x}_{\text{gen}}]_E$, since I has total canonization.

Proof For item (*i*) of the theorem, a reduced product of the Halpern–Läuchli forcing will be needed. Let $P \times_r P$ be the poset of all pairs $\langle T, U\rangle \in P \times P$ such that $a_T = a_U$. Similarly to the usual product, the reduced product adds points $\dot{x}_{\text{lgen}}, \dot{x}_{\text{rgen}} \in X$, which are the intersections of all trees on the left (or right, respectively) side of conditions in the generic filter. Still, the difference between $P \times_r P$ and $P \times P$ should be immediately apparent.

Claim 8.2 *Suppose that M is a countable elementary submodel of a large enough structure and $T \in M \cap P$ is a tree. There is a tree $U \in P$, $U \subset T$, such that every two distinct elements of U are M-generic for the reduced product.*

Proof Let M be a countable elementary submodel of a large enough structure, and let $\langle D_m : m \in \omega \rangle$ be an enumeration of all open dense subsets of $P \times_r P$ in the model M. By induction on $m \in \omega$ build conditions $\langle T_m, U_m \rangle \in M$ in the reduced product so that:

I1. $T_0 = T$, the conditions $T_m \in M \cap P$ form a decreasing sequence, and the first $m + 1$ splitting levels of T_m are equal to the first $m + 1$ splitting levels of T_{m+1}, and the same on the U side;
I2. for every choice of nodes $t \in T_{m+1}$, $u \in U_{m+1}$ just past the m-th splitting level, the condition $\langle T_{m+1} \restriction t, U_{m+1} \restriction u \rangle \in P \times_r P$ is in the set D_m.

The induction is elementary, at each step making a pass through all pairs $\langle t, u \rangle$ as in the second item to handle them all. In the end, the pair $\langle \bar{T} = \bigcap_m T_m, \bar{U} = \bigcap_n U_m \rangle \in P \times_r P$ is a condition such that every pair $\langle x, y \rangle \in [\bar{T}] \times [\bar{U}]$ is M-generic for the reduced product below the condition $\langle T, U \rangle$. □

The free set property of the σ-ideal I now immediately follows. Suppose that $T \in P$ is a tree and $A \subset [T] \times [T]$ is an analytic set with I-small vertical sections. Let M be a countable elementary submodel of a large structure containing T, A, and let $U \subset T$ be a tree such that $[U]$ consists of pairwise $P \times_r P$-generic points over the model M. Thinning out the tree U if necessary, we may assume that the intersections of U with Borel sets in M are either I-positive or empty. We claim that $A \cap [U]^2 \subset \text{id}$ as required. Suppose for contradiction that this fails and $x \neq y \in [U]$ are points such that $\langle x, y \rangle \in A$, then $M[x, y] \models \langle x, y \rangle \in A$ by analytic absoluteness between V and the transitive model $M[x, y]$. Since $M[x, y]$ is a $P \times_r P$-extension of M, there must be a condition $\langle S_0, S_1 \rangle \in M \cap P \times_r P$ such that $x \in [S_0]$, $y \in [S_1]$ and $\langle S_0, S_1 \rangle \Vdash \langle \dot{x}_{\text{lgen}}, \dot{x}_{\text{rgen}} \rangle \in \dot{A}$. The set $[U] \cap [S_1]$ is nonempty and therefore I-positive, and by the forcing theorem applied in the model M it is contained in the vertical section A_x. This contradicts the assumption that vertical sections of the set A are in I.

To prove the failure of the mutual generics property, suppose that M is a countable elementary submodel of a large structure, and suppose for contradiction that there is a tree $T \in P$ such that the points in $[T]$ are pairwise $P \times P$ over M. For every point $x \in [T]$ let $g_x \subset P \cap M$ be the filter generic over M, generated by x: $g_x = \{S \in P \cap M : x \in [S]\}$. Similarly, let $h_x \subset (\mathcal{P}(\omega)$ modulo finite$)\cap M$ be the filter generic over M generated by x: $h_x = \{a_S : S \in g_x\}$.

Claim 8.3 *The set* $B = \{x \in [T] : \exists S \in g_x\ a_T \not\subset a_S$ *modulo finite*$\}$ *is relatively meager in* $[T]$.

Proof If this failed, then by the Baire category theorem there would be a tree $S \in P \cap M$ and a node $t \in T$ such that for comeagerly many points $x \in [T]$ passing through t, we have $S \in g_x$ and $a_T \setminus a_S$ contains some elements above $|t|$. Take two extensions $t_0, t_1 \in T$ of the tree T that split at some level in $a_T \setminus a_S$ and points $x_0, x_1 \in [T]$ extending them respectively such that $S \in g_{x_0}, g_{x_1}$. However, this means that $t_0 \cap t_1$ is a splitnode of S, contradicting the choice of the two nodes t_0, t_1. □

Since the set B is meager, we can pick two distinct ponts $x_0, x_1 \in [T] \setminus B$ and observe that the set $a_T \subset \omega$ diagonalizes both of the filters h_{x_0}, h_{x_1}. In particular, h_{x_0} does not contain a set with finite intersection with some set in h_{x_1}, as would be the case if the filters g_{x_0}, g_{x_1} were generic over M for the product $P \times P$!

For item *(ii)* of the theorem, the σ-ideal I has total canonization as it satisfies the free set property. To show the failure of the Silver property, by Theorem 5.27 it is enough to argue for the failure of the selection property. Find a Borel injection $f : 2^\omega \to [\omega]^{\aleph_0}$ whose range consists of pairwise almost disjoint sets. Let $D \subset 2^\omega \times 2^\omega$ be the Borel set of all pairs $\langle x, y\rangle$ such that $y(n) = 0$ whenever $n \notin f(x)$. It is fairly obvious that the vertical sections of the set D are pairwise disjoint I-positive sets. We will show that D is the desired counterexample to the selection property.

Indeed, suppose that $T \in P$ is a tree such that the closed set $[T] \subset 2^\omega$ is covered by the sections of the set D, and visits each section in at most one point. There must be distinct points $y_0, y_1 \in [T]$ such that the set $a = \{n : y_0(n) = y_1(n) = 1\}$ is infinite. If $x \in 2^\omega$ is such that $y_0 \in D_x$, it must be the case that $a \subset f(x)$, and the same for y_1. However, since the sets in the range of f are pairwise almost disjoint, there can be only one such point $x \in 2^\omega$, and both y_0, y_1 must belong to D_x, contradicting the choice of the tree $T \in P$.

For item *(iii)*, we will show that the P extension has an intermediate extension generated by the algebra $\mathcal{P}(\omega)$ mod fin. Indeed, it is easy to see that if $G \subset P_I$ is a generic filter, then $H \subset \mathcal{P}(\omega)$ mod finite, $H = \{a_T : T \in G\} \subset \mathcal{P}(\omega)$ mod fin is a generic filter as well. The function $T \mapsto a_T$ is the associated pseudoprojection from P to $\mathcal{P}(\omega)$ mod fin. □

8.2 Silver cubes

In this section, we will consider the behavior of equivalence relations on Silver cubes. Recall that *Silver forcing* is the poset P whose elements p are just partial functions from ω to 2 with coinfinite domain. The ordering is that of reverse inclusion. For $p \in P$, write $[p] = \{x \in 2^\omega : p \subset x\}$; thus $[p] \subset 2^\omega$ is a compact set. A *Silver cube* is a set of the form $[p]$ for some $p \in P$.

The calculation of the associated σ-ideal gives complete information. Let $X = 2^\omega$ and let G be the Borel graph on X given by $x \; G \; y$ if and only if there is exactly one n such that $x(n) \neq y(n)$. A set $B \subset X$ is G-independent if no two elements of B are G-connected. Let I be the σ-ideal generated by Borel G-independent sets. We have:

Fact 8.4

(i) Whenever $A \subset X$ is an analytic set, either $A \in I$ or A contains a Silver cube, and these two options are mutually exclusive.
(ii) The ideal I is $\mathbf{\Pi}^1_1$ on $\mathbf{\Sigma}^1_1$.
(iii) The ideal I does not have the selection property.
(iv) The forcing P is proper, bounding, preserves Baire category, and adds an independent real.
(v) The map $p \mapsto [p]$ is an isomorphism between P and a dense subset of P_I.

Proof (*i*) comes from Zapletal (2004, lemma 2.3.37). (*ii*) follows from (*i*) and Zapletal (2008, theorem 3.8.9), or one can compute directly. (*iv*) is well-known; the independent real $\dot{a} \subset \omega$ is read off the generic real $\dot{x}_{\text{gen}} \in 2^\omega$ as $a = \{n \in \omega : \{m \in n : \dot{x}_{\text{gen}}(m) = 1\}$ has even size$\}$. Finally, the failure of the selection property is witnessed by the set $D \subset 2^\omega \times X$ given by $\langle y, x\rangle \in D$ if $x(3n) = x(3n+1) = y(n)$. □

The study of the canonization properties starts with the total canonization of smooth equivalence relations. This has been proved in a different language by Grigorieff (1971, corollary 5.5) in a rather indirect fashion. We provide a direct argument containing features useful later.

Theorem 8.5 *The ideal I has total canonization for smooth equivalences. Restated, every smooth equivalence on a Silver cube is equal to either the identity or the full equivalence relation on a Silver subcube.*

In other words, Silver forcing adds a minimal real degree. For the equivalence relations classifiable by countable structures, we have the slightly weaker:

Theorem 8.6 *If E is an equivalence on a Silver cube classifiable by countable structures, then there is a Silver subcube on which $E \subseteq E_0$ or E equals to the full equivalence relation. Restated,*

$$\text{classifiable by countable structures} \to_I \{\subseteq E_0, \text{ev}\}.$$

Note that E_0 and all of its Borel subequivalences are hyperfinite, and so the equivalence relations classifiable by countable structures simplify to hyperfinite ones. In related work, Doucha (2013) extended this result to equivalence relations Borel reducible to $=_J$ for some analytic P-ideal J. He also showed that no simple canonization theorem for subequivalences of E_0 is possible. The simplest example of an equivalence relation which is a subset of E_0 yet still does not simplify to E_0 or id on any Silver cube is the equivalence relation E connecting $x, y \in X$ if x, y differ on a finite set of entries, and the set is of even size.

On the anticanonization side, we are able to place E_0 and E_{K_σ} in the spectrum of the Silver ideal:

Theorem 8.7 *E_0 belongs to the spectrum of the Silver forcing. The poset $P_I^{E_0}$ is regularly embedded in P_I, it is $\aleph_0$-distributive, and it yields the $V[\dot{x}_{\text{gen}}]_{E_0}$ extension.*

The decomposition of Silver forcing into $P_I^{E_0}$ and the remainder forcing obtained from this theorem and Theorem 4.4 is different from the widely used Grigorieff decomposition (Grigorieff 1971).

Theorem 8.8 *E_{K_σ} is in the spectrum of I. It is witnessed by a certain Borel equivalence relation E on X such that the poset P_I^E is regular in P_I, it is $\aleph_0$-distributive and it yields the $V[\dot{x}_{\text{gen}}]_E$ extension.*

We now turn to the proofs. Fusion-type arguments will appear repeatedly below, and it will be useful to establish a fixed framework for them. A *fusion sequence* below a condition $p \in P$ is a sequence $\langle i_n, p_n \rangle$ such that $p = p_0 \geq p_1 \geq \cdots$ and $\langle i_m : m \in n \rangle$ is an increasing enumeration of the first n many points of $\omega \setminus \text{dom}(p_{n+1})$. It is clear that the conditions in a fusion sequence have a lower bound, the condition $q = \bigcup_n p_n$.

Proof of Theorem 8.5 The argument uses a claim of independent interest. Recall that if $x \in 2^\omega$ is an infinite binary sequence and $m \in \omega$ is a number, then x flip m is the binary sequence obtained from x by flipping its m-th bit.

Claim 8.9 *Whenever $p \in P$ and $\{f_j : j \in k\}$ is a finite collection of continuous functions from p to 2^ω such that preimages of singletons are I-small, then there is an arbitrarily large number $m \in \omega$ and a cube $q \leq p$ such that for every $x \in [q]$, both x, x flip $m \in [p]$ and $f_i(x) \neq f_i(x \text{ flip } m)$ for all $i \in k$.*

Proof This is proved by induction on k. The case $k = 0$ is trivial, since the collection of continuous functions in question is empty. Suppose that the claim has been established for some k and fails for some $\{f_j : j \in k+1\}$ and

$p \in P$. We will reach a contradiction by producing a cube $q \leq p$ such that f_k is constant on $[q]$.

We will build a fusion sequence $\langle i_n, p_n \rangle$ below p so that for every $n \in \omega$, the value of $f_k(x)$ does not depend on $x \restriction a_{n+1}$ for points $x \in [p_{n+1}]$, where $a_{n+1} = \{i_l : l \leq n\}$. The continuity of the function f_k implies that $f_k \restriction [q]$ must then be constant for any lower bound q of the fusion sequence.

The fusion sequence is built by induction on n. Suppose that p_n and i_{n-1} have been found. By the induction hypothesis at k, there is a subcube $p'_n \leq p_n$ and a number $i_n > i_{n-1}$ such that for every $x \in [p'_n]$ it is the case that x flip $i_n \in [p_n]$ and for all $j \in k$ it is the case that $f_j(i_n) \neq f_j(i_{n+1})$. The set $A = \{x \in [p'_n] : f_k(x) \neq f_k(x \text{ flip } i_{n+1})\}$ is relatively open in $[p'_n]$. If this set is nonempty, then it contains a Silver cube; however, this cannot happen since the claim is supposed to fail for $\{f_j : j \in k+1\}$ and p. Ergo, $A = 0$. Now let $p_{n+1} \in P$ be the function obtained from p'_n by omitting the numbers $i_0, \ldots, i_n$ from the domain of p'_n.

We claim that for $x \in [p_{n+1}]$, the value of $f_k(x)$ does not depend on $x \restriction a_n$. To verify the independence, argue that $f_k(x) = f_k(y)$ for every $y \in [p'_n]$ such that x agrees with y everywhere off the set a_{n+1}. Indeed, first amend y to y' by flipping the entry $y(i_n)$ if necessary to get an agreement with $x(i_n)$. As $y, y' \in [p'_n] \setminus A$, we still have $f_k(y') = f_k(y)$. Since $x, y' \in [p_n]$ and x, y' agree everywhere off a_n, the induction hypothesis on the fusion sequence shows that $f_k(x) = f_k(y') = f_k(y)$ as desired. This completes the induction step in the construction of the fusion sequence and the proof. □

Now suppose that $B \subset X$ is an I-positive Borel set and $f : B \to 2^\omega$ is a Borel function with I-small preimages of singletons; we must produce an I-positive Borel subset of B on which f is one-to-one. Use the bounding property of Silver forcing and Fact 2.52 to get a Silver cube $p \in P$ such that $[p] \subset B$ and $f \restriction [p]$ is continuous. Build a fusion sequence $\langle i_n, p_n \rangle$ below p_n such that, writing $a_n = \{i_l : l \in n\}$, every two points $x, y \in p_{n+1}$ that differ on the set a_n also have distinct f-images. It is clear that for any lower bound q of such a fusion sequence, the function $f \restriction [q]$ is injective as required.

To perform the inductive step of the construction, suppose that i_{n-1}, p_n have been found. For every $h \in 2^{a_n}$ let $f_h : [p_n] \to 2^\omega$ be the function defined by $f_h(x) = f(x \text{ rew } h)$ where $x_h \in p_n$ is the point obtained from x by rewriting the entries of $x \restriction a_n$ with h. These are continuous functions with I-small preimages of singletons. By Claim 8.9, there is a cube p'_n and a number $i_n > i_{n-1}$ such that $f_h(x) \neq f_h(x \text{ flip } i_n)$ for all $h \in 2^{a_n}$ and all $x \in [p'_n]$. By the continuity of the functions f_h, passing to a clopen subcube of p'_n if necessary we can find numbers $m_h \in \omega$ and bits $b_h \neq c_h$ such that for all

$x \in p'_n$, $f_h(x)(m_h) = b_h$ and $f_h(x \text{ flip } i_n)(m_h) = c_h$. Now let p_{n+1} be the subcube of p_n obtained from p'_n by omitting the numbers $i_0, \dots, i_n$ from its domain.

To verify the properties of the cube p_{n+1}, suppose that $x, y \in [p_{n+1}]$ are points that differ on the set a_{n+1}; we must show that they have different functional values. If they differ already on the set a_n, then we are done by the induction hypothesis. If $x \restriction a_n = y \restriction a_n = h$ for some $h \in 2^{a_n}$, then it must be the case that $x(i_n) \neq y(i_n)$. Find points $x', y' \in [p'_n]$ such that $x = x'$ rew h and $y = y'$ rew h and observe that $f(x)(m_h) = f_h(x')(m_h) \neq f_h(x' \text{ flip } i_n)(m_h) = f_h(y')(m_h) = f(y)(m_h)$. This completes the induction step and the proof. □

Proof of Theorem 8.6 Suppose that $B \subset 2^\omega$ is a Borel I-positive set and E is an equivalence on it classifiable by countable structures. Since the Silver forcing adds a minimal real degree, Theorem 4.9 implies that E simplifies to an essentially countable equivalence relation on a positive Borel set. Since the Silver forcing extension preserves Baire category, Theorem 4.7 shows that E restricted to a further I-positive subset is reducible to E_0. Now we can deal with this fairly simple situation separately.

Suppose that $p \in P$ and E is an equivalence relation on $[p]$ reducible to E_0 via a Borel function $f : [p] \to 2^\omega$. Passing to a subcube if necessary, we may assume that f is continuous on $[p]$. Suppose that E has no I-positive equivalence classes. We must produce a subcube $q \leq p$ such that $E \subseteq E_0$ on $[q]$.

By induction on $n \in \omega$, build a fusion sequence $\langle i_n, p_n : n \in \omega \rangle$ so that for every $x, y \in [p_{n+1}]$ such that $x(i_n) \neq y(i_n)$ there is a number $m > n$ such that $f(x)(m) \neq f(y)(m)$. It immediately follows that if q is a lower bound of such a fusion sequence, then $E \restriction [q] \subseteq E_0$. Suppose that p_n, i_{n-1} have been successfully found. Write $a_n = \{i_l : l \in n\}$ and for every $h \in 2^a$, let $f_h(x) = f(x \text{ rew } h)$. Use the σ-additivity of the σ-ideal I to find a condition $p'_n \leq p_n$ such that:

- for every $h, k \in 2^a$, either (a) for every $x \in [p'_n]$, $f_h(x)\ E_0\ f_k(x)$ holds or (b) for every $x \in [p'_n]$, $f_h(x)\ E_0\ f_k(x)$ fails;
- there is a number $m_0 > n$ such that if h, k are such that (a) above holds then $f_h(x) = f_k(x)$ on all entries above m_0 for all $x \in [p'_n]$; and for every $h \in 2^{a_n}$, for all $x \in [p'_n]$ the value of $f_h(x) \restriction m_0$ is the same;
- if h, k are such that (b) above holds then there is a number $m_{hk} > n$ and distinct bits $b_{hk} \neq c_{hk}$ such that for all $x \in [p'_n]$, $f_h(x)(m_{hk}) = b_{hk}$ and $f_k(x)(m_{hk}) = c_{hk}$.

Now let $p''_n \leq p'_n$ be a condition such that for some number $i_n > i_{n-1}$, for every $x \in [p''_n]$ and every $h \in 2^{a_n}$, it is the case that x flip $i_n \in [p'_n]$ and $f_h(x) \neq f_h(x \text{ flip } i_n)$. Such a subcube exists by Claim 8.9. Use the continuity of the function f to pass to a relatively clopen subcube if necessary and find numbers m_h and distinct bits $b_h \neq c_h$ such that for every $x \in [p''_n]$ and every $h \in 2^a$ it is the case that $f_h(x)(m_h) = b_h$ and $f_h(x \text{ flip } i_n)(m_h) = c_h$. Let p_{n+1} be the condition obtained from p''_n by removing the points $\{i_l : l \leq n\}$ from its domain. We claim that p_{n+1} works as required.

For this, let $x, y \in p_{n+1}$ be points such that $x(i_n) \neq y(i_n)$; we must show that $f(x)$ and $f(y)$ differ above n. Let $h = x \restriction a_n$ and $k = y \restriction a_n$, and find points $x', y' \in [p''_n]$ such that $x = x'$ rew h and $y = y'$ rew k; so $f(x) = f_h(x')$ and $f(y) = f_k(y')$. There are two distinct cases according to whether (a) or (b) above holds for h, k. If (b) holds then clearly $f_h(x')(m_{hk}) \neq f_k(y')(m_{hk})$ and the conclusion follows. If, on the other hand, clause (a) holds, then $f_h(x')(m_h) \neq f_h(x' \text{ flip } i_n)(m_h) = f_k(x' \text{ flip } i_n)(m_h) = f_k(y')(m_h)$ by the choice of the number m_h. This completes the proof. □

Proof of Theorem 8.7 Look at the equivalence relation E_0 on X. Whenever $p \in P$ is a Silver cube and $\pi : \omega \to \omega \setminus \text{dom}(p)$ is a bijection, then the map $g : X \to [p]$ defined by $g(x) = f \cup (x \circ \pi^{-1})$ is a continuous reduction of E_0 to $E_0 \restriction [p]$, and therefore E_0 is in the spectrum.

Theorem 4.4 will imply that $P_I^{E_0}$ is a regular subposet of P_I and generates the $V[\dot{x}_{\text{gen}}]_{E_0}$ extension as soon as we prove that E_0-saturations of sets in I are still in I. The easiest way to see that is to consider the action of the countable group of finite subsets of ω with symmetric difference on the space X where a finite set acts on an infinite binary sequence by flipping all entries of the sequence on the set. This action generates E_0 as its orbit equivalence relation, and it also preserves the graph G and with it the σ-ideal I. Thus, E_0-saturations of I-small sets are still I-small.

Finally, to argue for the $\aleph_0$-distributivity of $P_I^{E_0}$, we will use Proposition 3.23. For that, we must check that P_I has the continuous reading of names and E_0 is I-dense. The continuous reading of names follows from the bounding property of P_I and Fact 2.52. To see that E_0 is I-dense, just note that E_0-classes are dense in every Silver cube $[p]$ for $p \in P$. □

Proof of Theorem 8.8 Let $c : 2^\omega \to \omega^\omega$ be the continuous function defined by $c(x)(n) = |\{m \in n : x(m) = 1\}|$ and let E be the Borel equivalence relation on 2^ω connecting $x_0, x_1 \in 2^\omega$ if the function $|c(x_0) - c(x_1)| \in \omega^\omega$ is bounded. Clearly, the function h reduces the equivalence relation E to E_{K_σ}. To show that $E_{K_\sigma} \leq_B E$, choose successive finite intervals $I_n \subset \omega$ of length $2n + 1$ and for every function $y \in \omega^\omega$ below the identity consider the binary sequence $g(y) \in 2^\omega$ specified by the demand that $g(y) \restriction I_n$ begins with $y(n)$ many 1's, ends with $n - y(n)$ many 1's, and has zeroes in all other positions. It is not difficult to see that if $y_0, y_1 \in \omega^\omega$ below the identity are two functions then $|c(g(y_0)) - c(g(y_1)) \restriction I_n|$ is bounded by $|y_0(n) - y_1(n)|$ and the bound is actually attained at the midpoint of the interval I_n. Thus, the function g reduces E_{K_σ} to E. To show that $E \leq E \restriction [p]$ for every $p \in P$ let $\pi : \omega \to \omega \setminus \text{dom}(p)$ be the increasing bijection, and observe that the map $\bar{\pi} : 2^\omega \to [p]$ defined by $\bar{\pi}(x) = p \cup (x \circ \pi^{-1})$ reduces E to $E \restriction [p]$. Thus, the equivalence relation E shows that E_{K_σ} is in the spectrum.

The following definition will be helpful for the investigation of the poset P_I^E. Say that conditions p, q are *n-dovetailed* if between any two successive elements of $\omega \setminus \text{dom}(p)$ there is exactly one element of $\omega \setminus \text{dom}(q)$ and $\min(\omega \setminus \text{dom}(p)) < \min(\omega \setminus \text{dom}(q))$, and moreover, for every number $m \in \omega$, the difference $|\{k \in m : p(k) = 1\}| - |\{k \in m : q(k) = 1|$ is bounded in absolute value by n. Two conditions are *dovetailed* if they are n-dovetailed for some n.

Claim 8.10 *If p, q are dovetailed, then there is a continuous function $\pi : [p] \to [q]$ which preserves I and $\pi \subset E$.*

Proof Let $f : \omega \setminus \text{dom}(p) \to \omega \setminus \text{dom}(q)$ be the unique increasing bijection, and let $\pi : x \mapsto h \cup x \circ f^{-1}$. □

Claim 8.11 *If $p, q \in P$ are arbitrary conditions and $x \in [p]$ and $y \in [q]$ are E-equivalent points such that both $x \setminus p$, $y \setminus q$ take both* 0 *and* 1 *value infinitely many times, then there are dovetailed conditions $p' \leq p$ and $q' \leq q$.*

Proof Let $a, b \subset \omega$ be infinite subsets of $\omega \setminus \text{dom}(p)$ and $\omega \setminus \text{dom}(q)$ respectively such that between any two successive elements of a there is at most one element of b and vice versa, and $x \restriction a$ and $y \restriction b$ both return constantly zero value, and let $p' = x \restriction (\omega \setminus a)$ and $q' = y \restriction (\omega \setminus b)$. □

For the regularity of P_I^E, by Theorem 3.19 it will be enough to show that every Silver cube is an E-kernel. So let $[p]$ be a Silver cube for some $p \in P$ and let $A \subset [p]$ be an analytic set such that $[A]_E \notin I$; we must conclude that $[A]_E \cap [p] \notin I$. Choose a Silver cube $[q] \subset [A]_E$ and a point $y \in 2^\omega$ such that $q \subset y$ and on $\omega \setminus \text{dom}(q)$, y attains both values 0 and 1 infinitely many times. Find $x \in A$ such that $x \; E \; y$. Note that x must attain both values infinitely many times on $\omega \setminus p$: if it attained say 1 only finitely many times, there would be no E-equivalent point in $[p]$ to the point $y' \in [q]$ which extends q with only zeroes. Use Claim 8.11 to find dovetailed Silver cubes $p' \leq p$ and $q' \leq q$, observe that $[p'] \subset [q']_E$, and conclude that $[p'] \subset [p] \cap [q]_E$ as desired.

For the $\aleph_0$-distributivity of P_I^E, just note that E is I-dense and apply Proposition 3.23. To show that P_I^E yields the $V[x_{\mathrm{gen}}]_E$ extension, use Theorem 3.25. It is enough to prove that for all conditions $p, q \in Q$, either there is a condition $p' \leq p$ such that $[p']_E \cap [q] \in I$, or there is a condition $p' \leq p$ and an injective Borel map $f : [p'] \to [q]$ such that $f \subset E$ and f preserves the ideal I. So fix the conditions $p, q \in P$. If $[p]_E \cap [q] \in I$ then we are done; otherwise thin out q so that $[q] \subset [p]_E$ holds. By Claims 8.10 and 8.11, there will be dovetailed Silver cubes $p' \leq p$ and $q' \leq q$ and a homeomorphism $\pi : [p'] \to [q']$ whose graph is a subset of E and which preserves the ideal I. Thus, the assumptions of Theorem 3.25 are satisfied and P_I^E yields the $V[x_{\mathrm{gen}}]_E$ extension. □

Theorem 8.12 *There is a square coding function on Silver cubes.*

Proof Consider the function $f : (2^\omega)^2 \to 2^\omega$ defined by $f(x, y) = x \circ \pi^{-1}$ where π is the increasing enumeration of the set $\{n : x(n) \neq y(n)\}$ if this set is infinite, and $f(x, y) =$ trash if this set is finite. This function works.

Let $p \in P$ be a condition. Whenever $z \in 2^\omega$ is a binary sequence, let $x, y \in 2^\omega$ be the unique points such that $g \subset x, y$ and $z = x \circ \pi^{-1}$ and $1 - z = y \circ \pi^{-1}$ where π is the increasing enumeration of the complement of the domain of p. Clearly then, $f(x, y) = z$ and $f''[p]^2 = 2^\omega$. □

8.3 Ellentuck cubes

A central notion of Ramsey theory of Polish spaces is the following:

Definition 8.13 A *pure Ellentuck cube* is a set of the form $[b]^{\aleph_0}$ where $b \subset \omega$ is infinite. An *Ellentuck cube* is the set of the form $[a, b]$ where $a \subset \omega$ is finite, $b \subset \omega$ is infinite with $\min(b) > \max(a)$, and $[a, b] = \{a \cup c : c \in [b]^{\aleph_0}\}$.

The partition properties of Ellentuck cubes are governed by the Silver and Galvin–Prikry theorems:

Fact 8.14 (Silver (Silver 1970)) *For every partition of a pure Ellentuck cube into an analytic and a coanalytic set, one of the pieces contains a pure Ellentuck cube.*

(Galvin–Prikry (Galvin and Prikry 1973; Kechris 1994, theorem 19.11)) *For every Borel partition of an Ellentuck cube into countably many pieces, one of the pieces contains an Ellentuck cube.*

The *Mathias forcing* P is the poset of all Ellentuck cubes $p = [a_p, b_p]$ ordered by inclusion. The ordering is defined by by $q \leq p$ if $a_p \subset a_q$, $b_q \subset b_p$, and $a_q \setminus a_p \subset b_p$.

Let I be the σ-ideal on $X = [\omega]^{\aleph_0}$ consisting of *Ramsey null* sets: the sets $A \subset X$ which are nowhere dense in the quotient partial order $\mathcal{P}(\omega)$ modulo the ideal of finite sets.

Fact 8.15 *(i)* (Mathias 1977) *Every analytic or coanalytic subset of* $[\omega]^{\aleph_0}$ *is either in* I *or contains an Ellentuck cube.*

(ii) The map $p \mapsto [p]$ *is an isomorphism of* P *with a dense subset of* P_I.

(iii) (Sabok 2012b) *The* σ*-ideal* I *is* $\mathbf{\Pi}^1_2$ *on* $\mathbf{\Sigma}^1_1$ *and not* $\mathbf{\Delta}^1_2$ *on* $\mathbf{\Sigma}^1_1$.

(iv) The poset P *is proper.*

The canonization properties of Ellentuck cubes are complex. The total canonization for smooth equivalence relations fails badly, as the Mathias forcing adds many intermediate real degrees. Prömel and Voigt (1985) gave an explicit canonization formula for smooth equivalence relations which can be also derived from earlier work of Mathias (1977, theorem 6.1). For a function $\gamma : [\omega]^{<\aleph_0} \to 2$ let $\Gamma_\gamma : [\omega]^{\aleph_0} \to [\omega]^{\leq\aleph_0}$ be the function defined by $\Gamma_\gamma(a) = \{k \in a : \gamma(a \cap k) = 0\}$ and E_γ the smooth equivalence relation on $[\omega]^{\aleph_0}$ induced by the function Γ_γ.

Fact 8.16 (Mathias 1977; Prömel and Voigt 1985) *For every smooth equivalence relation* E *on* $[\omega]^{\aleph_0}$ *there is a function* $\gamma : [\omega]^{<\aleph_0} \to 2$ *and a pure Ellentuck cube on which* $E = E_\gamma$.

Here, we shall present a further canonization theorem, due to Mathias, regarding countable Borel equivalence relations. There is an obvious obstacle for canonization there, the relation E_0 as well as some of its subequivalences. It turns out that other countable equivalence relations can be simplified to these on Silver cubes.

Theorem 8.17 (Mathias) *For every countable Borel equivalence relation* E *on* $[\omega]^{\aleph_0}$*, there is a pure Ellentuck cube on which* $E \subseteq E_0$*. For every essentially countable equivalence relation* E *on* $[\omega]^{\aleph_0}$*, there is a pure Ellentuck cube on which* E *is reducible to* E_0*. In particular,*

$$\text{countable} \to_I \text{hyperfinite, essentially countable} \to_{I \leq_B} E_0.$$

Note the parallel of the first arrow with the Connes–Feldman–Weiss theorem (Connes *et al.* 1981).

Proof The proof of Theorem 8.17 starts with three claims of independent interest.

Claim 8.18 *Let* $c \subset \omega$ *be an infinite set and* $f : [c]^{\aleph_0} \to \mathcal{P}(\omega)$ *be a Borel function such that for every* $e \in [c]^{\aleph_0}$*,* $f(e) \setminus e$ *is infinite. Then there exists an infinite set* $d \subset c$ *such that for every* $e \in [d]^{\aleph_0}$*,* $f(e) \setminus d$ *is infinite.*

Proof For an infinite set $e \subset c$, write e_{ev} for the set of its elements that are indexed by even numbers in its increasing enumeration, and e_{od} for $e \setminus e_{\text{ev}}$. Partition $[c]^{\aleph_0}$ into two Borel parts: $B = \{e : f(e_{\text{ev}}) \cap e_{\text{od}}$ is infinite$\}$, and $C = \{f(e_{\text{ev}}) \cap e_{\text{od}}$ is finite$\}$. By the Galvin–Prikry theorem (Fact 8.14), there must be a set $d' \subset c$ such that $[d']^{\aleph_0}$ is either wholly contained in B or C. Let $d = d'_{ev}$; we claim that the set d works as required.

Case 1 Suppose first that $[d']^{\aleph_0} \subset B$. Then for every infinite set $e \subset d$ there is $e' \subset d'$ such that $e'_{\text{ev}} = e$ and $e'_{\text{od}} \subset d'_{\text{od}}$. Now, $f(e) \setminus d \supset f(e) \cap d'_{\text{od}} = f(e'_{\text{ev}}) \cap d'_{\text{od}} \supset f(e'_{\text{ev}}) \cap e'_{\text{od}}$. The rightmost set in this chain of inclusions is infinite by the homogeneity assumption, and so is the leftmost set. In this case, we are done.

Case 2 Now suppose that $[d']^{\aleph_0} \subset C$. We claim that for every infinite set $e \subset d$, it must be the case that $f(e) \cap d'$ is finite. If this set was infinite, find a set $e' \subset d'$ such that $e'_{\text{ev}} = e$ and for any two successive elements of the set e, if there is an element of the set $f(e) \cap d'$ between them then the unique element of e' between them belongs to $f(e) \cap d'$. Then clearly $f(e'_{\text{ev}}) \cap e'_{\text{od}}$ is infinite, contradicting the case assumption. Thus, the set $f(e) \setminus d'$ must be infinite, as well as its superset $f(e) \setminus d$. The claim follows. □

Claim 8.19 *Let $f : [\omega]^{\aleph_0} \to [\omega]^{\aleph_0}$ be a Borel function. The Mathias forcing forces the following: either $\dot{f}(\dot{x}_{\text{gen}}) \subset \dot{x}_{\text{gen}}$ up to finitely many numbers, or there is a ground model superset $d \supset \dot{x}_{\text{gen}}$ such that $f(\dot{x}_{\text{gen}}) \setminus d$ is infinite.*

Proof This is an elaboration of Claim 8.18. Suppose $B \in P_I$ is a condition forcing $\dot{f}(\dot{x}_{\text{gen}}) \setminus \dot{x}_{\text{gen}}$ is infinite. The set $C = \{x \in B : f(x) \setminus x$ is infinite$\}$ is still I-positive and Borel, so thinning it out if necessary we may assume that it is an Ellentuck cube. Using the first item, we can thin out this cube still further to find $d \subset \omega$ such that all sets in the cube are subsets of d and their f-images have infinitely many elements outside of d. This condition forces the "or" clause and the claim follows. □

Claim 8.20 *Let M be a countable elementary submodel of a large structure and let $b \subset \omega$ be an M-generic real for Mathias forcing. If $e_0, e_1 \subset b$ are infinite sets then $M[e_0] = M[e_1]$ if and only if the symmetric difference $e_0 \Delta e_1$ is finite.*

Proof An infinite subset of a Mathias generic real is again Mathias generic (Mathias 1977), so it is the case that both e_0, e_1 are Mathias generic reals over the model M. Clearly, if their symmetric difference is finite then they construct each other and the models $M[e_0]$ and $M[e_1]$ are the same. For the opposite implication, suppose that $e_0 \setminus e_1$ is infinite and argue that $e_0 \in M[e_1]$ is impossible. Indeed, if $e_0 \in M[e_1]$, then $e_0 = f(e_1)$ for some Borel function $f \in M$. Claim 8.19 shows that there is an infinite set $d \in M$ such that $e_1 \subset d$ and $f(e_1) \setminus d$ is infinite. Since b is Mathias generic over M with infinite intersection with d, all but finitely many of its elements belong to d. But then, the set $f(e_1) \setminus b$ must be infinite, and therefore $f(e_1)$ cannot be equal to the set e_0 which is a subset of b. □

Towards the proof of the theorem, suppose first that the Borel equivalence relation E has in fact countable classes; it is then generated as an orbit equivalence relation of a Borel action of some countable group G. Let M be a countable elementary submodel of a large structure containing the group action, and let $b \subset a$ be a Mathias generic real over M. If $e \subset b$ is an infinite set, then the model $M[e]$ contains all E-equivalents of e, since they are simply the images of e under the Borel action of G. Claim 8.20 then shows that the E-class of e is a subset of the E_0-class of e, and as $e \subset b$ was arbitrary, $E \restriction [b]^{\aleph_0} \subseteq E_0$.

The case of the general essentially countable equivalence relation uses the canonization of smooth equivalence relations, as in Fact 8.16. Suppose that $a \subset \omega$ is an infinite set, F is a countable Borel equivalence relation on a Polish space Y, $f : [a]^{\aleph_0} \to Y$ is a Borel function, and E is the pullback of F by the function f. Use the canonization of smooth equivalence relations (Fact 8.16), to thin out the set a if necessary to find a function $\gamma : [\omega]^{<\aleph_0} \to 2$ such that the smooth equivalence relations associated with f and Γ_γ coincide at $[a]^{\aleph_0}$. This allows us to replace the general case with the situation where $f = \Gamma_\gamma$ and F is a Borel equivalence relation with countable classes on $[\omega]^{\aleph_0}$. Let $C = \Gamma''_\gamma [a]^{\aleph_0}$ and note that the set C is analytic. By Silver's theorem (Fact 8.14), there is an infinite set $c \subset a$ such that $[c]^{\aleph_0}$ is either a subset of C or disjoint from C. The latter case cannot happen, since the set $\Gamma_\gamma(c) \subset c$ would contradict the homogeneity; thus, the former case rules. By the previous paragraph, there must be an infinite set $d \subset c$ such that $F \restriction [d]^{\aleph_0}$ is Borel reducible to E_0. It follows that $E \restriction [d]^{\aleph_0}$ is also reducible to E_0, since the function Γ_γ Borel reduces it to $F \restriction [d]^{\aleph_0}$. □

Beyond the realm of essentially countable equivalence relations, there is a rich and chaotic spectrum of Borel equivalences on $[\omega]^{\aleph_0}$ which maintain their complexity on every Ellentuck cube.

Theorem 8.21 *E_0 is in the spectrum of the Ramsey null ideal I. The poset $P_I^{E_0}$ is regularly embedded in P_I, it is $\aleph_0$-distributive and it yields the $V[\dot{x}_{\mathrm{gen}}]_{E_0}$ model.*

Proof Identify subsets of ω with their characteristic functions and use this identification to transfer E_0 to the space X. The theorem follows from Theorem 4.4 once we prove that E_0 saturations of Borel I-small sets are I-small. However, it is clear that E_0-saturation of a set nowhere dense in the algebra $\mathcal{P}(\omega)$ modulo finite is again nowhere dense, because the saturation does not change the place of the elements of the set in this algebra. The $\aleph_0$-distributivity follows from Proposition 3.23, since the E_0 orbit of any element in an Ellentuck cube is dense in the cube. □

In fact, it is not difficult to see that the decomposition of Mathias forcing associated with the E_0 equivalence is just the standard decomposition into $\mathcal{P}(\omega)$ modulo finite followed with a poset shooting a set through a Ramsey ultrafilter. Note that in the Mathias extension, there are many E_0-classes which are subset of every E_0-invariant set in the generic filter; every infinite subset of the generic real will serve as an example. It turns out though that E_0 is the only countable equivalence relation in the spectrum of Mathias forcing.

Theorem 8.22 *The spectrum of the σ-ideal I is cofinal among Borel equivalence relations in $\leq_B$; it includes E_{K_σ}.*

Proof Let K be a Borel ideal on ω containing all finite sets and let E_K be the equivalence relation on the space Y of increasing functions in ω^ω defined by $y\ E_K\ z$ if $\{n \in \omega : y(n) \neq z(n)\} \in K$. For the cofinality of the spectrum, by Fact 2.21, it is enough to show that E_K is in the spectrum of Mathias forcing. Look at E, the relation on $[\omega]^{\aleph_0}$ connecting two points if their increasing enumerations are E_K-equivalent. Clearly, this is an equivalence relation Borel reducible to E_K. On the other hand, if $B \subset X$ is a Borel I-positive set, then E_K is Borel reducible to $E \restriction B$. Just find an Ellentuck cube $[a, b]$ inside the set B and let $\pi : \omega \to a \cup b$ be the increasing enumeration. To each point $y \in Y$ associate the set $b_y = a \cup \mathrm{rng}((\pi \circ y) \restriction \omega \setminus |a|)$ and observe that the map $y \mapsto b_y$ reduces E_K to $E \restriction B$. If K is an F_σ-ideal on ω such that $=_K$ is bireducible with E_{K_σ} as in Proposition 2.20, then also E_K is bireducible with E_{K_σ}, showing that E_{K_σ} is in the spectrum of I. □

Theorem 8.23 *F_2 is in the spectrum of the Ramsey null ideal.*

Proof The argument follows closely Case 4 of the proofs of Theorems 6.24 and 6.67. List the rational numbers as $\{q_m : m \in \omega\}$. Divide ω into successive intervals I_n for $n \in \omega$, each of respective length n, and find a function $\chi : \omega \to \mathbb{Q}$ such that $\chi \restriction I_n$ is an order isomorphism between I_n and $\{q_m : m \in n\}$. Finally, for every $x \in [\omega]^{\aleph_0}$ and every $q \in \mathbb{Q}$, let $x^q \subset x$ be the image of the set $\{m : \chi(m) = q\}$ under the increasing enumeration of

x, and write $f(x) = \{x^q : q \in \mathbb{Q}\} \subset [\omega]^{\aleph_0}$. Clearly, the collection $f(x)$ consists of pairwise almost disjoint subsets of x. Define the equivalence relation E on $[\omega]^{\aleph_0}$ by setting $x \mathrel{E} y$ if the sets $f(x)$ and $f(y)$ have the same E_0-saturation. It is clear that $E \leq_B F_2$; the proof will be complete once we show that $F_2 \leq_B E \restriction C$ for every Ellentuck cube $C \subset [\omega]^{\aleph_0}$. For simplicity we will only consider the case of a pure Ellentuck cube C, $C = [x]^{\aleph_0}$ for some $x \subset \omega$.

Let M be a countable elementary submodel of a large enough structure.

Claim 8.24 *There is a continuous injection $h : 2^\omega \to [\omega]^{\aleph_0}$ such that its range consists of pairwise almost disjoint sets, and for every countable dense set $a \subset 2^\omega$ consisting of irrational points there is a set $y \subset x$ Mathias-generic over M such that the sets $h''a$ and $f(y)$ have the same E_0-saturation.*

Proof Choose a set $\bar{x} \subset x$ which is Mathias-generic over M. Choose long successive intervals $\{J_n : n \in \omega\}$ of natural numbers so that for each n there is an injection $h_n : 2^n \to J_n \cap \bar{x}$ which preserves the lexicographic order, such that between any two distinct elements of the range of h_n and/or endpoints of the interval J_n there are at least n-many elements of the set $\bar{x}$. Define $h : 2^\omega \to [\omega]^{\aleph_0}$ by $h(z) = \{h_n(z \restriction n) : n \in \omega\}$. We claim that this function works.

It is clear that the range of the function h consists of pairwise almost disjoint sets. Now suppose that $a \subset 2^\omega$ is a countable dense set consisting of irrational points. We will produce a set $y \subset \bar{x}$ such that the sets $f(y)$ and $h''a$ have the same E_0-saturation. Since $\bar{x}$ is Mathias-generic over M, its infinite subset y must be again Mathias-generic (Mathias 1977); this will complete the proof.

Use the density of the set a in the lexicographic order to build an order-preserving bijection $\phi : \mathbb{Q} \to a$. We will find the point $y \subset \bar{x}$ so that for every rational $q \in \mathbb{Q}$, $y^q = h(\phi(q))$ up to a finite modification. For every $n \in \omega$ find a number $k_n \in \omega$ such that the lexicographic order of the points $\{\phi(q_m) : m \in n\} \subset a$ is apparent already from their initial segments of length k_n. Now use the initial choice of the intervals I_n, J_n and the function χ to find the set $y \subset \bar{x}$ so that for its increasing enumeration e the following holds:

- $e''I_n \subset J_n \cap \bar{x}$;
- for every $n \in \omega$, every $m \in n$, every $k > k_n$ and every $l \in I_k$, $\chi(l) = q_m$ iff $e(l) = h_k(\phi(q_m) \restriction k)$.

Such a point will clearly satisfy $y^{q_m} = h(\phi(q_m))$ above $\min(J_{k_{m+1}})$ for every rational q_m. □

Extend the definition of h to $h : (2^\omega)^\omega \to ([x]^{\aleph_0})^\omega$ by $h(z) = \langle h(z(n)) : n \in \omega \rangle$. Let F be the equivalence relation on $([\omega]^{\aleph_0})^\omega \cup [\omega]^{\aleph_0}$ connecting $z \in ([\omega]^{\aleph_0})^\omega$ and $y \in [\omega]^{\aleph_0}$ if the sets $f(y)$ and $\mathrm{rng}(z)$ have the same

E_0-saturations. The equivalence relation F is clearly Borel reducible to F_2. Theorem 2.29 shows that the set $C = \{z \in ([\omega]^{\aleph_0})^\omega : \exists y \subset x$ Mathias-generic over M which is F-related to $z\}$ is Borel; it also gives a Borel function $k : C \to [x]^{\aleph_0}$ such that for every $z \in C$ the set $k(z) \subset x$ is Mathias-generic over M and the sets $f(k(z))$ and $\mathrm{rng}(z)$ have the same E_0-saturation. Claim 8.24 shows that if $z \in (2^\omega)^\omega$ enumerates a dense subset of 2^ω consisting of irrational points, then $h(z) \in C$. Now, let $g_0 : 2^\omega \to (2^\omega)^\omega$ be any Borel function such that for $y \neq z \in 2^\omega$, $g_0(y)$ and $g_0(z)$ enumerate disjoint dense subsets of 2^ω consisting of irrational points. Let $g_1 : (2^\omega)^\omega \to (2^\omega)^\omega$ be any Borel function such that for every $z \in (2^\omega)^\omega$, $g_1(z)$ enumerates the set $\bigcup_n \mathrm{rng}(g_0(z(n)))$. The Borel function $k \circ h \circ g_1$ is the required reduction of F_2 to $E \restriction [x]^{\aleph_0}$. □

8.4 Milliken cubes

Let $(\omega)^\omega$ be the space of all infinite sequences a of finite sets of natural numbers with the property that $\min(a_{n+1}) > \max(a_n)$ for every number $n \in \omega$. Define a partial *Milliken ordering* $\leq$ on this space by $b \leq a$ if every set on b is a union of several sets on a. The combinatorics of $(\omega)^\omega$ is governed by the following:

Fact 8.25 (Bergelson–Blass–Hindman (Todorcevic 2010, corollary 4.49)) *For every partition $(\omega)^\omega = B_0 \cup B_1$ into Borel sets there is $a \in (\omega)^\omega$ such that the set $\{b \in (\omega)^\omega : b \leq a\}$ is wholly included in one of the pieces.*

There is also an associated canonization theorem:

Fact 8.26 (Taylor (Taylor 1976)) *For every equivalence relation E on $[\omega]^{<\aleph_0}$, there is a sequence $a \in (\omega)^\omega$ such that on the set of all finite unions of entries of a, the equivalence E simplifies to one of the following forms:*

(i) $u \mathrel{E} v \leftrightarrow u = v$;
(ii) $u \mathrel{E} v \leftrightarrow \min(u) = \min(v)$;
(iii) $u \mathrel{E} v \leftrightarrow \max(u) = \max(v)$;
(iv) $u \mathrel{E} v \leftrightarrow \min(u) = \min(v)$ *and* $\max(u) = \max(v)$;
(v) every two sets are E-related.

The purpose of this section is to prove a related canonization theorem on infinite sequences in $(\omega)^\omega$.

Definition 8.27 A *pure Milliken cube* is a set of the form $\{c \in (\omega)^\omega : c \leq a\}$ for a sequence $a \in (\omega)^\omega$. A *Milliken cube* is a set of the form $\{t^\frown c : c \leq a\}$ where t is a finite increasing sequence of finite subsets of ω and $a \in (\omega)^\omega$ is such that $\max \bigcup t < \min(a(0))$.

Regarding the canonization properties of Milliken cubes for smooth equivalence relations, Klein and Spinas (2005) proved a parallel of the Prömel–Voigt canonization theorem. They identified a well-organized collection Γ of continuous maps from $(\omega)^\omega$ to $V_{\omega+1}$ such that, writing E_γ for the equivalence relation $\{\langle x, y\rangle \in (\omega)^\omega \times (\omega)^\omega : \gamma(x) = \gamma(y)\}$, the following holds: for every $a \in (\omega)^\omega$ and every smooth equivalence relation E on $(a)^\omega$ there is $b \leq a$ and $\gamma \in \Gamma$ such that $E \restriction (b)^\omega = E_\gamma$.

We will prove a canonization theorem for countable equivalence relations on the space $(\omega)^\omega$. Among these, the relation E_{tail} stands out. It connects two sequences a, b if they have the same tail, that is, if there is an integer n such that for all but finitely many $m \in \omega$, $a(m) = b(m + n)$. Our canonization theorem elaborates on the central position of E_{tail}.

Theorem 8.28 *For every $a \in (\omega)^\omega$ and every countable equivalence relation E on $(a)^\omega$ there is $b \leq a$ such that $E \restriction (b)^\omega \subseteq E_{\mathrm{tail}}$.*

There is a σ-ideal on $(\omega)^\omega$ and a poset associated with the notion of a Milliken cube. Let P, the *Milliken forcing*, be the poset of all pairs $p = \langle t_p, a_p\rangle$ such that t_p is an increasing finite sequence of nonempty finite subsets of ω and $a_p \in (\omega)^\omega$ is a sequence of sets above t_p. The ordering is defined by $q \leq p$ if $t_p \subset t_q$, $a_q \leq a_p$, and all sets on the sequence $t_q \setminus t_p$ are unions of some sets on a_p. For a condition $p \in P$ write $[p]$ to denote the cube obtained from t_p, a_p. Quite obviously, the poset P adds a generic sequence $\dot{x}_{\mathrm{gen}} \in (\omega)^\omega$ which is the union of the first coordinates of conditions in the generic filter. To define the associated σ-ideal, let I be the collection of Borel subsets of $(\omega)^\omega$ which do not contain a Milliken cube. The following seems to be folklore knowledge:

Theorem 8.29 *The poset P is proper. The collection I is a σ-ideal and the map $p \mapsto [p]$ is an isomorphism between P and a dense subset of P_I.*

Proof The argument is best channeled through the following useful fact:

Claim 8.30 *Whenever M is a countable elementary submodel of a large enough structure, $a \in (\omega)^\omega$ is a P-generic sequence over M and $b \leq a$, then b is a P-generic sequence over M as well.*

Proof We will show that for every condition $p \in P$ and every open dense set $D \subset P$, there is a condition $q \leq p$ with $t_q = t_p$ such that for every $c \leq t_q{}^\frown a_q$ there is a number n such that $\langle (c \restriction n), a_q \setminus n\rangle \in D$. The lemma will follow: if $D \in M$ then the sequence a must pass though a condition $q \in M$ as above by genericity, and then every sequence $c \leq a$ passes through a condition in $D \cap M$ by the properties of q.

To find the condition $q \leq p$, assume for simplicity that $t_p = 0$. Use a simple fusion argument to find a sequence $a'_p \leq a_p$ such that for every $n \in \omega$ and every sequence $u \leq a'_p \restriction n$, either (a) $\langle u, a'_p \setminus n\rangle \in D$ or (b) there is no sequence $c \leq a'_p$ such that $\langle u, b\rangle \in D$. Now let $(a'_p)^\omega = B_0 \cup B_1$ be the partition into two Borel sets defined by $c \in B_0$ if for some $n \in \omega$, (a) above occurs for $u = c \restriction n$. The Milliken theorem 8.25 shows that one of the sets B_0, B_1 contains a pure strong Milliken cube. If $a \leq a'_p$ is such that $(a)^\omega \in B_0$ then the condition $q = \langle t_p, a\rangle$ is as required. On the other hand, the set B_1 cannot contain a strong Milliken cube: if $a \leq a'_p$ is such that $(a)^\omega \in B_1$ and $r \leq \langle t_p, a\rangle$ is a condition in the open dense set D, then (a) must have occurred for $u = t_r$, contradicting the assumption that $(a)^\omega \subset B_1$. □

To finish the proof of Theorem 8.29, let J be the collection of those Borel sets B such that $P \Vdash \dot{x}_{\text{gen}} \notin \dot{B}$. The collection J is a σ-ideal; we will show that $J = I$. It is clear that $J \subseteq I$: if $B \notin I$ then there is a condition $p \in P$ such that $[p] \subset B$, thus $p \Vdash \dot{x}_{\text{gen}} \in B$ and $B \notin J$. To show that $I \subseteq J$, look at an arbitrary Borel set $B \notin J$ and a condition $p \in P$ forcing $\dot{x}_{\text{gen}} \in \dot{B}$. Let M be a countable elementary submodel of a large enough structure and let b be a P-generic sequence over M meeting the condition p. Let $q \leq p$ be the condition defined by $t_q = t_p$ and $a_q = b \setminus t_p$; we will show that $[q] \subset B$ and therefore $B \notin I$. Indeed, by Claim 8.30 every sequence $c \leq b$ is generic over M, so every sequence $c \in [q]$ is generic over M meeting the condition p, so $M[c] \models c \in B$ by the forcing theorem and $c \in B$ by the analytic absoluteness between the models $M[c]$ and V.

Thus, I is a σ-ideal; the map $p \mapsto [p]$ is a bijection between P and a dense subset of P_I that is also an isomorphism of $\leq$ and $\subseteq$. Finally, the properness of the poset P also follows abstractly from Claim 8.30. If $B \in P_I$ is an I-positive Borel set and M is a countable elementary submodel of a large structure containing the set B, we must show that the set $\{x \in B : x \text{ is } P_I\text{-generic over } M\}$ is I-positive. Find a condition $p \in M$ such that $[p] \subset B$, and find a sequence $x \in (\omega)^\omega$ generic over M which meets p. Let $q = \langle t_p, x \setminus t_p\rangle$. The claim shows that $[q]$ consists only of points generic over M and meeting the condition p. As $[q] \notin I$, the proof is complete. □

Proof of Theorem 8.28 The whole argument follows closely the proof of the first sentence of Theorem 8.17. For $b, c \in (\omega)^\omega$ write $c \leq^* b$ if for some tail c' of c, we have $c' \leq b$.

Claim 8.31 *Whenever $c \in (\omega)^\omega$ and $f : (c)^\omega \to (\omega)^\omega$ is a Borel function such that for every $e \leq c$, $f(e) \not\leq^* e$, then there is a sequence $d \leq c$ such that for every $e \leq d$, we have $f(e) \not\leq^* d$.*

Proof First, divide $(c)^\omega$ to two Borel classes: $e \in B_0$ if $\bigcup_n f(e)(n) \setminus \bigcup_n e(n)$ is a finite set, and $e \in B_1$ otherwise. The Milliken theorem (Fact 8.25) divides the treatment into two cases:

Case 1 There is $c' \leq c$ such that $(c')^\omega \subset B_1$. In this case, for every $e \leq c'$, write e_{de} for the sequence $\langle e(4n) \cup e(4n+2) : n \in \omega \rangle$ partition $(c')^\omega$ into two Borel classes: $e \in C_0$ if $\bigcup_n f(e_{\mathrm{de}})(n) \cap \bigcup_n e(2n+1)$ is finite and $e \in C_1$ otherwise. Find a sequence $d' \leq c'$ homogeneous with respect to this partition, and let $d = d'_{\mathrm{de}}$. We claim that d works as required. The verification splits into two subcases:

Case 1a Either, $(d')^\omega \subset C_1$. Then, for every sequence $e \leq d$ there is a sequence $e' \leq d'$ such that $e'_{\mathrm{de}} = e$ and for every odd $n \in \omega$ there is odd $k \in \omega$ such that $e'(n) = d'(k)$. The Case 1a assumption shows that $\bigcup_n f(e)(n) \cap \bigcup_n d'(2n+1)$ is infinite, and so $f(e) \not\leq^* d$ as desired.

Case 1b Or, $(d')^\omega \subset C_0$. We claim that for every sequence $e \leq d$, it must be the case that the set $\bigcup_n f(e)(n) \cap (\bigcup_n d'(n) \setminus \bigcup_n e(n))$ is finite. Otherwise, it would be easy to find a sequence $e' \leq d'$ such that, for every $n \in \omega$:

- $e'(4n) \cup e'(4n+2) = e(n)$;
- if there is some $i \in [\min(e(n), \max(e(n))] \cap (\bigcup_m d'_m \setminus e)$ then there is such an i in $e'(4n+1)$;
- if there is some $i \in (\max(e(n)), \min(e(n+1))) \cap \bigcup_m d'_m$ then there is such an i in $e'(4n+3)$.

Then clearly $\bigcup_n f(e'_{\mathrm{de}})(n) \cap \bigcup_n e'(2n+1)$ is infinite, contradicting the Case 1b assumption. Thus, for every sequence $e \leq d$, the set $\bigcup_n f(e)(n) \cap (\bigcup_n d'(n) \setminus \bigcup_n e(n))$ is finite. By the Case 1 assumption, the set $\bigcup_n f(e)(n) \setminus \bigcup_n d'(n)$ must be infinite, and therefore $f(e) \not\leq^* d$ as desired.

Case 2 There is $c' \leq c$ such that $(c')^\omega \subset B_0$. In this case, for every $e \leq c'$ write e_{co} for the sequence $\langle e(2n) \cup e(2n+1) : n \in \omega \rangle$. Since $f(e_{\mathrm{co}}) \not\leq^* e_{\mathrm{co}}$ and Case 2 holds, it must be the case that for infinitely many n, $u = f(e_{\mathrm{co}})(n)$ is a finite set that splits some $v = e_{\mathrm{co}}(m_n)$ (meaning that v has nonempty intersection with both u and its complement), and this can happen in one of four ways. The Milliken theorem 8.25 provides a split into four corresponding parallel subcases:

Case 2a There is $d' \leq c'$ such that for every $e \leq d'$ there are infinitely many n for which there exists m_n such that $e(2m_n) \subset f(e_{\mathrm{co}})(n)$ and $e(2m_n+1) \cap f(e_{\mathrm{co}})(n) = 0$. Let $d = d'_{\mathrm{co}}$; we claim that d works as required in the statement of the claim. Indeed, if $e \leq d$, then write $\bar{e} \leq d'$ for the sequence defined by

$\bar{e}(2m+1) = d'(k)$ for the largest k such that $d'(k) \subset e(m)$, and $\bar{e}(2m) = e(m) \setminus \bar{e}(2m+1)$; note that m must be odd as $e \leq d$. By the case assumption, there are infinitely many n for which there is m_n such that $\bar{e}(2m_n) \subset f(e)(n)$ and $\bar{e}(2m_n+1) \cap f(e)(n) = 0$. A review of the definitions will show that $f(e)(n)$ then splits the set $d(\frac{m-1}{2})$. Since this happens for infinitely many n, we conclude that $f(e) \not\leq^* d$ as required.

Case 2b There is $d' \leq c'$ such that for every $e \leq d'$ there are infinitely many n for which there exists m_n such that $e(2m_n+1) \subset f(e_{\mathrm{co}})(n)$ and $e(2m_n) \cap f(e_{\mathrm{co}})(n) = 0$.

Case 2c The set $f(e_{\mathrm{co}})(n) = 0$ splits $e(2m_n)$.

Case 2d The set $f(e_{\mathrm{co}})(n) = 0$ splits $e(2m_n+1)$.

These three last subcases are treated in a fashion parallel to Case 2a. □

Claim 8.32 *Let $f : (\omega)^\omega \to (\omega)^\omega$ be a Borel function. The Milliken forcing forces that either $\dot{f}(\dot{x}_{\mathrm{gen}}) \leq^* \dot{x}_{\mathrm{gen}}$ or there is a ground model $d \in (\omega)^\omega$ such that $\dot{x}_{\mathrm{gen}} \leq^* d$ and $\dot{f}(\dot{x}_{\mathrm{gen}}) \not\leq^* d$.*

Proof Let I be the Milliken ideal. Let $B \in P_I$ be any Borel I-positive set. There are two options: either the Borel set $C = \{x \in B : f(x) \leq^* x\}$ is I-positive, in which case it is a condition forcing the "either" clause of the claim; or Theorem 8.29 shows that there is a Milliken cube $[t, c] \subset B \setminus C$. Use Claim 8.31 to thin it down to $[t, d]$ such that for every $e \in [t, d]$, $f(e) \not\leq^* d$. The condition $[t, d] \leq B$ forces the "or" clause of the claim. □

Claim 8.33 *Let M be a countable elementary submodel of a large structure and $a \in (\omega)^\omega$ a P-generic sequence over M. For any two sequences $b, c \leq a$, $M[b] = M[c] \leftrightarrow b \ E_{\mathrm{tail}} \ c$.*

Proof Both b, c are P-generic reals over the model M by Claim 8.30. Clearly, if b, c share the same tail, then they construct each other and the models $M[b]$ and $M[c]$ are the same. For the opposite implication, suppose that $b \not\leq^* c$ and argue that $b \in M[c]$ is impossible. Indeed, if $b \in M[c]$, then $b = f(c)$ for some Borel function $f \in M$. Use Claim 8.32 to find a sequence $d \in M$ such that $c \leq d$ and $b \not\leq^* d$. Since a is P-generic over M and a, d are compatible in $\leq$ (their lower bound is c), a simple genericity argument shows that $a \leq d$ has to hold. Then $b \leq a \leq d$, contradicting the properties of the sequence d. □

Finally, Theorem 8.28 follows. Let E is a countable Borel equivalence relation on $(\omega)^\omega$. Use the Feldman–Moore theorem (Fact 2.13) to find a Borel action of a countable group G on $(\omega)^\omega$ such that E is its orbit equivalence relation. Let M be a countable elementary submodel of a large enough structure containing the group action. Let $a \in (\omega)^\omega$ be a P-generic sequence over M. Whenever $b\ E\ c \leq a$ are sequences, then $M[b] = M[c]$ since b and c are images of each other of the group action, which is coded in M. By Claim 8.33, it follows that $b\ E_{\text{tail}}\ c$, in other words $E \restriction (a)^\omega \subset E_{\text{tail}}$ as desired. □

Question 8.34 Let E be an essentially countable Borel equivalence relation on $(\omega)^\omega$. Is there a Milliken cube such that on it E is Borel reducible to E_0?

8.5 Matet cubes

Definition 8.35 The symbol $(\omega)^\omega$ stands for the collection of all sequences a of finite sets of natural numbers such that for every $n \in \omega$, $\max(a(n)) < \min(a(n+1))$. A *pure Matet cube* is a set of the form $[a] = \{\bigcup_{n \in b} a(n) : b \in [\omega]^{\aleph_0}\}$ where $a \in (\omega)^\omega$. A *Matet cube* is a set of the form $\{c \cup b : b \in [a]\}$ for some finite set $c \subset \omega$ and $a \in (\omega)^\omega$.

The partition theory of Matet cubes is governed by Fact 8.25, which immediately implies the following:

Fact 8.36 *For every Borel partition of a pure Matet cube into finitely many pieces, there is a pure Matet cube wholly contained in one of the pieces of the partition. For every Borel partition of a Matet cube into countably many pieces, there is a Matet cube wholly contained in one piece of the partition.*

Regarding the canonization properties of Matet cubes, one can immediately identify two obstacles to total canonization of Borel equivalence relations: the equivalence $E_{\min}$ which connects two sets of natural numbers if they have the same minimum, and the equivalence E_0. It turns out that these two obstacles are to a great extent unique. Write I for the collection of analytic sets which do not contain a Matet cube. We have:

Theorem 8.37 *Let E be an equivalence relation on $[\omega]^{\aleph_0}$ in one of the following classes:*

(i) Borel reducible to E_{K_σ};
(ii) classifiable by countable structures;
(iii) Borel reducible to $=_J$ for some analytic P-ideal J on ω.

Then there is a pure Matet cube C such that $E \restriction C$ is equal to one of the following: id, ev, E_0, $E_{\min}$, $E_0 \cap E_{\min}$. *For each of these classes* $\mathbf{E}$ *of equivalence relations, we have*

$$\mathbf{E} \to_I \{\text{id}, \text{ev}, E_0\}.$$

Note that if we are permitted to pass to a Matet cube instead of the pure Matet cube, the equivalence relation $E_{\min}$ disappears from the canonization basis. This is easy to achieve by first canonizing on a pure cube and then passing to a nonpure subcube, all of whose elements will have the same minimum.

Before we turn to the proof of the above canonization theorem, we take the opportunity to introduce a proper forcing associated with Matet cubes and its combinatorial presentation. Define the *Matet forcing* P (Matet 1988; Blass 1989) to consist of all pairs $p = \langle t_p, a_p \rangle$ where $t_p \subset \omega$ is a finite set and a_p is an infinite sequence of finite subsets of natural numbers such that $\max(a_p(n)) < \min(a_p(n+1))$ for all $n \in \omega$. Thus, every condition $p \in P$ is associated with a Matet cube $[p] = \{t_p \cup \bigcup_{n \in b} a_p(n) : b \in [\omega]^{\aleph_0}\}$. The ordering is defined by $q \leq p$ if $t_p \subset t_q$ and t_q is the union of t_p with some sets on the sequence a_p, and all sets on a_q are unions of some sets on a_p. Let I, the *Matet ideal*, be the collection of all analytic subsets of $[\omega]^{\aleph_0}$ containing no Matet cube as a subset. We have:

Theorem 8.38 *(i) The collection I is a σ-ideal and the map $p \mapsto [p]$ is an isomorphism of P with a dense subset of P_I.*

(ii) I is $\mathbf{\Pi}^1_1$ on $\mathbf{\Sigma}^1_1$.

(iii) I has the continuous reading of names: for every Borel function $f : [\omega]^{\aleph_0} \to \omega^\omega$ there is a pure Matet cube on which f is continuous.

(iv) I fails to have the selection property.

(v) (Blass 1989) *The forcing P is proper, preserves Baire category and outer Lebesgue measure, and adds no independent reals. It adds an unbounded real.*

Before turning to the proofs, we will introduce a standard fusion-type terminology for the forcing P. $q \leq p$ is a *direct extension* of p if $t_q = t_p$. If $a \in (\omega)^\omega$ then a *fusion sequence* below a is a sequence $\langle u_i, b_i : i \in \omega \rangle$ such that $u_0 = 0$, $b_0 = a$, and for every $i \in \omega$, u_i is a finite union of sets on the sequence b_i and $b_{i+1} \leq b_i$ is a member of $(\omega)^\omega$ with $\min(b_{i+1}(0)) > \max(u_{i+1})$. If $a \in (\omega)^\omega$ is a sequence and $m \in \omega$ then $a \setminus m$ is the sequence in $(\omega)^\omega$ obtained by erasing the first m entries on the sequence a.

Proof of Theorem 8.38 The proof starts with the identification of basic forcing properties of Matet forcing P.

Claim 8.39 *Let $O \subset P$ be an open dense set and $p \in P$ a condition. There is a direct extension $q \leq p$ such that for every $n \in \omega$, the condition $\langle t_q \cup a_q(n), a_q \setminus n\rangle$ belongs to the set O.*

Proof Set up a simple fusion process to find a sequence $b \leq a$ such that for every finite set $u \subset \omega$, either $\langle t_p \cup \bigcup_{n \in u} a(n), b \setminus \max(u)\rangle \in O$ or else none of the direct extensions of this condition belong to O. Applying Fact 8.25 to the partition $\pi : (\omega)^\omega \to 2$ given by $\pi(d) = 0$ if the former case holds for $d(0)$, we get a homogeneous sequence $c \subset b$. The homogeneous color cannot be 1 since the condition $\langle t_p, c\rangle$ has an extension in the open dense set O. If c is homogeneous in color 0 then it confirms the statement of the claim with $q = \langle t_p, c\rangle$. □

For the proof of properness of the poset P, let M be a countable elementary submodel of a large enough structure, let $p \in P \cap M$ be a condition. We must produce a master condition $q \leq p$. Enumerate the open dense sets of P in the model M by $\langle O_n : n \in \omega\rangle$ and use Claim 8.39 and the elementarity of the model M repeatedly to build a sequence $p = p_0, p_1, p_2, \ldots$ of direct extensions of p in the model M such that, for every $n \in \omega$, $a_{p_n} \restriction n = a_{p_{n+1}} \restriction n$ and for every finite set $u \subset \omega$ with $\max(u) > n$, it is the case that the condition $\langle t_p \cup \bigcup_{m \in u} a_{p_{n+1}}(m), a_{p_{n+1}} \setminus \max(u)\rangle \in M$ belongs to the open dense set O_n. The limit $q \leq p$ of the sequence of conditions $\langle p_n : n \in \omega\rangle$ will be the required master condition.

In order to show that I is a σ-ideal, suppose that $\{B_n : n \in \omega\}$ are Borel sets in the collection I and suppose for contradiction that there is a condition $p \in P$ such that $[p] \subset \bigcup_n B_n$. Clearly $p \Vdash \dot{x}_{\text{gen}} \in \bigcup_n B_n$ and so, strengthening p if necessary, we may find a single natural number $n \in \omega$ such that $p \Vdash \dot{x}_{\text{gen}} \in \dot{B}_n$. Let M be a countable elementary submodel of a large enough structure, and let $q \leq p$ be an M-generic condition described in the previous proof. It is immediate that for every point $x \in [q]$ the conditions $\langle t_q \cup (q \cap \bigcup_{i \in n} a_q(i)), a_q \setminus n\rangle : n \in \omega\rangle$ generate an M-generic filter $g_x \subset P \cap M$. By the forcing theorem applied in the model M, $M[x] \models x \in B_n$, and by Borel absoluteness, $x \in B_n$. Thus $[q] \subset B_n$, contradicting the assumption that $B_n \in I$.

In order to show that $p \mapsto [p]$ is an isomorphism between P and a dense subset of P_I, it is necessary to show that if $p, q \in P$ are conditions such that $[q] \subset [p]$ then in fact $q \leq p$. To simplify the discussion a bit, assume that $t_p = t_q = 0$. Suppose that $q \leq p$ fails, and find a number $n \in \omega$ such that $a_q(n)$ is not a union of sets on the sequence a_p. In such a case, either there is an element $i \in a_q(n)$ such that $i \notin \bigcup \text{rng}(a)$, in which case no set $x \in [q]$ containing i can be in $[p]$; or else there is a number m such that $a_p(m) \cap a_q(n) \neq 0$ and $a_p(m) \setminus a_q(n) \neq 0$. In the latter case find a point

$x \in [q]$ containing $a_q(n)$ as a subset and disjoint from the set $a_p(m) \setminus a_q(n)$; such a point cannot belong to $p]$ either. In both cases we have a contradiction with the assumption that $[q] \subset [p]$!

To show that the ideal I is $\mathbf{\Pi}^1_1$ on $\mathbf{\Sigma}^1_1$, let $A \subset 2^\omega \times [\omega]^{\aleph_0}$ be an analytic set, a projection of a closed set $C \subset 2^\omega \times [\omega]^{\aleph_0} \times \omega^\omega$. We must show that the set $\{y \in 2^\omega : A_y \in I\}$ is coanalytic. Fix a point $y \in 2^\omega$ and suppose that $A_y \notin I$. This means that there is a condition $p \in P$ such that $[p] \subset A_y$, by the Shoenfield absoluteness this will be true in the P-forcing extension as well, and so there must be a P-name τ for a point in ω^ω such that $p \Vdash \langle \check{y}, \dot{x}_{\text{gen}}, \tau \rangle \in \dot{C}$. By a fusion argument, there is a condition $r \leq p$ and a function $g : [\omega]^{<\aleph_0} \to \omega^{<\omega}$ such that the following holds: for every $b \in [\omega]^{<\aleph_0}$, write $r_b \leq r$ to be that condition whose trunk consists of t_r and the union of i-th elements of the sequence a_r for all $i \in b$, and $a_{r_b} = a_r \setminus \max(b) + 1$; we will have that for every $b \in [\omega]^{<\aleph_0}$, $\text{dom}(g(b)) = \max(b) + 1$ and $r_b \Vdash g(b) \subset \tau$. In particular, the following formula $\phi(y, r, g)$ holds: "the function g respects inclusion, $\text{dom}(g(b)) = \max(b) + 1$ and the basic open neighborhood of the space $2^\omega \times [\omega]^{\aleph_0} \times \omega^\omega$ given by the first $\max(b)$ elements of y in the first coordinate, the trunk t_{r_b} in the second coordinate, and the sequence $g(b)$ in the third coordinate has nonempty intersection with the closed set C." It is also clear that $\phi(y, r, g)$ implies that $[r] \subset A_y$ since the function g yields the requisite points in the set C witnessing that elements of the set $[r]$ belong to A_y. We conclude that $A_y \notin I$ if and only if there is $r \in P$ and a function $g : [\omega]^{<\aleph_0} \to \omega^{<\omega}$ such that $\phi(y, r, g)$ holds, and this is an analytic statement.

For the continuous reading of names, let $p \in P$ and let $f : [p] \to \omega^\omega$ be a Borel function. Use Fact 8.36 to find a direct extension $q \leq p$ such that for every number $n \in \omega$ and every set $u \subset n + 1$ with $n \in u$ there is a sequence $s \in \omega^n$ such that for every $x \in [t_p \cup \bigcup_{i \in u} a_q(i), a_q \setminus n+1]$ we have $s \subset f(x)$. Clearly, $f \restriction [q]$ is a continuous function.

For the preservation of Baire category, there is a simple fusion-type proof. A more elegant argument will show that the ideal I is the intersection of a collection of meager ideals corresponding to Polish topologies on the Polish space $[\omega]^{\aleph_0}$ that yield the same Borel structure, and use the Kuratowski–Ulam theorem as in Zapletal (2008, corollary 3.5.5). This is the same as to show that for every condition $p \in P$ there is a Polish topology on $[p]$ such that all sets in the ideal I are meager with respect to it.

We will deal with the case of the largest condition $p \in P$; there, $t_p = 0$ and $a_p(n) = \{n\}$, so $[p] = [\omega]^{\aleph_0}$. We will show that every Borel nonmeager set in the usual topology on the space $[\omega]^{\aleph_0}$ is I-positive. Suppose that $B \subset [\omega]^{\aleph_0}$ is such a nonmeager set, containing the intersection $O \cap \bigcap_{n \in \omega} O_n$, where O is a nonempty basic open set and $\{O_n : n \in \omega\}$ are open dense sets. Let $r_0 \in 2^{<\omega}$

be a finite sequence such that whenever the characteristic function of a point $x \in [\omega]^{\aleph_0}$ contains r_0 then $x \in O$. By a simple induction on $n \in \omega$ build finite sequences $r_0 \subset r_1 \subset r_2 \subset \cdots$ such that for every point $x \in [\omega]^{\aleph_0}$ whose characteristic function contains $r_{n+1} \setminus r_n$, $x \in \bigcap_{m \in n} O_m$ holds. In the end, let $t = \{i \in \mathrm{dom}(r_0) : r_0(i) = 1\}$ and let $a \in (\omega)^\omega$ be the sequence defined by $a(n) = \{i \in \mathrm{dom}(r_{n+1} \setminus r_n) : r_{n+1}(i) = 1\}$, and observe that $[t, a] \subset B$ and so $B \notin I$ as required.

For the preservation of outer Lebesgue measure, suppose that $A \subset 2^\omega$ is a set of outer Lebesgue mass $> \varepsilon + \delta$ and $p \in P$ is a condition forcing $\dot{O}$ to be a subset of 2^ω of mass $< \varepsilon$. We must find a condition $q \leq p$ and a point $z \in A$ such that $q \Vdash \check{z} \notin \dot{O}$.

Find positive real numbers $\delta_u : u \in [\omega]^{<\aleph_0}$ that sum up to less than δ, and find names $\dot{O}_j : j \in \omega$ for basic open sets whose union $\dot{O}$ is. Using Claim 8.39 and a simple fusion process, find a direct extension $q \leq p$ such that for every nonempty set $u \in [\omega]^{<\aleph_0}$, there is an open set P_u such that the condition $t_p \cup \bigcup_{i \in u} a_q(i) \Vdash \bigcup_{j \in \max(u)} \dot{O}_j \subset P_u \subset \dot{O}$ and the mass of $\dot{O} \setminus P_u$ is less than δ_u. Clearly, the set $R_0 = \bigcup\{P_u \setminus P_{u \cap \max(u)} : u \in [\omega]^{<\aleph_0}, |u| > 1\}$ has mass at most δ, and the set $R_1 = \limsup\{P_{\{n\}} : n \in \omega\}$ has size at most ε. Find a point $z \in A \setminus (R_0 \cup R_1)$, thin out the sequence q if necessary to make sure that $z \notin P_{\{n\}}$ for any $n \in \omega$, and conclude that $q \Vdash \check{z} \notin \dot{O}$ as desired.

Not adding independent reals is always a somewhat more delicate issue. Suppose that $p \in P$ is a condition and $\dot{y}$ a name for an infinite binary sequence; for simplicity assume that $t_p = 0$. Passing to a direct extension of p if necessary, we may assume that $p \Vdash \dot{y} = \dot{f}(\dot{x}_{\mathrm{gen}})$ for some continuous function $f : [p] \to 2^\omega$. Define a partition $\pi : (a_p)^\omega \to 2$ by setting $\pi(c) = 0$ if $f(\bigcup_{n \neq 1} c(n))(\min(c(1)) = 0$. This is clearly a clopen partition and therefore there is a homogeneous sequence $d \in (\omega)^\omega$ by Fact 8.25. Let us say that the homogeneous color is 0. Consider the condition $q \leq p$ given by $t_q = d(0)$, and $a_q(j) = d(2j + 2)$. It is not difficult to use the homogeneity of the sequence d to show that the condition q forces $\dot{y} \restriction \{\min(b_{\min(d(2j+1))}) : j \in \omega\} = 0$.

Finally, P certainly adds an unbounded real: the increasing enumeration of the generic subset of ω cannot be bounded by any ground model function. The failure of the selection property of the ideal I is also simple. Let $C \subset \mathcal{P}(\omega)$ be a perfect set consisting of pairwise almost disjoint sets, and consider the set $D \subset C \times [\omega]^{\aleph_0}$ consisting of all pairs $\langle c, x\rangle$ such that $c \in C$ and $x \subset c$. It is clear that D is a Borel set with I-positive vertical sections. If $B \subset [\omega]^{\aleph_0}$ is a Borel set covered by the vertical sections of the set D, visiting each vertical section in at most one point, it must be the case that the elements of B are pairwise almost disjoint, therefore B cannot contain a subset of the form $[p]$ for any $p \in P$, and so $B \in I$. □

Proof of Theorem 8.37(i) Recall that E_{K_σ} is the equivalence relation whose domain consists of all functions in ω^ω below the identity equipped with the distance $\delta(x, y) = \sup_n |x(n) - y(n)|$; E_{K_σ} then connects the points x, y if $\delta(x, y)$ is finite. Let $a \in (\omega)^\omega$ be a sequence, and let E be an equivalence relation reducible to E_{K_σ} by a Borel function $f : [a] \to \text{dom}(E_{K_\sigma})$. Thinning out the sequence a we may assume that the function f is in fact continuous. Consider the partition $(a)^\omega = B \cup C$ placing a sequence $b \leq a$ into the set B just in case that for every i there are points $x, y \in [b(0), b \setminus 1]$ such that $\delta(f(x), f(y)) \geq i$. Using the Milliken theorem 8.25 we may thin out the sequence a so that $(a)^\omega \subset B$ or $(a)^\omega \subset C$. There is an appropriate split into cases:

Case 1 Assume first that $(a)^\omega \subset B$. In this case, we will first thin out a to c that $E \restriction [c] \subseteq E_0$, and then proceed with another partition argument. To perform the thinning, by induction on $i \in \omega$ build a fusion sequence $\langle u_i, c_i : i \in \omega \rangle$ so that whenever $j \in \omega$ and $t, s \subset j$ are two sets, then for every $x \in [\bigcup_{i \in s} u_i \cup u_j, c_{j+1}]$ and every $y \in [\bigcup_{i \in t} u_i, c_{j+1}]$ it is the case that $\delta(f(x), f(y)) \geq j$. It is clear that the limit c of the fusion sequence will satisfy $E \restriction [c] \subseteq E_0$.

Suppose that $u_i : i \in j$ and c_j have been constructed. Enumerate the set $\mathcal{P}(j) \times \mathcal{P}(j)$ by $\langle \langle s_k, t_k \rangle : k < 2^{2j} \rangle$, and by induction on k build finite sets u_j^k and Milliken cubes c_j^k so that:

I1. $u_j^0 = 0$ and the set u_j^{k+1} is a union of u_j^k with some sets on the sequence c_j^k;
I2. $c_j = c_j^0 \geq c_j^1 \geq \cdots$;
I3. for every k there are numbers m_k, l_k^s, l_k^t so that for every $x \in [\bigcup_{i \in s_k} u_i \cup u_j^{k+1}, c_j^{k+1}]$ and every $y \in [\bigcup_{i \in t_k} u_i, c_j^{k+1}]$ it is the case that $f(x)(m_k) = l_k^s$, $f(y)(m_k) = l_k^t$, and $|l_k^s - l_k^t| > j$.

This is not difficult to do using the case assumption. Suppose that c_j^k, u_j^k have been found. The case assumption yields $x_0, x_1 \in [\bigcup_{i \in s_k} u_i \cup u_j^{k+1}, c_j^k]$ such that $d(f(x_0), f(x_1)) > 2j$. Find m_k such that $|f(x_0)(m_k) - f(x_1)(m_k)| > j$. Find a sequence $c_j^{k+1} \leq c_j^k$ such that for every $y \in [\bigcup_{i \in t} u_i, c_j^k]$ the value $f(y)(m_k)$ is the same, equal to some r_k. Then either $f(x_0)(m_k)$ or $f(x_1)(m_k)$ is farther than j from $f(y)(m_k)$, say that it is x_0, and put $l_k = f(x_0)(m_k)$. Use the continuity of the function f to find a set u_j^{k+1} such that $x_0 \in [\bigcup_{i \in s} u_i \cup$

$u^k_{j+1}, c^k_j]$ and for all points x in this dented cube it is the case that $f(x)(m_k) = l_k$. This concludes the induction step.

After the induction on k has been performed, let u_j, c_j be the last set and the last sequence arrived at. This completes the construction of the fusion sequence, and we now have $E \restriction (c)^\omega \subset E_0$, where c is the limit of the fusion sequence. Let $(c)^\omega = B' \cup C'$ where $b \in B'$ if $\bigcup_{i\in\omega} b(i)$ is equivalent to $\bigcup_{i\in\omega, i\neq 1} b(i)$. The Milliken theorem 8.25 can again be used to thin out c so that either $(c)^\omega \subset B'$ or $(c)^\omega \subset C'$. There are the corresponding two subcases:

Case 1a Suppose that $(c)^\omega \subset B'$. In this case, first observe that $E \restriction [c]$ connects every pair of E_0-equivalent points with the same minimum. To verify the connection, use the case assumption to argue that for every infinite set $e \subset \omega$ and every $j > \min(e)$ it is the case that $\bigcup_{i\in e} c(i)$ is E-related to $\bigcup_{i\in e\cup\{j\}} c(i)$. Then, whenever $x, y \in [c]$ are E_0-related points with the same minimum, it is possible to transform one into another in finitely many steps described in the previous sentence, and the transitivity of the relation E shows that x and y are E-related.

Consider the partition of $(c)^\omega$ into two Borel pieces B'' and C'', putting $b \in B''$ if $\bigcup_{i\in\omega} b(i)$ is E-related to $\bigcup_{i\in\omega, i\neq 0} b(i)$. The Milliken theorem 8.25 again provides a split into cases:

Case 1aa Either there is $d \leq c$ such that $(d)^\omega \subset B''$. In such a case, $E \restriction [d] = E_0$: whenever $x, y \in [d]$ are E_0-related, if their minimums are the same then they are E-related by the preceding paragraph. If their minimums are different, say $\min(x) < \min(y)$, then there is a sequence $b \leq d$ such that $\min(b_0) = \min(x)$, $\min(b_1) = \min(y)$, and $x \; E_0 \; \bigcup_i b(i)$. The homogeneity assumption applied to b shows that $x \; E \; \bigcup_{i\in\omega} b(i) \; E \; \bigcup_{i\in\omega, i\neq 0} b(i) \; E \; y$ as desired.

Case 1ab Or, there is $d \leq c$ such that $(d)^\omega \subset C''$. In such a case, $E \restriction [d] = E_0 \cap E_{\min}$ by a similar homogeneity argument as in Case 1aa.

Case 1b Suppose now that $(c)^\omega \subset C'$. In this case, we will thin out c to d such that $E \restriction [d] = \text{id}$. By induction on $i \in \omega$ build a fusion sequence $\langle v_i, d_i : i \in \omega \rangle$ so that whenever $j \in \omega$ and $s, t \subset j$ are two distinct nonempty sets then for every $x \in [\bigcup_{i\in s} v_i, d_j]$ and every $y \in [\bigcup_{i\in t} v_i, d_j]$ such that $x \setminus \min(d_j) = y \setminus \min(d_j)$ it is the case that $x \; E \; y$ fails. The fusion d of this fusion sequence will certainly have the required property that $E \restriction [d] = \text{id}$: if $x, y \in [d]$ are not E_0-related then they are not E-related by the initial work in Case 1, and if they are E_0-related and distinct then they are not E-related by the assumption on the fusion sequence past the last difference of the two points x, y.

Suppose that $\{v_i : i \in j\}$ and d_j have been constructed. For every pair s, t of distinct nonempty subsets of j define the partition of $(d_j)^\omega$ into two Borel pieces $B_{s,t}$ and $C_{s,t}$ by placing $b \leq d_j$ into $B_{s,t}$ if $\bigcup_{i \in s} v_i \cup \bigcup_{i \in \omega} b(i)$ is E-related to $\bigcup_{i \in t} v_i \cup \bigcup_{i \in \omega, i \neq 0} b(i)$. Observe that there can be no sequence $e \leq d_j$ such that $(e)^\omega \subset B_{s,t}$: this would mean that $\bigcup_{i \in s} v_i \cup \bigcup_{i \in \omega} e(i)$ is equivalent to both $\bigcup_{i \in t} v_i \cup \bigcup_{i \in \omega, i \neq 0} b(i)$ and $\bigcup_{i \in t} v_i \cup \bigcup_{i \in \omega, i \neq 0,1} b(i)$, but the equivalence between the last two points would contradict the fact that $(c)^\omega \subset C'$. Therefore, the Milliken theorem 8.25 applied repeatedly yields a sequence $e \leq d_j$ such that $(e)^\omega \subset C_{s,t}$ for all pairs of distinct nonempty sets $s, t \subset j$. Let $v_j = e(0)$ and $d_{j+1} = e \restriction (0, \omega)$. This completes the induction step and the proof of Case 1b.

Case 2 Suppose now that $(a)^\omega \subset C$. In this case, first observe that every two points in $[a]$ with the same minimum are E-related: if $\min(x) = \min(y)$ then the homogeneity assumption applied to the sequence $a \setminus \min(x)$ shows that the distance $d(f(x), f(y))$ must be finite. Now let F be the equivalence relation on ω relating i and j if for every $x, y \in [c]$ such that $\min(x) = \min(c(i))$ and $\min(y) = \min(c(j))$, $x \; E \; y$ holds. There are two possibilities:

Case 2a Either F has an infinite class, say $e \subset \omega$. Let $d \leq c$ be the sequence consisting only of those entries of c indexed by elements of the set $e \subset \omega$. It is clear that $E \restriction [d] = \mathrm{ev}$.

Case 2b Or, F has an infinite set consisting of pairwise inequivalent elements, say $e \subset \omega$. Again, let $d \leq c$ be the sequence consisting only of those entries of c indexed by elements of the set $e \subset \omega$. It is clear that $E \restriction [d] = E_{\min}$.

This concludes the case of equivalence relations reducible to E_{K_σ}. □

Proof of Theorem 8.37(ii) Observe that the canonization of smooth equivalence relations proved in (*i*) shows that the Matet forcing P adds a minimal real: for smooth equivalence relation it is impossible to canonize to the non-smooth E_0, and therefore it will be equal to either id or ev on a Matet cube. Thus, we will be able to use Theorem 4.9 to simplify any equivalence relation classifiable by countable structures to an essentially countable one on a Matet cube, and then use the canonization of equivalence relations reducible to E_{K_σ} to canonize the essentially countable relation further. The only obstacle is that we want to end up with a pure Matet cube as opposed to just a Matet cube.

Suppose that E is an equivalence relation on $[\omega]^{\aleph_0}$ classifiable by countable structures. Using Theorem 4.9 repeatedly, a simple fusion process yields a pure Matet cube $[0, a]$ for some $a \in (\omega)^\omega$ such that for every $n \in \omega$, the equivalence relation $E \restriction [a(n), a \setminus n+1]$ is essentially countable, and the Borel function reducing it to a Borel countable equivalence relation is either one-to-one or constant. In other words, the equivalence relation $E \restriction [a(n), a \setminus n+1)$ is Borel, and it either has countable classes or it is equal to ev. There are two cases:

Case 1 The set $u = \{n \in \omega : E \restriction [a(n), a \setminus n+1] = \text{ev}\}$ is infinite. Then $E \restriction [0, \langle a(n) : n \in u\rangle]$ has countably many classes, and we can employ Theorem 8.37(i) to canonize it on a pure subcube to either ev or $E_{\min}$.

Case 2 The set $v = \{n \in \omega : E \restriction [a(n), a \setminus n+1] \neq \text{ev}\}$ is infinite. Then $E \restriction [0, \langle a(n) : n \in v\rangle]$ is a Borel countable equivalence relation, therefore Borel reducible to E_{K_σ}, and again Theorem 8.37(i) can be used to canonize it on a pure subcube to either id, E_0 or to $E_0 \cap E_{\min}$. □

Proof of Theorem 8.37(iii) Now assume that $a \in (\omega)^\omega$ and E is an equivalence relation on $[a]$ reducible to $=_J$ for some analytic P-ideal J on ω. Use the theorem of Solecki (1999) to find a submeasure lower semicontinuous submeasure ϕ on ω such that $J = \{e \subset \omega : \lim_i \phi(e \setminus i) = 0\}$. Let $f : [a] \to \mathcal{P}(\omega)$ be a Borel function reducing E to $=_J$; thinning out the sequence a we may assume that the function f is continuous. For every sequence $b \leq a$ and every $j \in \omega$ let $\varepsilon(j, b)$ be the supremum of all numbers of the form $\phi(f(x)\Delta\phi(y))$ as x, y vary over all points in $[\bigcup_{i\in j} b(i), b \setminus j]$. Note that for a fixed sequence b, the numbers $\varepsilon(j, b)$ cannot increase as j increases, and so they converge to some real number $\varepsilon(b) \geq 0$. Define the partition $(a)^\omega$ into two Borel pieces B and C by placing $b \in B$ if $\varepsilon(b) = 0$. The Milliken theorem 8.25 gives us the first split into subcases:

Case 1 It is possible to thin out the sequence a so that $(a)^\omega \subset B$. In this case, we will find $c \leq a$ so that $E \restriction [c]$ becomes a relatively closed equivalence relation. Closed equivalence relations are smooth by Gao (2009, proposition 5.4.7), and so a reference to the previously treated case of equivalence relations reducible to E_{K_σ} will complete the argument here. To find c, by induction on $i \in \omega$ find nonempty finite sets u_i such that $\min(u_{i+1}) > \max(u_i)$, each u_i is a union of finitely many entries on a, and for every $j \in \omega$, every set $s \subset j$ and every $x, y \in [a]$ such that the set $\bigcup_{i\in s} u_i \cup u_j$ is an initial segment of both x and y it is the case that $\phi(f(x)\Delta f(y)) < 2^{-j}$. This is easily possible using the case assumption repeatedly. Let $c = \langle u_i : i \in \omega\rangle$. To see that the complement of E is relatively open in $[c]$, let $x, y \in [c]$ be E-unrelated points, and $j \in \omega$ be so large that $\lim_i \phi((f(x)\Delta f(y)) \setminus i) > 2^{-j}$. Let s, t be initial segments of x, y such that for some numbers $k_s, k_t > j$ it is the case that $u_{k_s} \subset s$ and $u_{k_t} \subset t$. We claim that the relatively open sets $[s]$ and $[t]$ inside $[c]$ are E-unrelated. Well, if $x' \in [s] \cap [c]$ and $y' \in [t] \cap [c]$, then $\phi(f(x)\Delta f(x')) < 2^{k_s} < 2^{j+1}$ and $\phi(f(y)\Delta f(y')) < 2^{k_t} < 2^{j+1}$. Since $\lim_i \phi((f(x)\Delta f(y))\setminus i) > 2^{-j}$, this means that $\lim_i \phi((f(x')\Delta f(y'))\setminus i) > 0$ and $\neg x' \; E \; y'$ as required.

Case 2 It is possible to thin out the sequence a so that $(a)^\omega \subset C$. In this case, we will thin out a further so that $E \restriction [a] \subset E_0$. The treatment will then hook up into Case 1 of the equivalence relations reducible to E_{K_σ}. To perform the thinning, by induction on $i \in \omega$ build a fusion sequence $\langle u_i, c_i : i \in \omega\rangle$ so that whenever $j \in \omega$ and $t, s \subset j+1$ are two sets then for every $x \in [\bigcup_{i \in s} u_i \cup u_j, c_{j+1}]$ and every $y \in [\bigcup_{i \in t} u_i, c_{j+1}]$ it is the case that $\phi((f(x)\Delta f(y)) \setminus j) > \frac{1}{2}\varepsilon(\langle u_i : i \in t\rangle^\frown c_j)$. Once such a construction is performed, writing $c \in (\omega)^\omega$ for its limit, it will satisfy $E \restriction [c] \subset E_0$. To verify the inclusion, let $x, y \in [c]$ be E_0-unrelated points. One of them, say x, must have infinitely many elements that do not occur in the other. Let $d \leq c$ be the sequence enumerating all j such that $u_j \subset x$ and note that $\varepsilon(d) > 0$ by the homogeneity assumption. Now there must be infinitely many $j \in \omega$ such that $u_j \subset x$ and $u_j \cap y = 0$, and for each such $j \in \omega$, writing $s = \{i < j : u_i \subset x\}$ and $t = \{i < j : u_i \subset y\}$, the construction yields $\phi(f(x)\Delta f(y)) \setminus j) > \varepsilon(d)$. Therefore $f(x)\Delta f(y)$ is not in the ideal J and x, y are not E-related.

Suppose that $\langle u_i : i \in j\rangle$ and c_j have been constructed. Thin out c_j so that for every set $s \subset j$, for every $x \in [\bigcup_{i \in s} u_i, c_j]$ the set $f(x) \cap j$ is the same. Enumerate the set $\mathcal{P}(j) \times \mathcal{P}(j)$ by $\langle\langle s_k, t_k\rangle : k \in 2^{2j}\rangle$, and by induction on k build finite sets u^k_j and Milliken cubes c^k_j so that:

I1. $u^0_j = 0$ and the set u^{k+1}_j is a union of u^k_j with some sets on the sequence c^k_j;
I2. $c_j = c^0_j \geq c^1_j \geq \cdots$;
I3. for every k there is a number m_k and sets $w^s_k, w^t_k \subset m_k$ so that for every $x \in [\bigcup_{i \in s_k} u_i \cup u^{k+1}_j, c^{k+1}_j]$ and every $y \in [\bigcup_{i \in t_k} u_i, c^{k+1}_j]$ it is the case that $f(x) \cap m_k = w^s_k$, $f(y) \cap v_k = w^s_k$, and $w^s_k \Delta w^t_k > \frac{1}{2}\varepsilon(\langle u_i : i \in t\rangle^\frown c^k_j)$.

This is not difficult to do using the case assumption. Suppose that c^k_j, u^k_j have been found. Let $d = \langle u_i : i \in s\rangle^\frown \langle u^k_j\rangle^\frown c^k_j$. The definition of $\varepsilon(d)$ yields points $x_0, x_1 \in [\bigcup_{i \in s_k \cup u^{k+1}_j, c^k_j}]$ such that $\phi(f(x_0)\Delta f(x_1)) > \frac{1}{2}\varepsilon(d)$. Note that the set $f(x_0)\Delta f(x_1)$ contains no elements smaller than j. Let $m_k \in \omega$ be a number such that $\phi((f(x_0)\Delta f(x_1)) \cap m_k) > \frac{1}{2}\varepsilon(d)$. Find a sequence $c^{k+1}_j \leq c^k_j$ such that for every $y \in [\bigcup_{i \in t_k} u_i, c^k_j]$ the set $f(y) \cap m_k$ is the same, equal to some W^t_k. Then either $f(x_0)(m_k)$ or $f(x_1)(m_k)$ is farther than j from $f(y)(m_k)$, say that it is x_0, and put $w^s_k = f(x_0) \cap m_k$. Use the continuity of the function f to find a set u^{k+1}_j such that $x_0 \in [\bigcup_{i \in s} u_i \cup u^{j+1}_i, c^k_j]$ and for all points x in this dented cube it is the case that $f(x) \cap m_k = w^s_k$. This concludes the induction step.

After the induction on k has been performed, let u_j, c_j be the last set and the last sequence arrived at. This completes the construction of the fusion sequence and the proof. □

We now show that E_0 is indeed an unavoidable element of the canonization basis, and that it belongs to the spectrum of the Matet ideal.

Theorem 8.40 *E_0 is in the spectrum of the σ-ideal I. The poset $P^{E_0}_I$ is regularly embedded in P_I, it is $\aleph_0$-distributive, and it yields the model $V[\dot{x}_{\text{gen}}]_{E_0}$.*

Proof This is an immediate corollary of Theorem 4.4 and the simple observation that E_0-saturations of I-small sets are still I-small. This observation in turn follows from the fact that a rational shift of a Matet cube is again a Matet cube. For the $\aleph_0$-distributivity note that the poset P_I has the continuous reading of names, E_0 classes are dense in every Matet cube, and apply Proposition 3.23. □

As a final remark in this section, canonization of objects more complicated than equivalence relations must fail badly in the case of Matet cubes due to the following simple proposition:

Proposition 8.41 *The σ-ideal I has a square coding function.*

Proof Let $x, y \in X$ be two points, not equal modulo finite. Consider all sets $u \subset x \cap y$ which are an interval inside x and inside y, and are maximal with respect to this property; list them in an increasing order as $\langle u_n : n \in \omega \rangle$. Define $f(x, y) \in 2^\omega$ by setting $f(x, y)(n) = 1$ if $\max((x \cup y) \setminus \min(u_n)) \in x$ and $f(x, y)(n) = 0$ otherwise.

Suppose now that $B \subset X$ is an I-positive set; it contains a pure Matet cube of the form $[t, a]$ for some nonempty finite set $t \subset \omega$ and $a \in (\omega)^\omega$. Let $z \in 2^\omega$ be a prescribed value. Let x be the union of t the sets $a(2n + 1)$ for all $n \in \omega$ and the sets $a(2n)$ for all n with $z(2n) = 1$. Let y be the union of t the sets $a(2n + 1)$ for all $n \in \omega$ and the sets $a(2n)$ for all n with $z(2n) = 0$. It is immediate that $f(x, y) = z$. □

Question 8.42 Does Theorem 8.37 remain in force for the class of all analytic equivalence relations?

9

Product-type ideals

9.1 Finite products and the covering property

Products of ideals are a demanding field of study, laden with Ramsey-type considerations. In this section, we analyze a class of σ-ideals fairly frequent in practice, for which the product ideals possess a great degree of regularity as well as strong canonization properties. It is the class of σ-ideals σ-generated by coanalytic collection of compact sets, with the covering property:

Definition 9.1 A σ-ideal I has the *covering property* if every I-positive analytic set has a compact I-positive subset.

Since the σ-ideals σ-generated by closed sets have the continuous reading of names, and their quotient forcings are proper by Fact 6.1, in the context of this class of σ-ideals the covering property is equivalent to the bounding property of the quotient forcing by Fact 2.52.

Before the statement of the canonization theorem, recall the definition of the *box product ideal* from Fact 2.56. Let $\{X_i : i \in n\}$ be Polish spaces with respective σ-ideals I_i on each. The box product $\prod_{i\in n} I_n$ is the collection of all Borel sets $B \subset \prod_{i\in n} X_i$ containing no subset of the form $\prod_{i\in n} B_n$ where for each $i \in n$, $B_i \subset X_i$ is Borel and I_i-positive. Fact 2.56 has the following immediate corollary:

Corollary 9.2 *If the σ-ideals in the collection $\{I_i : i \in n\}$ are σ-generated by a coanalytic collection of closed sets and have the covering property, then $\prod_{i\in n} I_i$ is a σ-ideal.*

Note that the ideals in the assumption of the corollary are $\mathbf{\Pi}^1_1$ on $\mathbf{\Sigma}^1_1$ and their quotient forcings are proper and preserve Baire category by Fact 6.1. They also have the continuous reading of names, so the covering property implies the bounding property and the assumptions of Fact 2.56 are satisfied.

Theorem 9.3 *Let $n \in \omega$ and $\{I_i : i \in n\}$ be σ-ideals on the respective Polish spaces $\{X_i : i \in n\}$ σ-generated by a coanalytic collection of closed sets, with the covering property. For every sequence $\langle B_i : i \in n\rangle$ of respectively I_i-positive Borel sets and every analytic equivalence E on $\prod_i B_i$ there are Borel I_i-positive sets $C_i \subset B_i$ such that $E \restriction \prod_i C_i = \prod_i F_i$ where each F_i is either* id *or* ev. *Restated,*

$$\text{analytic} \to_{\prod_i I_i} \{\prod_i F_i : \forall i\ F_i = \text{id} \vee F_i = \text{ev}\}.$$

Proof Let P be the product forcing $\prod_{i\in n} P_{I_i}$. Let $a \subset n$ be a set, and let E_a be the equivalence on $\prod_{i\in n} X_i$ connecting two sequences if they have the same entries on the set a. The first step is to define the *reduced product* associated with the equivalence relation E_a. Let P_a be the product of P_{I_i} for $i \in a$ with $P_{I_i} \times P_{I_i}$ for $i \notin a$. A condition $p \in P_a$ will be identified with a sequence $\langle C_j : j \in n \times 2\rangle$ where $C_{i,0}, C_{i,1} \in P_{I_i}$ for all $i \in n$ and $C_{i,0} = C_{i,1}$ for $i \in a$. We will organize the P_a-generic sequence into two, $\vec{x}_{\text{lgen}} \in (2^\omega)^n$ and $\vec{x}_{\text{rgen}} \in (2^\omega)^n$ which use the P_{I_i}-generic point for $i \in a$, and the left (in the case of $\vec{x}_{\text{lgen}}$) or right (in the case of $\vec{x}_{\text{rgen}}$) P_{I_i}-generic point for $i \notin a$; thus, $\vec{x}_{\text{lgen}}\ E_a\ \vec{x}_{\text{rgen}}$.

Lemma 9.4 *Let M be a countable elementary submodel of a large structure, and let $\{B_i : i \in n\}$ be I_i-positive Borel sets in the model M. Let $a \subset n$ be a set. There are I_i-positive Borel subsets $\{C_i \subset B_i : i \in n\}$ such that the product $\Pi_{i\in n} C_i$ consists of pairwise reduced product generic sequences for the model M.*

This is to say that any two sequences $\vec{x}, \vec{y} \in \Pi_{i\in n} C_i$ satisfying $x(i) = y(i) \leftrightarrow i \notin a$, are P_a-generic for the model M.

Proof We must first introduce the game-theoretic characterization of the forcing properties of the quotient posets P_I as described in Zapletal (2008, section 3.10.10). For a fixed σ-ideal I on a Polish space, the Boolean game G_I associated with I between Players I and II starts with Player II indicating an initial I-positive Borel set B_{ini}; then the game has ω many megarounds, and each of the megarounds consists of finitely many rounds, in each of which Player I plays an I-positive Borel set and Player II answers with its Borel I-positive subset. At some point of his choice, Player I decides to end the megaround and pass to a next one. A play p which ended with Player I ending its last megaround will be called *regular*. For a regular play p write $[p]$ for the intersection of B_{ini} with all the sets obtained as unions of all Player II's answers in a fixed megaround of p; similar usage will prevail for infinite plays p. Player I wins an infinite play p if he ended infinitely many megarounds and

$[p] \notin I$. In what follows, the letters $p, q, \ldots$ will denote either infinite plays of this game or finite plays at the end of which Player I decided to end the current megaround.

Fact 9.5 (Zapletal 2008, theorem 3.10.24) *Suppose that the σ-ideal I is $\mathbf{\Pi}^1_1$ on $\mathbf{\Sigma}^1_1$ and the quotient poset is proper. The following are equivalent:*

(i) P_I preserves Baire category and is bounding;
(ii) Player I has a winning strategy in the game G.

Since the σ-ideals I_i are $\mathbf{\Pi}^1_1$ on $\mathbf{\Sigma}^1_1$ and their quotient forcings are proper, bounding, and preserve Baire category by Fact 6.1, we conclude that for every $i \in n$ Player I has a winning strategy σ_i in the game G_{I_i}. This provides ammunition for various fusion arguments like the following claim:

Claim 9.6 *Let $\langle p_i : i \in n \rangle$ be finite regular plays according to the respective strategies σ_i, and let D be an open dense subset of $\prod_i P_{I_i}$. Then there are regular extensions p'_i of p_i such that the set $\prod_i [p'_i]$ is covered by finitely many elements of D.*

Proof For simplicity, we will treat the case $n = 2$. First argue that for every I_1-positive Borel set B and every finite regular play p according to the strategy σ_0 there is a Borel I_1-positive set $C \subset B$ and a regular extension q of the play p such that $[q] \times C$ is covered by finitely many elements of the open dense set D. To find the set C, by induction on $j \in \omega$ build sets A_j, A'_j, B_j such that:

I1. A_j, A'_j are Borel I_0-positive sets and the play $p, A_0, A'_0, A_1, A'_1, \ldots$ follows the strategy σ_0;
I2. $B \supset B_0 \supset B_1 \supset \cdots$ are Borel I_1-positive sets;
I3. $\langle A'_j, B_j \rangle \in D$.

The induction is trivial using the density of the set D. At some finite stage k of the construction, the strategy σ_0 will call for an end of the next megaround past p. Let $q = p, A_0, A'_0, A_1, A'_1, \ldots, A_{k-1}, A'_{k-1}$ and $C = B_{k-1}$. Then q, C work as required.

To get the statement of the claim for $n = 2$, fix p_0, p_1 and by induction on $j \in \omega$ build regular plays q_j and sets B_j, C_j so that:

I1. B_j, C_j are Borel I_1-positive set and the play $p_1, B_0, C_0, B_1, C_1, \ldots$ follows the strategy σ_1;
I2. $p_0 \subset q_0 \subset q_1 \subset q_2 \subset \cdots$ are plays according to strategy σ_0;
I3. $[q_j] \times C_j$ is covered by finitely many elements of the set D.

The induction is trivial using the previous paragraph in the induction step. At some finite stage k of the construction, the strategy σ_1 will call for an end of the next megaround past p_1. Let $p'_0 = q_{k-1}$ and $p'_1 = p_1, B_0, C_0, \dots, B_{k-1}, C_{k-1}$. The regular plays p'_0, p'_1 clearly satisfy the demands of the claim. □

Now, write $n \times \omega^{<\omega} = \bigcup_m T_m$ as an increasing union of finite sets T_m such that for every m, i the set $\{t : \langle i, t\rangle \in T_m\}$ is closed under initial segments. By induction on m build functions f_m with the domain T_m such that:

I1. for every $\langle i, t\rangle \in T_m$ the value $f_m(i, t)$ is a finite regular play in the model M respecting the strategy σ_i, and $f_m(i, 0)$ starts with B_i;
I2. for $m < m'$ and $\langle i, t\rangle \in T_m$, the play $f_{m'}(i, t)$ extends $f_m(i, t)$;
I3. if $\langle i, t\rangle \in T_m$ and $\langle i, t^\frown j\rangle \in T_{m+1} \setminus T_m$ and $[f_m(i, t)]$ has nonempty intersection with the j-th open set in some fixed enumeration of a basis for the space X_i, then $[f_{m+1}(i, t^\frown j)]$ is a subset of this intersection;
I4. whenever $a \subset n$ is a set, and $\{\langle i, t_i\rangle : i \in a\} \cup \{\langle i, t_i^0\rangle, \langle i, t_i^1\rangle : i \notin a\}$ are nodes of T_m such that $t_i^0 \neq t_i^1$ for all $i \notin a$, then the product $\prod[f_m(i, t_i)] \times \prod_{i \notin a}[f_m(i, t_i^0)] \times \prod_{i \in a}[f_m(i, t_i^1)]$ is covered by finitely many conditions in the m-th dense subset of P_a in the model M under some fixed enumeration.

The induction process is trivial, using Claim 9.6 repeatedly to assure that I4 holds. In the end, for every $i \in n$ and every $y \in \omega^\omega$ look at the filter generated by the conditions $\{[f_m(i, y \restriction k)] : m$ is the least such that $\langle i, y \restriction k\rangle \in T_m\}$. This is an M-generic filter for P_I and so there is a unique point $g_i(y) \in B_i$ in it. We will first argue that $\operatorname{rng}(g_i)$ is an I_i-positive set for every $i \in n$.

Indeed, let $\{A_k : k \in \omega\}$ be closed sets in the σ-ideal I_i; we must produce $y \in \omega^\omega$ such that $g_i(y) \notin \bigcup_k A_k$. By induction build an increasing sequence of nodes $t_k \in \omega^{<\omega}$ of length k such that $[f_m(i, t_k)] \cap A_k = 0$, where m is the the least such that $\langle i, t_k\rangle \in T_m$. The induction step is arranged by noting that $[\bigcup_m f_m(i, t_k)]$ is an I_i-positive set and therefore has I_i-positive intersection with some basic open set disjoint from A_k (this is the only place where we use the assumption that the generators of I_i are closed). If this basic open set is enumerated by some j then the induction hypothesis at $k+1$ with $t_{k+1} = t_k^\frown j$ will hold. In the end, let $y = \bigcup_k t_k$ and observe that $g_i(y) \notin \bigcup_k A_k$ as desired.

Now use Solecki's result (Solecki 1994) to find a Borel I_i-positive set $C_i \subset \operatorname{rng}(g_i)$ for each $i \in n$. The induction hypothesis I4 above immediately implies that the sets C_i have the required mutual genericity property. □

Towards the proof of the theorem, let E be an analytic equivalence relation on the space $\prod_i B_i$. Let p_0 be the condition in the reduced products given by

$p(i, 0) = p(i, 1) = B_i$ for every $i \in n$. Find an inclusion minimal set $a \subset n$ such that some condition $p \in P_a$ below p_0 forces $\vec{x}_{\text{lgen}}\ E\ \vec{x}_{\text{rgen}}$. Consider the *symmetrization* of the condition p, the condition q defined by $q(i, 0) = q(i, 1) = p(i, 0)$ for all $i \in n$. We claim that q also forces $\vec{x}_{\text{lgen}}\ E\ \vec{x}_{\text{rgen}}$.

To show this, let $\vec{x}_{\text{lgen}}, \vec{x}_{\text{rgen}}$ are P_a-generic sequences meeting the condition q. In a further generic extension one can find a sequence $\vec{y}_{\text{gen}}$ such that the pairs $\vec{x}_{\text{lgen}}, \vec{y}_{\text{gen}}$ and $\vec{x}_{\text{rgen}}, \vec{y}_{\text{gen}}$ are P_a-generic sequences meeting the condition p. These two pairs consist of E-equivalent sequences by the forcing theorem, and consequently $\vec{x}_{\text{lgen}}\ E\ \vec{x}_{\text{rgen}}$ by the transitivity of the relation E.

Now let M be a countable elementary submodel of a large structure and use the claim to find sets $C_i \subset B_i : i \in n$ whose product consists of pairwise P_a-generic sequences for the model M. We claim that $E \restriction \Pi_{i \in n} C_i = E_a$.

To show this, suppose that $\vec{x}, \vec{y} \in \Pi_{i \in n} C_i$ are sequences and let $b = \{i \in n : \vec{x}(i) = \vec{y}(i)\}$. Find sequences $\vec{x}_0, \vec{y}_0$ in the product $\prod_{i \in n} C_i$ such that $a = \{i \in n : \vec{x}(i) = \vec{x}_0(i)\} = \{i \in n : \vec{y}(i) = \vec{y}_0(i)\}$ and $a \cap b = \{i \in n : \vec{x}_0(i) = \vec{y}_0(i)\}$. The forcing theorem applied to the reduced product P_a together with Borel absoluteness implies that $\vec{x}\ E\ \vec{x}_0$ and $\vec{y}\ E\ \vec{y}_0$. Now if $a \cap b = a$ then the forcing theorem also implies that $\vec{x}_0\ E\ \vec{y}_0$ and so $\vec{x}\ E\ \vec{y}$. On the other hand, if $a \cap b \neq a$ then the minimal choice of the set a together with the forcing theorem applied to $P_{a \cap b}$ give $\neg \vec{x}_0\ E\ \vec{y}_0$ and $\neg \vec{x}\ E\ \vec{y}$! □

9.2 Finite products of the E_0 ideal

In this section, we provide a stronger, multidimensional version of the E_0 canonization theorem (Theorem 7.1). Recall the definition of the E_0-ideal I on 2^ω as generated by the Borel sets on which E_0 is smooth. We will first describe the associated σ-ideal. Let $n \in \omega$ and $X_n = (2^\omega)^n$.

Definition 9.7 The σ-ideal I_n on X_n is σ-generated by all analytic sets $A \subset X$ such that there is $i \in n$ such that each section of A in the i-th coordinate consists of pairwise E_0-inequivalent elements of 2^ω.

Fact 9.8 *Let $n \in \omega$. Then:*

(i) *every analytic subset of X_n either is in I or contains a product of Borel sets $\prod_{i \in n} B_i$ such that $B_i \notin I$, and these two options are mutually exclusive;*

(ii) *the quotient poset P_{I_n} is proper, bounding and preserves Baire category;*

(iii) *I_n is a $\mathbf{\Pi}^1_1$ on $\mathbf{\Sigma}^1_1$ σ-ideal consisting of meager sets.*

In particular, the quotient poset P_{I_n} has a dense subset naturally isomorphic to the product of n copies of the poset P_I.

Proof (*i*) was proved in Zapletal (2004, lemma 2.3.58). (*ii*) follows from the fact that P_{I_n} is a product of n many copies of the poset P_I, that poset is proper, bounding, and preserves Baire category by Fact 7.2, and the conjunction of these properties is preserved under products by Fact 2.56. A direct argument for (*ii*) is of course possible as well. The complexity part of (*iii*) follows from Fact 2.56 again, and the rest of (*iii*) follows from the Kuratowski–Ulam theorem and the fact that Borel E_0-independent subsets of 2^ω are meager. □

Theorem 9.9 *Let $n \in \omega$, let $B_i \subset 2^\omega$ for $i \in n$ be Borel sets on which E_0 is not smooth, and let E be an analytic equivalence relation on $\prod_{i\in n} B_i$. There are Borel sets $C_i \subset B_i$ for $i \in n$ on which E_0 is still not smooth, and $E \restriction \prod_{i\in n} C_i$ is equal to $\prod_{i\in n} F_i$ where each F_i is either* id, ev, *or E_0 equivalence relation on 2^ω. In other words,*

$$\text{analytic} \to_{I_n} \{\prod_{i\in n} F_i : \forall i\ F_i = \text{id} \vee F_i = \text{ev} \vee F_i = E_0\}.$$

It will be helpful to observe that the Borel I_n-positive sets come in a canonical form. Recall the combinatorial presentation P of the poset P_I from Definition 7.3.

Definition 9.10 A set $B \subset X_n$ is an E_0*-box* if it is of the form $\prod_{i\in n}[p_i]$ where each p_i is a condition in the poset P. If this is the case, we write $B(i) = [p_i]$.

Note that if $p \in P$ is a condition then there is a natural E_0-preserving homeomorphism of 2^ω and $[p]$, assigning each point $x \in 2^\omega$ the concatenation of the trunk of p and $x(i)$-th element of the pair p_i for all $i \in \omega$. If B is an E_0-box then there is a natural homeomorphism of X_n with B – namely the product of the natural E_0-preserving homeomorphisms between 2^ω and $B(i)$ for each $i \in n$. This homeomorphism preserves the E_0 equivalence in each coordinate. It will allow us to reduce work on an arbitrary E_0-box to work on the whole space X_n by simply pulling back the context to X_n along the homeomorphism.

The present proof of Theorem 9.9 depends on a strong partition theorem. Suppose that $\langle a_n, \phi_n : n \in \omega\rangle$ is a sequence of finite sets and submeasures on them such that the numbers $\phi_n(a_n)$ increase to infinity very fast.

Fact 9.11 (Shelah and Zapletal 2011) *For every partition $\Pi_n a_n \times \omega = B_0 \cup B_1$ into two Borel pieces, one of the pieces contains a product $\Pi_n b_n \times c$ where each $b_n \subset a_n$ is a set of ϕ_n mass at least* 1 *and $c \subset \omega$ is infinite.*

Proof of Theorem 9.9 The strategy of the proof can be described as follows. First, we treat a couple of special cases: the smooth equivalence relations, and the equivalence relations that are a subset of the product of E_0's (this uses a partition theorem). The conjunction of these two cases can be cranked up to cover the case of essentially countable equivalence relations. After that, a brief induction argument on the dimension of the product reduces the general case to the case of an essentially countable equivalence relation, concluding the proof.

The following remnant of the free set property will be useful at two distinct places in the proof:

Claim 9.12 *Let $n \in \omega$ be a number.*

(i) Let $R \subset (2^\omega)^n \times 2^\omega$ be a meager relation. Then there is an E_0-box $C \subset (2^\omega)^n$ such that for every $x \in C$ and every $y \in C(0)$, $\langle x, y\rangle \in R$ implies $x(0)\ E_0\ y$.

(ii) Let $B \subset (2^\omega)^n$ be an E_0-box. Let $R \subset B \times B(0)$ be a relatively meager relation. Then there is a E_0-box $C \subset B$ such that for every $x \in C$ and every $y \in C(0)$, $\langle x, y\rangle \in R$ implies $x(0)\ E_0\ y$.

Proof We will start with the first item. Let $\{O_k : k \in \omega\}$ be a collection of open dense subsets of $(2^\omega)^n \times 2^\omega$ such that their intersection is disjoint from R. By induction on $l \in \omega$ build finite binary sequences $t(0, i, l)$ and $t(1, i, l)$ for $i \in n$ so that:

I1. for a fixed number l the sequences are pairwise distinct and have all the same nonzero length;

I2. whenever $b : n \times l \to 2$ and $c : l \to 2$ are functions such that $c(l-1) \neq b(0, l-1)$, then writing $u(i)$ for the concatenation of $\langle t(b(i,k), i, k) : k \in l\rangle$ for each $i \in n$ and v for the concatenation of $\langle t(c(k), 0, k) : k \in l\rangle$, it is the case that $\prod_{i\in n}[u(i)] \times [v] \subset \bigcap_{k\in l} O_k$ holds.

This is easy to do by repeatedly extending the candidates for $t(0, i, l)$ and $t(1, i, l)$ in order to satisfy item I2 for all of the finitely many possibilities of functions b, c. In the end, let $C_i = \{x \in 2^\omega : x = s_0^\frown s_1^\frown \cdots$ where $\forall l \in \omega\ s_l \in \{t(0, i, l), t(1, i, l)\}\}$. It is not difficult to see that $C_i \subset 2^\omega$ is an I-positive compact set and for $p_i \in P$ the E_0-box $C = \prod_{i\in n} C_i$ has the required properties.

The second item can be immediately reduced to the first. Let p_i for $i \in n$ be the conditions in P such that $[p_i] = B(i)$, let $f_i : 2^\omega \to [p_i]$ be the natural E_0-preserving homeomorphism, and apply the first item to the relation $(f_0 \times f_1 \times \cdots \times f_n \times f_0)^{-1} R \subset (2^\omega)^n \times 2^\omega$. □

Lemma 9.13 *Let $n \in \omega$ be a number, let $B \subset X_n$ be an I_n-positive Borel set, and let E be a smooth equivalence relation on B. There is a set $a \subset n$ and a Borel I_n-positive set $C \subset B$ such that $E \restriction C = \prod_{i\in n} F_i$ where $F_i = \mathrm{id}$ for $i \in a$ and $F_i = \mathrm{ev}$ for $i \notin a$.*

Proof The proof proceeds by induction on $n \in \omega$. The basis case $n = 0$ is trivial. Let $f : B \to 2^\omega$ be a Borel function reducing the equivalence relation

E to the identity. Thinning out the set B if necessary, we may assume that the function f is continuous – note that the poset P_{I_n} is proper and bounding and so has the continuous reading of names by Fact 2.51. Inscribing an E_0-box and pulling back along a canonical homeomorphism of X_n with the box, we can assume that the set B is in fact equal to the whole space X_n. There are two cases.

Case 1 There is a Borel I_n-positive set $B' \subset B$ and an index $i \in n$ such that whenever $x, y \in B'$ are two sequences whose entries are the same except possibly at the entry i, then $f(x) = f(y)$; without loss of generality assume that $i = n - 1$. In such a case, the functional value of f on a point $x \in B'$ depends only on $x \restriction n - 1$, and so f induces a smooth equivalence relation E' on the projection of B' into the first $n - 1$ many coordinates. Then, use the induction hypothesis to thin out the first $n - 1$ many coordinates of B' so as to find an I_n-positive Borel set $C \subset B'$ and a set $a \subset n - 1$ such that $E' \restriction \prod_{i \in n-1} C(i) = \prod_{i \in n-1} F_i$ where $F_i = \text{id}$ if $i \in a$ and $F_i = \text{ev}$ if $i \notin a$. Then $E \restriction C = \prod_{i \in n} F_i$ where $F_i = \text{id}$ if $i \in a$ and $F_i = \text{ev}$ if $i \notin a$, as required.

Case 2 If Case 1 fails, then we will produce an E_0-box $C \subset B$ such that $E \restriction C = \text{id}$. A claim will be useful:

Claim 9.14 *For every two distinct n-tuples $\langle t_i : i \in n \rangle$, $\langle s_i : i \in n \rangle$ of finite binary strings of the same length there are n-tuples $\langle t'_i : i \in n \rangle$, $\langle s'_i : i \in n \rangle$ of finite binary strings such that for every $i \in n$, $t_i \subset t'_i$, $s_i \subset s'_i$, and $t_i = s_i \rightarrow t'_i = s'_i$ all hold, and the sets $f'' \prod_i [t'_i]$ and $f'' \prod [s'_i]$ are disjoint.*

Proof Let $j \in n$ be such that $s_j \neq t_j$. Use the case assumption to find $x, y \in \prod_i [t_i]$ such that $x(i) = y(i)$ holds for all $i \neq j$ and $f(x) \neq f(y)$. Find $z \in \prod_i [s_i]$ such that for all $i \in n$, $s_i = t_i$ implies $x(i) = z(i) = y(i)$. It must be the case that either $f(x) \neq f(z)$ or $f(y) \neq f(z)$; for definiteness assume the former case and pick $m \in \omega$ such that $f(x) \restriction m \neq f(y) \restriction m$. Use the continuity of the function f to find finite binary strings t'_i, s'_i for $i \in n$, all of the same length, such that $t'_i \subset x(i)$, $s'_i \subset z(i)$, and such that $f'' \prod_i [t'_i] \subset [f(x) \restriction m]$ and $f'' \prod_i [s'_i] \subset [f(z) \restriction m]$. This completes the proof. □

By induction on $l \in \omega$ build finite binary sequences $t(0, i, l)$ and $t(1, i, l)$ for $i \in n$ so that:

I1. the sequences $t(0, i, l)$ and $t(1, i, l)$ are distinct of the same length;

I2. writing $d(i, l)$ for the set $\{u \in 2^{<\omega} :$ for some $b \in 2^l$, u is the concatenation of $\langle t(b(k), i, k) : k \in l \rangle\}$, the sets $f'' \prod_i [t_i]$ for $\langle t_i : i \in n \rangle \in \prod_i d(i, l)$ are pairwise disjoint.

The induction step is easy, using Claim 9.14 repeatedly. In the end, let $C_i = \{x \in 2^\omega : x = s_0^\frown s_1^\frown \cdots$ where $\forall l\ s_l \in \{t(0,i,l), t(1,i,l)\}$. It is immediate that $C_i \subset 2^\omega$ is a compact I-positive set. Let $C = \prod_{i \in n} C_i$ and observe that I2 implies $E \restriction C = \mathrm{id}$, as desired. □

Lemma 9.15 *Let $n \in \omega$ be a number, and let $E \subset E_0^n$ be a Borel equivalence relation on an I_n-positive Borel set $B \subset (2^\omega)^n$. There is a set $a \subset n$ and an I_n-positive Borel set $C \subset B$ such that $E \restriction C = \prod_{i \in n} F_i$, where $F_i = \mathrm{id}$ if $i \in a$, and $F_i = E_0$ if $i \notin a$.*

Proof Most of the work necessary for the proof is performed in an abstract and well-known context of canonization on products of finite sets. We record the finite results first. To compactify the notation, the following nonstandard usage will prevail. The number $n \in \omega$ denotes the dimension of the product. n-sequences will be adorned with an arrow (such as $\vec{a}$) and simple set theoretic operations and relations will be applied coordinatewise. Thus $\vec{c} \subset \vec{b}$ means that for each $i \in n$, $\vec{c}(i) \subset \vec{b}(i)$, $\vec{c} \in [\vec{b}]^3$ means that $\forall i \in n\ \vec{c}(i) \in [\vec{b}(i)]^3$, etc. If $\vec{b}$ is a sequence of sets, then $\vec{x} \in \vec{b}$ means that $\forall i \in n\ \vec{x}(i) \in \vec{b}(i)$; in other words the sequence $\vec{b}$ is identified with the product $\prod_{i \in n} \vec{b}(i)$. $\sum \vec{k}$ denotes the sum of all numbers on the sequence $\vec{k}$. For a set $a \subset n$, E_a is the equivalence relation on n-sequences connecting $\vec{c}$ and $\vec{d}$ if for all $i \in a$, $\vec{c}(i) = \vec{d}(i)$.

Claim 9.16 *For every sequence $\vec{k} \in \omega^n$ there is a sequence $\vec{l} \in \omega^n$ such that for every sequence $\vec{b}$ of finite sets of size $\vec{l}$ and every partition of the set $[\vec{b}]^3$ into two parts there is a sequence $\vec{c} \subset \vec{b}$ of size $\vec{k}$ such that the set $[\vec{c}]^3$ is a subset of one piece of the partition.*

Proof Use the Ramsey theorem to find a number $l \in \omega$ such that $l \to (\Sigma\vec{k})^{3n}_2$. We claim that the sequence $\vec{l}$ repeating n many times the number l, works as required. Suppose that $p : [\vec{l}]^3 \to 2$ is a partition. Let $\bar{p} : [l]^{3n} \to 2$ be the partition defined by $\bar{p}(x) = p(\vec{u})$ where $\vec{u}(i)$ is the triple consisting of $2i$-th, $2i+1$-th, and $2i+2$-th elements of x. Let $a \subset l$ be a set of size $\Sigma\vec{k}$ such that the set $[d]^{3n}$ is a subset of one piece of the partition $\bar{p}$. Decompose d as $\bigcup_{i \in n} \vec{c}(i)$ where $\vec{c}(i) = \vec{k}$ and all numbers in $\vec{c}(i)$ are smaller than numbers in $\vec{c}(i+1)$. Clearly, the set $[\vec{c}]^3$ is a subset of one piece of the partition. □

Claim 9.17 *For every sequence $\vec{k} \in \omega^n$ there is a sequence $\vec{l} \in \omega^n$ such that for every sequence $\vec{b}$ of finite sets of size $\vec{l}$ and every equivalence relation E on $\vec{b}$, there is a sequence $\vec{c} \subset \vec{b}$ of size $\vec{k}$ and a set $a \subset n$ such that $E \restriction \vec{c} = E_a$.*

Proof Use the Erdős–Rado theorem (Todorcevic 2010, corollary 1.9) to find a number l such that for every equivalence relation E on $[l]^n$ there is a set $d \subset l$ of size $\Sigma\vec{k}$ and a set $a \subset n$ such that for every $x, y \in [d]^n$, $x\ E\ y$ iff for every $i \in a$ the i-th element of x in increasing enumeration is the same as the i-th element of y in the increasing enumeration. The sequence $\vec{l}$ repeating the number l n times works as required; the proof is the same as in Claim 9.16. □

Claim 9.18 *For every sequence $\vec{k} \in \omega^n$ there is a sequence $\vec{l} \in \omega^n$ such that for every sequence $\vec{b}$ of finite sets of size $\vec{l}$ and every equivalence relation E on $2 \times \vec{b}$ there is a sequence $\vec{c} \subset \vec{b}$ of size $\vec{k}$ such that:*

(i) either, for every $\vec{x} \in \vec{c}$, $\langle 0, \vec{x}\rangle \; E \; \langle 1, \vec{x}\rangle$ holds;
(ii) or, for every $\vec{x}, \vec{y} \in \vec{c}$, $\langle 0, \vec{x}\rangle \; E \; \langle 1, \vec{y}\rangle$ fails.

Proof Let $\vec{l} \in \omega^n$ be large enough sequence. Suppose that E is an equivalence relation on $2 \times \vec{l}$. First, use Claim 9.16 to thin down the sequence $\vec{l}$ so that one of the following occurs:

Case 1 Either for each $\vec{x} \in \vec{l}$ $\langle 0, \vec{x}\rangle \; E \; \langle 1, \vec{x}\rangle$ holds. In this case, we are in the first item of the claim.

Case 2 Or, for each $\vec{x} \in \vec{l}$ $\langle 0, \vec{x}\rangle \; E \; \langle 1, \vec{x}\rangle$ fails. Use Claim 9.17 to find sets $a_0, a_1 \subset n$ and a sequence $\vec{b}$ such that $E \restriction \{0\} \times \vec{b} = E_{a_0}$ and $E \restriction \{1\} \times \vec{b} = E_{a_1}$. For definiteness suppose that a_1 is not a proper subset of a_0. For a sequence $\vec{d}$ consisting of unordered triples of natural numbers, write $\mathrm{mid}(\vec{d})$ for the sequence consisting of the numbers in the middle of each triple in $\vec{d}$. Recall the sequence notation, and define a partition of $[\vec{b}]^3$ by $p(\vec{d}) = 0$ if $\{0\}^\frown \mathrm{mid}(\vec{d})$ is E-related to some element of $\{1\} \times \vec{d}$. Use Claim 9.16 to further thin out the sequence $\vec{b}$ so that it is large enough and homogeneous for the partition p. We claim that the homogeneous color must be 1. Then, let $\vec{c}$ be the sequence obtained from $\vec{b}$ by removing the smallest and largest element from $\vec{b}(i)$ for all $i \in n$, and observe that $\vec{c}$ satisfies *(ii)* of the claim.

Suppose for contradiction that the homogeneous color is equal to 0. It is not difficult to produce two sequences $\vec{d}, \vec{e} \in [\vec{b}]^3$ such that for all $i \in a_0$, $\vec{d}(i) \cap \vec{e}(i)$ contains just the midpoint of these two sets, and for all $i \notin a_0$, $\vec{d}(i) \cap \vec{e}(i) = 0$. Clearly, $\{0\}^\frown \mathrm{mid}(\vec{d}) \; E \; \{0\}^\frown \mathrm{mid}(\vec{e})$.

Case 2a Suppose that $a_0 \neq a_1$. Since a_1 is not a proper subset of a_0, the sets $\{1\} \times \vec{d}$ and $\{1\} \times \vec{e}$ are E-unrelated.This makes it impossible for $p(\vec{d})$ and $p(\vec{e})$ to be both equal to 0, contradiction.

Case 2b Suppose that $a_0 = a_1$. Again, the sets $\{1\} \times \vec{d}$ and $\{1\} \times \vec{e}$ are E-unrelated except for the single equivalence class $1^\frown \mathrm{mid}(\vec{d})$. By the assumption of Case 2, this class does not contain $0^\frown \mathrm{mid}(\vec{d})$, and so again $p(\vec{d})$ and $p(\vec{e})$ cannot be both equal to 0, a contradiction. □

Now back to the proof of Lemma 9.15. Let $B \subset (2^\omega)^n$ be an I_n-positive Borel set, and let E be a Borel equivalence relation which is a subset of the product of E_0's. Let $g : B \to 2^{(\omega^{<\omega})^n}$ be the function defined by $g(\vec{x})(\vec{t}) = 0$ if $\vec{x}$ rew $\vec{t}\ E\ \vec{x}$. Shrinking the set B if necessary we may assume that the function g is continuous. First, inscribing an E_0-box into B and pulling back through the natural homeomorphism between $(2^\omega)^n$ and the box if necessary we may assume that in fact the set B is the whole space $(2^\omega)^n$. To cut the notational clutter, below we identify 2^ω with $\mathcal{P}(\omega)$ via the characteristic functions.

By induction on $j \in \omega$ build n-sequences $\vec{b}_j$ and $\vec{e}_j$ of finite subsets of ω and numbers $m_j, \bar{m}_j$ so that:

I1. $m_0 < \bar{m}_0 < m_1 < \bar{m}_1 < \cdots$ the sets on the sequence $\vec{b}_j$ are all subsets of the interval $(m_j, \bar{m}_j)$ and the sets on the sequence $\vec{e}_j$ are all subsets of the interval $(\bar{m}_j, m_{j+1})$;

I2. for any pair $\vec{c}, \vec{d}$ of sequences of sets below $\bar{m}_j$,

   (a) either for every $\vec{x} \in \mathcal{P}(\omega)^n$ such that each set on the sequence $\vec{e}_j$ is an initial segment of the corresponding set on $\vec{x}$, $\vec{c} \cup \vec{x}\ E\ \vec{d} \cup \vec{x}$ holds,

   (b) or for every such $\vec{x}$, $\vec{c} \cup \vec{x}\ E\ \vec{d} \cup \vec{x}$ fails;

I3. the sizes of the sets $\vec{b}_j$ grow sufficiently fast as indicated by the needs of the argument to follow.

For every $j \in \omega$, write F_j for the binary relation on $\mathcal{P}(\bar{m}_j)^n$, connecting $\vec{c}$ and $\vec{d}$ if I2(a) above occurs. I2 shows that F_j is an equivalence relation. For every sequence $\vec{d}$ of sets below m_j, let $E_{\vec{d},j}$ be the equivalence relation on $\prod \vec{b}_j$ defined by $\vec{u}\ E_{\vec{d},j}\ \vec{v}$ if $\vec{d} \cup \{\vec{u}\}\ F_j\ \vec{d} \cup \{\vec{v}\}$. By a repetitive use of Claims 9.17 and 9.18 at every $j \in \omega$, shrink the sets on the sequences $\vec{b}_j$ if necessary so that:

I4. for every sequence $\vec{d}$ whose entries are below m_j there is a set $a_{\vec{d},j} \subset n$ such that $E_{\vec{d},j} \restriction \prod \vec{b}_i = E_{a_{\vec{d},j}}$;

I5. for sequences $\vec{c}, \vec{d}$ below m_j,

   (a) either for every $\vec{u} \in b_j$ $\vec{c} \cup \{\vec{u}\}\ F_j\ \vec{d} \cup \{\vec{u}\}$ holds,

   (b) or for every $\vec{u}, \vec{v} \in b_j$ $\vec{c} \cup \{\vec{u}\}\ F_j\ \vec{d} \cup \{\vec{v}\}$ fails.

For every $j \in \omega$, let μ_j be the submeasure on the set $[\vec{b}_j]^2$ defined by $\mu_j(h) = 0$ if $h = 0$, $\mu_j(h) = 1$ if $h \neq 0$ contains no subset of the form $[\vec{c}_j]^2$ for a sequence of sets $\vec{c}_j \subset \vec{b}_j$ each of size at least 2, and $\mu_j(h) \leq r$ if h can be decomposed into r many sets of μ_j-mass ≤ 1. The sequences $\vec{b}_j$ must be chosen to be growing so fast that:

I6. the numbers $\mu_i([\vec{b}_j]^2)$ tend to infinity fast enough so that the partition theorem (Fact 9.11) may be applied for partitions into 2^n many pieces.

This ends the list of inductive requirements. The induction itself is easy to perform: I2 uses the continuity of the function g, I4 uses Claim 9.17, and I5 uses Claim 9.18.

Consider the partition $p : \prod_j [\vec{b}_j]^2 \times \omega \to \mathcal{P}(n)$ defined by $p(\langle \vec{d}_j : j \in \omega \rangle, k) = a_{\vec{d},k}$ where $\vec{d}$ is the sequence of sets defined as $\{\max(\vec{d}_0)\} \cup \vec{e}_0 \cup \{\max(\vec{d}_1)\} \cup \cdots \cup \vec{e}_{j-1}$. The partition theorem (Fact 9.11) yields sequences $\langle \vec{c}_j : j \in \omega \rangle$ such that $\vec{c}_j \subset \vec{b}_j$ is a sequence of sets of size at least three, an infinite set $w \subset \omega$ and a set $a \subset n$ such that whenever $\vec{d}_j$ for $j \in \omega$ are sequences of pairs with $\vec{d}_j \in [\vec{c}_j]^2$, and $k \in w$, then $p(\langle \vec{d}_j : j \in \omega \rangle, k) = a$.

Finally, we are in a position to construct the sets C_i for $i \in n$ such that each $C_i \subset \mathcal{P}(\omega)$ is a closed set on which the equivalence relation E_0 is nonsmooth, and such that $E \restriction \prod_i C_i$ connects pairs $\vec{x}$ and $\vec{y}$ just in case that $\forall i \in a\ \vec{x}(i) = \vec{y}(i)$ and $\forall i \notin a\ \vec{x}(i)\ E_0\ \vec{y}(i)$ holds. Let C_i consist of all sets of the form $\{k_0\} \cup e_0 \cup \{k_1\} \cup e_1 \cup \cdots$ where k_j is the largest element of the set $\vec{c}_j(i)$ if $j \notin w$, and k_j is one of the two largest elements of the set $\vec{c}_j(i)$ if $j \in w$. We must verify the advertised properties of the compact sets C_i for $i \in n$. The following three claims complete the proof:

Claim 9.19 *For every $i \in n$ the equivalence relation E_0 is not smooth on the set C_i.*

Proof Consider the map $h : 2^\omega \to C_i$ defined by $h(z) =$ the unique element of the set B such that for every $l \in \omega$, if j is l-th element of the set $w \subset \omega$, $h(z)$ contains the largest element of the set $\vec{c}_j(i)$ if and only if $x(l) = 1$. This map reduces E_0 to $E_0 \restriction C_i$, and therefore $E_0 \restriction C_i$ is not smooth. □

Claim 9.20 *For every sequence $\vec{x} \in \prod_i C_i$ and every $k \in w$, $a_{\vec{x} \cap m_j, j} = a$ holds.*

Proof For every $j \in \omega$ let $\vec{d}_j$ be the n-tuple of pairs defined by $\vec{d}_j(i) = \{\min(\vec{c}_j(i)), m\}$ where m is the unique number in $\vec{x}(i) \cap \vec{c}_j(i)$. Observe that since the set $\vec{c}_j(i)$ has size at least three and m is one of its two largest elements, it must be the case that $\min(\vec{c}_j(i)) < m$. The claim then follows from the homogeneity, since $p(\langle \vec{d}_j : j \in \omega \rangle, k) = a$ for every $k \in w$. □

Claim 9.21 *If $\vec{x}, \vec{y} \in \prod_i C_i$ are such that for all $i \in n$ $\vec{x}(i)\ E_0\ \vec{y}(i)$ and for all $i \in a$ $\vec{x}(i) = \vec{y}(i)$, then $\vec{x}\ E\ \vec{y}$ holds.*

Proof Fix $\vec{x}, \vec{y} \in \prod_i C_i$ satisfying the assumptions. A simple induction on $j \in w$ together with Claim 9.20 shows that for all $j \in \omega$, $\vec{x} \cap \bar{m}_j\ F_j\ \vec{y} \cap \bar{m}_j$

holds. Let j be large enough so that all differences between the entries of $\vec{x}$ and $\vec{y}$ occur below $\bar{m}_j$, and note that for such j, $\vec{x} \cap \bar{m}_j \; F_j \; \vec{y} \cap \bar{m}_j$ implies $\vec{x} \; E \; \vec{y}$ by the definition of F_j. □

Claim 9.22 *If $\vec{x}, \vec{y} \in \prod_i C_i$ are such that for some $i \in a$ $\vec{x}(i) \neq \vec{y}(i)$, then $\vec{x} \; E \; \vec{y}$ fails.*

Proof Since $E \subset E_0^n$, we can assume that for all $i \in n$ $\vec{x}(i) \; E_0 \; \vec{y}(i)$ holds. Let $k \in \omega$ be the least number such that $\vec{x}(i) \cap \bar{m}_k \neq \vec{y}(i) \cap \bar{m}_j$ for some $i \in a$. As in the previous claim, for all $j < k$ it is the case that $\vec{x} \cap \bar{m}_j \; F_j \; \vec{y} \cap \bar{m}_j$ holds. By induction on $j \geq k$ argue that $\vec{x} \cap \bar{m}_j \; F_j \; \vec{y} \cap \bar{m}_j$ fails. The base case $j = k$ holds by the minimality of k. To perform the induction step from j to $j+1$, note that inductive item I5(a) cannot be satisfied for $\vec{c} = \vec{x} \cap m_j$ and $\vec{d} = \vec{y} \cap m_j$ by the induction hypothesis at j, and I5(b) confirms the induction hypothesis at $j+1$. Let $j > k$ be large enough so that all differences between the entries of $\vec{x}$ and $\vec{y}$ occur below $\bar{m}_j$, and note that for such j, the failure of $\vec{x} \cap \bar{m}_j \; F_k \; \vec{y} \cap \bar{m}_j$ implies the failure of $\vec{x} \; E \; \vec{y}$ by the definition of F_j. □

This completes the proof of Lemma 9.15. □

The previous lemmas can be easily cranked up to cover the general case of analytic equivalence relations with countable classes, and then to the essentially countable equivalence relations.

Lemma 9.23 *Let $n \in \omega$ be a number, and let E be an analytic equivalence relation with countable classes on an I_n-positive Borel set $B \subset (2^\omega)^n$. There is a set $a \subset n$ and an I-positive Borel set $C \subset B$ such that $E \restriction C = \prod_{i \in n} F_i$, where $F_i = \mathrm{id}$ if $i \in a$, and $F_i = E_0$ if $i \notin a$.*

Proof For every $i \in n$ let $R_i \subset (2^\omega)^n \times 2^\omega$ be the relation consisting of all pairs $\langle x, y \rangle$ such that $x \in B$ and there is $x' \in B$ E-related to x such that $x'(i) = y$. These are analytic relations with countable vertical sections, and therefore relatively meager in every E_0-box in X_{n+1}. A repeated application of Claim 9.12 then yields an E_0-box $B' \subset B$ such that for every pair $x, x' \in B'$ and every $i \in n$, $x \; E \; x'$ and $\langle x, x'(i) \rangle \in R$ implies $x(i) \; E_0 \; x'(i)$. In other words, the equivalence E is a subset of the product of E_0's on B'. Lemma 9.15 then yields the final I_n-positive Borel set $C \subset B'$. □

Lemma 9.24 *Let $n \in \omega$ be a number, and let E be an essentially countable equivalence relation on an I_n-positive Borel set $B \subset (2^\omega)^n$. There is a partition $n = a \cup b \cup c$ and an I-positive Borel set $C \subset B$ such that $E \restriction C = \prod_{i \in n} F_i$, where $F_i = \mathrm{id}$ if $i \in a$, $F_i = E_0$ if $i \in b$, and $F_i = \mathrm{ev}$ if $i \in c$.*

Proof Let $f : B \to Y$ be a Borel function reducing E to some Borel equivalence relation F on a Polish space Y with countable classes. First apply Lemma 9.13 to the smooth equivalence E' induced by f to find a Borel I_n-positive set $B' \subset B$ and a set $c \subset n$ such that $E' \restriction B' = \prod_{i \in n} F_i$ where $F_i = \text{id}$ if $i \notin c$ and $F_i = \text{ev}$ if $i \in c$. This is to say that the values $f(x)$ depend only on $x \restriction (n \setminus c)$ and that in an injective way. Ergo, the equivalence relation F induces a Borel equivalence relation with countable classes on the projection of B' into the coordinates outside the set c. An application of Lemma 9.23 concludes the proof. □

Now we are ready to complete the proof of Theorem 9.9. By induction on $n \in \omega$ we will prove the following statement:

($\star$) for every $m \in \omega$, for every analytic equivalence relation E on a Borel I_{m+n}-positive set $B \subset (2^\omega)^m \times (2^\omega)^n$, there is a Borel I_{m+n}-positive subset $C \subset B$ such that for every $z \in (2^\omega)^m$, $E \restriction C_z$ is essentially countable.

The theorem will follow from the case $m = 0$ and Lemma 9.24 above.

Case $n = 0$ is trivial. Suppose that m, n are given. Use the induction hypothesis to thin out B so that for every index $i \in n$ every $z \in (2^\omega)^m$, and every $y \in 2^\omega$, the equivalence relation $E \restriction \{x \in B : x \restriction m = z \wedge x(m+i) = y\}$ is essentially countable. Thin out B further to an E_0-box. Write $B_0 = B$ and consider the relation R_0 on $B_0 \times B_0(m)$ defined by $\langle x, y\rangle \in R_0$ if there is $x' \in B_0$ such that $x' \mathrel{E} x$, $x' \restriction m = x \restriction m$ and $x'(m) = y$.

Case 1 If the relation R_0 is not meager in B_0, then use the Kuratowski–Ulam theorem to find $y \in B_0(m)$ such that the horizontal section R_0^y is nonmeager in B_0 – thus, $R_0^y \subset B_0$ is an analytic I_{m+n}-positive set. Use uniformization (Fact 2.51(*iii*)) to find a Borel I_{m+n}-positive set $C \subset R_0^y$ and a Borel function $f : C \to B_0$ such that for every $x \in C$, we have $f(x) \restriction m = x \restriction m$, $f(x)(m) = y$, and $f(x) \mathrel{E} x$. Now, for every $z \in (2^\omega)^m$, the function f reduces $E \restriction C_z$ to the equivalence relation $E \restriction \{x \in B : x \restriction m = z, x(m) = y\}$. This latter equivalence relation is essentially countable and in this case we are done.

Case 2 If the relation R_0 is meager in B_0, then by Claim 9.12 there is a box $B_1 \subset B_0$ such that for every $x \in B_1$ and every $y \in B_1(m)$, $\langle x, y\rangle \in R_0$ implies $x(m) \mathrel{E_0} y$. Now repeat the whole argument at the coordinate $m+1$ with the relation $R_1 \subset B_1 \times B_1(m+1)$ defined by $\langle x, y\rangle \in R_1$ if there is $x' \in B_1$ such that $x' \mathrel{E} x$, $x' \restriction m = x \restriction m$ and $x'(m+1) = y$, and by induction on $i \in n$ repeat it at every coordinate i. If at any stage of the induction we end up in Case 1, then we are done. Otherwise, we end up with a box $C = B_n \subset B$ such that for every $i \in n$ and every $x_0, x_1 \in B_n$ with $x_0 \restriction m = x_1 \restriction m$ and $x_0 \mathrel{E} x_1$ it is the case that $x_0(m+i) \mathrel{E_0} x_1(m+i)$. This means that for every $z \in (2^\omega)^m$

the equivalence relation $E \restriction C_z$ is a subset of product of E_0's. The classes of $E \restriction C_z$ are already countable, but we still have to deal with the possibility that $E \restriction C_z$ is analytic and not Borel. An application of Lemma 9.23 shows that we can thin out the box C even further so that for every $z \in (2^\omega)^m$, $E \restriction C_z$ takes one of the finitely many prescribed Borel forms, and they all have countable equivalence classes. The proof is complete. □

9.3 Infinite products of perfect sets

In this section, we will provide several canonization and anticanonization results for infinite products of perfect sets on the underlying Polish space $X = (2^\omega)^\omega$. The associated poset is the countable support product of Sacks forcing: the poset P of all ω-sequences p of perfect binary trees, ordered by coordinatewise inclusion. The computation of the associated ideal yields complete information. For a condition $p \in P$ write $[p] = \{x \in X : \text{for all } n \in \omega, x(n) \text{ is a branch of the tree } p(n)\}$. Let I be the collection of those Borel subsets of $X = (2^\omega)^\omega$ which do not contain a subset of the form $[p]$ for some $p \in P$. The following fact is a conjunction of the rectangular Ramsey property of Sacks forcing (Fact 2.56) and Zapletal (2008, proposition 2.1.6).

Fact 9.25

(i) I is a $\mathbf{\Pi}^1_1$ on $\mathbf{\Sigma}^1_1$ σ-ideal of Borel sets.

(ii) Every positive analytic set contains a compact positive subset.

(iii) The map $p \mapsto [p]$ is an isomorphism between P and a dense subset of the poset P_I.

(iv) The poset P is proper, bounding and preserves Baire category.

We provide two canonization theorems for infinite products of perfect sets. We avoid any attempt at organizing the very complicated class of smooth equivalence relations, and prove:

Theorem 9.26 *For every equivalence relation on X classifiable by countable structures there is an infinite product of perfect sets on which E is smooth. In other words,*

$$\text{classifiable by countable structures} \to_I \text{smooth}.$$

Theorem 9.27 *Whenever J is an analytic P-ideal on ω, then for every equivalence relation E on X Borel reducible to $=_J$ there is an infinite product of perfect sets such that E is smooth on it. In other words,*

$$\leq_B =_J \to_I \text{smooth}.$$

Theorem 9.28 *E_1 is in the spectrum of I. Moreover, the poset $P_I^{E_1}$ is regular in P_I, it is $\aleph_0$-distributive, and its extension is equal to $V[\dot{x}_{\mathrm{gen}}]_{E_1}$.*

Theorem 9.29 *The spectrum of I is cofinal among the Borel equivalence relations under $\leq_B$. It contains E_{K_σ}.*

These results leave an obvious gap, which we state as an open question:

Question 9.30 Is it true that for every equivalence relation E on $(2^\omega)^\omega$, reducible to the orbit equivalence of a Polish group action, there are perfect sets $P_n \subset 2^\omega$ for each $n \in \omega$ such that E restricted to $\Pi_n P_n$ is smooth?

For the proofs, first set up a fixed framework for fusion arguments. Let $p \in P$. A *fusion sequence* below p is a sequence $\langle u_n, p_n : n \in \omega\rangle$ such that u_n is a sequence of length n whose entries are finite binary trees with n splitting levels, p_n is a condition in P below p, and for every $m \in \omega$ the tree $p_n(m)$ end-extends $u_n(m)$. Moreover, we require that $p_{n+1} \leq p_n$ and for every $m \in n$, the tree $u_{n+1}(m)$ end-extends the tree $u_n(m)$. It is clear that the conditions $\{p_n : n \in \omega\}$ in a fusion sequence share a lower bound, the condition $q \in P$ such that for every $m \in \omega$, $q(m) = \bigcup_{m<n} u_n(m)$. In this context, we will write $[u_n]$ for the set of all sequences t of length n such that $t(m)$ is an endnode of $u_n(m)$. For every $t \in [u_n]$ the condition $p_n \restriction t$ is defined as that sequence r such that for all $m \in n$, $r(m)$ is the set of all nodes of $p_n(m)$ compatible with $t(m)$, and for all $m \geq n$, $r(m) = p_n(m)$. For every $x \in [p_n]$ let $x \restriction u_n$ be the unique element of $[u_n]$ such that $x \in p_n \restriction t$. For a finite sequence $\langle t_i : i \in m\rangle$ of finite binary strings, write $[t_i : i \in m] \subset X$ for the clopen subset of X consisting of all points $x \in X$ such that for every $i \in m$, $t_i \subset x(i)$ holds.

As a useful fusion warm-up, we will prove:

Proposition 9.31 *The σ-ideal I consists of sets which are both meager and null in the sense of product topology and the usual product measure on X.*

Proof Let us first deal with the measure case. Let μ be the usual product measure on X, and let $A \subset X$ be an analytic set of positive μ-mass. We must inscribe an infinite product of perfect sets into A. Thinning out A if necessary, we may assume that in fact A is closed.

By induction on $n \in \omega$ build a fusion sequence $\langle u_n, p_n\rangle$ according to the following demands:

I1. $[p_n] \subset X$ is a clopen set;
I2. for every $t \in [u_n]$, the set A has relative μ-mass $> 1 - (2^{-n^2})$ in the set $[p_n \restriction t]$.

Once this is done, the lower bound q of the fusion sequence will satisfy $[q] \subset A$; this will prove the measure part of the proposition. The induction step is not difficult to perform as follows. To construct u_{n+1} and p_{n+1} from u_n and p_n, fix some $t \in [u_n]$ and let $\pi_s : [p_n \restriction s] \to [p_n \restriction t]$ be the map defined by $\pi_s(x)(i) = x(i)$ rew $t(i)$ for all $i \in n$. It is not difficult to see that each map π_s preserves the measure μ. By item I2, the set $C_n = \bigcap_{s \in [u_n]} \pi_s'' A$ has positive μ-mass. By the Lebesgue density theorem find a number $m > n$ and finite binary strings $t'(i)$ for $i \in m$ such that $[t'(i) : i \in m] \subset [p_n \restriction t$ and C_n has very large relative mass in the clopen set $[t'(i) : i \in m]$. Now, whenever $t''(i)$ for $i \in m$ are binary strings such that for $i \in n+1$ $t''(i)$ is a one-step extension of $t'(i)$ and for $i \geq n+1$ $t''(i) = t'(i)$, the set C_n has relative mass at least $1 - 2^{-(n+1)^2}$ in the set $[t''(i) : i \in m]$. Let u_{n+1} be the sequence of trees such that for every $i \in n$, $u_{n+1}(i)$ has endnodes $(t'(i)^\frown 0)$ rew s and $(t'(i)^\frown 1)$ rew s for every endnode $s \in u_n(i)$, and $u_{n+1}(n)$ has endnodes $t'(n)^\frown 0$ and $t'(n)^\frown 1$. Let $p_{n+1} \leq p_n$ be the condition such that $[p_{n+1}] = \{x \in X : \forall i \in n\ t'(i)$ rew $s \subset x(i)$ for some endnode $s \in u_n(i)$ and $\forall i > n\ t'(i) \subset x(i)\}$. This completes the inductive step.

For the category case, assume that $A \subset X$ is an analytic nonmeager set. We must inscribe a product of perfect sets into A. Thinning out A if necessary, we may assume that A is G_δ, and write A as a decreasing intersection $\bigcap_n O_n$ of open sets. By induction on $n \in \omega$ build a fusion sequence $\langle u_n, p_n \rangle$ so that:

I1. $[p_n] \subset X$ is a clopen subset of O_n;
I2. for every $t \in [u_n]$, the set A is comeager in the set $[p_n \restriction t]$.

The inductive procedure is the same as in the measure case and we leave it to the reader. □

Proof of Theorem 9.26 We will start with a simple claim:

Claim 9.32 *Let $p \in P$ be a condition and $\{f, g_m : m \in \omega\}$ be Borel functions from $[p]$ to 2^ω and let $k \in \omega$. At least one of the following two options holds:*

(i) either there is a condition $q \leq p$ such that for every $x, y \in [q]$, $(\forall n \neq k\ x(n) = y(n))$ implies $f(x) = f(y)$;
(ii) or there is a condition $q \leq p$ such that for every $x, y \in [q]$, $x(k) \neq y(k)$ implies $f(x) \neq g_m(y)$ for every $m \in \omega$.

Proof Without loss of generality assume that $k = 0$. Thinning the condition p if necessary, we may assume that the functions f, g_m are continuous. Suppose that the first item fails. Build a fusion sequence $\langle u_n, p_n : n \in \omega \rangle$ below p such that for every $n \in \omega$ and every $t, s \in [u_n]$, if $t(0) \neq s(0)$ then $f''[p_n \restriction t] \cap g_m''[p_n \restriction s] = 0$ for all $m \in n$. It is clear that in the end, the lower bound q of the fusion sequence will be the condition required in the second item.

To perform the induction step, find any sequence u_{n+1} of finite trees below p_n that can serve as the next entry of the fusion sequence. We must show how to thin out p_n to p_{n+1} and preserve the induction hypothesis.

Suppose that $t, s \in [u_{n+1}]$ are two sequences such that $t(0) \neq s(0)$ and choose a number $m \in n$. Use the assumption to find points $x_0, x_1 \in [p_n \restriction t]$ which differ at the first coordinate only, such that $f(x_0) \neq f(x_1)$; perhaps $f(x_0)(i) = 0$ while $f(x_1)(i) = 1$ for some $i \in \omega$. Use the continuity of the function f to find finite sequences $\hat{t}_0, \hat{t}_1$ of the same length that differ only at 0-th coordinate, $x_0 \in [p_n \restriction \hat{t}_0]$, every $x \in [p_n \restriction \hat{t}_0]$ has $f(x)(i) = 0$, and similarly for subscript 1. Consider the sequence $\hat{s}$ which is defined by $\hat{s}(j) = s(j)$ if $t(j) \neq s(j)$ and $\hat{s}(j) = \hat{t}(j)$ otherwise; so $\hat{s} \geq s$. The continuity of the function g_m on $[p_n]$ shows that strengthening $\hat{s}$ if necessary, it must be the case that either $g_m(x)(i) = 0$ for all $y \in [p_n \restriction \hat{s}]$, or $g_m(x)(i) = 1$ for all $y \in [p_n \restriction \hat{s}]$. Suppose the former case holds. Thinning out p_n below s so that $p_n \restriction s = p_n \restriction \hat{s}$ and below t so that $p_n \restriction t \leq p_n \restriction \hat{t}_1$ (which is possible to do simultaneously) we achieve that $f''[p_n \restriction t] \cap g''_m[p_n \restriction s] = 0$.

The condition $p_{n+1} \leq p_n$ is now obtained by repeating the thinning procedure in the previous paragraph for all pairs of sequences $t, s \in [u_{n+1}]$ which differ at the first coordinate and all numbers $m \in n$. The induction hypothesis continues to hold. □

The canonization of equivalence relations classifiable by countable structures is achieved via the separation property of Definition 4.21 and Theorem 4.22. To prove the separation property, suppose that $p \in P$ is a condition forcing $\dot{z} \in 2^\omega$ and $\dot{a} \subset 2^\omega$ is a countable set not containing $\dot{z}$. We must produce a condition $q \leq p$ and a compact set $C \subset 2^\omega$ such that $q \Vdash \dot{z} \in \dot{C}$ and $\dot{a} \cap \dot{C} = 0$. Thinning out the condition p if necessary, we can find continuous functions $f, g_n : [p] \to 2^\omega$ for $n \in \omega$ such that $p \Vdash$"$\dot{z} = \dot{f}(\dot{x}_{\text{gen}})$ and $\dot{a} = \{\dot{g}_n(\dot{x}_{\text{gen}}) : n \in \omega\}$" and for every $x \in [p]$, $f(x) \notin \{g_n(x) : n \in \omega\}$. It will be enough to find a condition $q \leq p$ such that $f''[q]$ is disjoint from $\bigcup_n g''_n[q]$; then, q together with the set $C = f''[q]$ will have the required properties.

To find the condition q, build a fusion sequence $\langle u_n, p_n : n \in \omega\rangle$ below p such that:

I1. for every $t \in [u_n]$, if there is a condition $r \leq p_{n+1}$ such that $r \restriction t \neq 0$ and the value of $f(x)$ does not depend on $x(n)$ for $x \in [r \restriction t]$, then p_{n+1} is such a condition;

I2. for every $t \in [u_n]$, if there is a condition $r \leq p_{n+1}$ such that for all $s \in [u_n]$, $r \restriction s \neq 0$ and for all $x \in [r \restriction t]$ and every $y \in [r]$, $x(n) \neq y(n)$ implies $f(x) \neq g_m(y)$ for all $m \in \omega$, then p_{n+1} is such a condition.

The induction process is trivial. Let $q \in P$ be a lower bound of the fusion sequence, and let $x, y \in [q]$. We must prove that $f(x) \neq g_m(y)$ for all $m \in \omega$. There are two distinct cases:

Case 1 There is no number $n \in \omega$ such that, writing $t = x \restriction u_n$, it is the case that $x(n) \neq y(n)$ and the assumption of item I1 fails for t and n. In this case, write $z_n \in [q]$ for the sequence obtained from y by replacing its first n entries with the corresponding entries of x and argue that the value $f(z_n)$ does not depend on n. To show that $f(z_n) = f(z_{n+1})$, discern two subcases. If $x(n) = y(n)$ then $z_n = z_{n+1}$ and $f(z_n) = f(z_{n+1})$ certainly holds. On the other hand, if $x(n) \neq y(n)$, the case assumption shows that the assumption of item I1 holds and implies exactly that $f(z_n) = f(z_{n+1})$. Now note that $z_0 = y$ and $\lim_n z_n = x$. Since the function f is continuous, it follows that $f(y) = f(x)$ and $f(x) = f(y) \neq g_m(y)$ for all $m \in \omega$ by the assumptions on the functions $f, g_m : m \in \omega$.

Case 2 If Case 1 fails then there must be a witnessing number n. Write $t = x \restriction u_n$. Since the assumption of item I1 fails for t, the first item of Claim 9.32 must fail for $k = n$. Thus, the second item holds, and therefore the assumption of item I2 holds. Since $x(n) \neq y(n)$, we conclude that $f(x) \neq g_m(y)$ for all $m \in \omega$ as required. The proof is complete. □

Proof of Theorem 9.27 Use a result of Solecki (1999) to find a lower semi-continuous measure ϕ on ω such that $J = \{a \subset \omega : \lim_n \phi(a \setminus n) = 0\}$. The proof of the following claim is parallel to Claim 9.32:

Claim 9.33 *Let $p \in P$, let $f, g : [p] \to 2^\omega$ be Borel functions and let $a \subset \omega$ be a finite set. At least one of the following happens:*

(i) *either for every $\varepsilon > 0$ there is a condition $q \leq p$ such that for every pair $x, y \in [q]$ with $\forall n \notin a\ x(n) = y(n)$ it is the case that $\phi(f(x)\Delta f(y) < \varepsilon$;*
(ii) *or there is a condition $q \leq p$ and $\varepsilon > 0$ such that for every pair $x, y \in [q]$ with $\forall n \in a\ x(n) \neq y(n)$ it is the case that $\neg f(x) =_J f(y)$.*

Let $p \in P$ and let $f : [p] \to 2^\omega$ be a Borel function reducing $E \restriction [p]$ to $=_J$. Thinning out the condition p we may assume that the function f is in fact continuous. We will construct a fusion sequence $\langle u_n, p_n : n \in \omega\rangle$ below p subject to a fairly nasty induction hypothesis. The aim of the construction is to find equivalence relations F_n on $[u_n]$ for every $n \in \omega$ such that E restricted to the lower bound q of the fusion sequence is Borel reducible to the equivalence relation $F = \prod_n F_n/\text{fin}$ on the set $\prod_n [u_n]$. The equivalence relation is a modulo finite product of equivalence relations on finite sets, and as such it is Borel reducible to E_0. The previous theorem can then be applied to show that thinning out q further if necessary, $E \restriction q$ is smooth. The reduction of $E \restriction q$ to F will be attained through the natural continuous one-to-one function $g : [q] \to \prod_n [u_n]$ defined by $g(x)(n) = x \restriction u_n$.

It is time to expose the induction hypothesis on the fusion sequence. For every $n \in \omega$ there will be a number $k_n > n$ and a real number $\varepsilon_n > 0$ so that:

I1. the numbers k_n are increasing, the numbers ε_n decrease towards 0 with $\varepsilon_{n+1} < \varepsilon_n/8$;

I2. for every pair $s, t \in [u_n]$, writing $a = \{m \in n : t(m) = s(m)\}$,

(a) either for all $x \in p_n \restriction t$ and $y \in p_n \restriction s$ such that $x \restriction a \cup [n, \omega) = y \restriction a \cup [n, \omega)$ $x \; E \; y$ fails;

(b) or for all such x, y it is the case that $x \; E \; y$ holds;

I3. for every pair $s, t \in [u_n]$ such that clause (I2)(a) holds, for all x, y as above $\limsup_k \phi(f(x) \Delta f(y) \setminus k) > \varepsilon_n$ holds;

I4. for every pair $s, t \in [u_n]$ such that clause (I2)(b) holds, for all such x, y $\phi(f(x) \Delta f(y) \setminus k_n) < \varepsilon_n/8$ holds;

I5. for all $t \in [u_n]$ and all $a \subset n$, if there is a condition $r \leq p_n \restriction t$ such that for every $x, y \in [r]$, $x \restriction a \cup [n, \omega) = y \restriction a \cup [n, \omega)$ implies $\phi(f(x) \Delta f(y)) < \varepsilon_n/8$ then $p_n \restriction t$ is such a condition;

I6. for all $t \in [u_n]$ and all $a \subset n$, if there is a condition $r \leq p_n$ such that for all $s \in [u_n]$, $r \restriction s \neq 0$ and such that for every $x \in [r \restriction t]$, $y \in [r]$, $\forall m \in n \setminus a \; x(m) \neq y(m)$ implies $\neg x \; E \; y$, then $p_n \restriction t$ is such a condition;

I7. for all $t \in [u_n]$ and all $a \subset n$, if there is a condition $r \leq p_{n+1} \restriction t$ such that for every $x, y \in [r]$, $x \restriction a \cup [n+1, \omega) = y \restriction a \cup [n+1, \omega)$ implies $\phi(f(x) \Delta f(y)) < \varepsilon_n/8$ then $p_{n+1} \restriction t$ is such a condition;

I8. for all $t \in [u_n]$ and all $a \subset n$, if there is a condition $r \leq p_{n+1}$ such that for all $s \in [u_n]$, $r \restriction s \neq 0$ and such that for every $x \in [r \restriction t]$, $y \in [r]$, $\forall m \in (n+1) \setminus a \; x(m) \neq y(m)$ implies $\neg x \; E \; y$, then $p_{n+1} \restriction t$ is such a condition.

While the list of induction hypotheses may sound daunting, it is in fact very easy to get. Suppose that $u_n, p_n, k_n, \varepsilon_n$ have been obtained. First, strengthen p_n to $p_{n+1} \leq p_n$ satisfying the last two items. Let u_{n+1} to be the sequence of the finite trees recording $p_{n+1}(m)$ up to $n+1$-st splitting level, for $m \in n+1$. Now run through the list of all pairs $s, t \in [u_{n+1}]$, using the σ-additivity of the σ-ideal I each time to gradually strengthen p_{n+1} so as to satisfy item I2. Run through the list again, gradually thinning out p_{n+1} again and using the σ-additivity of the σ-ideal I to find real numbers $\varepsilon_{n+1}(s, t) > 0$ such that if I2(a) holds for s, t then item I3 is satisfied with $\varepsilon_{n+1}(s, t)$ in place of ε_n. Let

$\varepsilon_{n+1} > 0$ be some real number smaller than all the numbers thus obtained. Run through the list again, thinning out p_{n+1} and using σ-additivity of the ideal I again to find numbers $k_{n+1}(s, t)$ such that if the clause I2(b) holds for s, t then I4 is satisfied with $k_{n+1}(s, t)$ in place of k_n. Let k_{n+1} be some number larger than all numbers thus obtained. Finally, run through the list of all $t \in [u_{n+1}]$ and $a \subset n + 1$, gradually strengthening the condition p_{n+1} again so as to satisfy I5 and I6. This completes the induction step.

Now suppose that the fusion has been performed. For each $n \in \omega$, let F_n be the relation on $[u_n]$ defined by $s \; F_n \; t$ if I2(b) is satisfied for s, t. A brief check is in order:

Claim 9.34 *F_n is an equivalence relation.*

Proof The only nontrivial point is the transitivity. Let $t_0 \; F_n \; t_1 \; F_n \; t_2$; we want to show that $t_0 \; F_n \; t_2$. Write $a_{01} = \{m \in n : t_0(m) = t_1(m)$ and similarly for a_{02} and a_{12}. Choose points $x_0 \in [p_n \restriction t_0]$, $x_1 \in [p_n \restriction t_1]$ and $x_2 \in [p_n \restriction t_2]$ such that $x_0 \restriction a_{01} \cup [n, \omega) = x_1 \restriction a_{01} \cup [n, \omega)$ and $x_1 \restriction a_{12} \cup [n, \omega) = x_2 \restriction a_{12} \cup [n, \omega)$. By the assumptions, we have $x_0 \; E \; x_1 \; E \; x_2$. Now amend x_0 on $a_{02} \setminus a_{01}$ so that $x_0 = x_2$ on this set; the assumptions of I2(a) are still satisfied and show that $x_0 \; E \; x_1$. Similarly, amend x_2 on $a_{02} \setminus a_{12}$ so that $x_0 = x_2$ on this set; it will be still the case that $x_1 \; E \; x_2$. In conclusion, $x_0 \; E \; x_2$, and since $x_0 \restriction a_{02} \cup [n, \omega) = x_2 \restriction a_{02} \cup [n, \omega)$, the pair x_0, x_2 contradicts I2(a) for t_0, t_2. It follows that I2(b) for t_0, t_2 must occur, in other words $x_0 \; F_n \; x_1$ holds. □

Let q be a lower bound of the fusion sequence and suppose $x, y \in [q]$; we must show that $x \; E \; y$ if and only if $g(x) \; F \; g(y)$. There are several cases:

Case 1 There is a number n such that, writing $t = g(x)(n)$ and $a = \{m \in n : x(m) = y(m)\}$, the assumption of I5 of the induction hypothesis fails for t, a. In this case, we will conclude that both $x \; E \; y$ and $x \; F \; y$ fail. Indeed, Claim 9.33 shows that then the assumption of I6 is satisfied and therefore $x \; E \; y$ must fail. Also, for every sufficiently large number $\bar{n} \in \omega$, writing $\bar{a} = \{m \in \bar{n} : g(x)(\bar{n})(m) = g(y)(\bar{n})(m)\}$, it will be the case that $(n \setminus a) \cap \bar{a} = 0$, and so the clause I2(a) is satisfied for $\bar{t} = g(x)(\bar{n})$ and $\bar{s} = g(y)(\bar{n})$ by the conclusion of (I6) at n, t, a. Thus, both $g(x)(\bar{n}) \; F_{\bar{n}} \; g(y)(\bar{n})$ and $x \; F \; y$ must fail.

Case 2 Same as Case 1, except about item I7 of the induction hypothesis. The argument and the conclusion are the same as in Case 1.

Case 3 If both Case 1 and Case 2 fail, then for every n let $z_n \in [q]$ be the sequence obtained by replacing the first n entries of x with the first n entries of y, and observe that $\phi(f(z_n) \Delta f(z_{n+1})) < \varepsilon_n/2$. To verify the inequality, discern two cases: if $x(n) = y(n)$ then $z_n = z_{n+1}$ and the inequality certainly holds. If $x(n) \neq y(n)$ then, since Case 2 failed at n, it must be that item I7 of the induction hypothesis occurred for $g(y)(n)$ and the inequality again follows. Note that since $\lim_n z_n = y$, the function f is continuous, and the submeasure ϕ is lower semicontinuous, it follows that $\phi(f(z_n)\Delta f(y)) < \Sigma_{m\geq n}\varepsilon_m/2 < \varepsilon_n$.

We will now show that if $g(x)(n)\ F_n\ g(y)(n)$ for infinitely many n, then $x\ E\ y$, and if $\neg g(x)(n)\ F_n\ g(y)(n)$ for some n then $\neg x\ E\ y$. This will prove the reduction property of the function g on the pair x, y.

Case 3a Suppose first that n is such that $t = g(x)(n)\ F_n\ g(y)(n) = s$. Writing $a = \{m \in n : x(m) = y(m)\}$ and $b = \{m \in n : g(x)(m) = g(y)(m)\}$, it is obvious that $a \subseteq b$. Write z'_n for the sequence obtained from x by replacing the entries in $b \setminus a$ with the corresponding entries of y. Since Case 1 fails, item I5 of the induction hypothesis must hold for t, a, and we have $\phi(f(x)\Delta f(z'_n)) < \varepsilon_n/4$. Since $t\ F_n\ s$, clause I2(b) shows that $\phi((f(z'_n)\Delta f(z_n)) \setminus k_n) < \varepsilon_n/4$. Moving from $f(x)$ to $f(z'_n)$ to $f(z_n)$ to $f(y)$, we have that $\phi((f(x)\Delta f(y)) \setminus k_n < \varepsilon_n/4 + \varepsilon_n/4 + \varepsilon_n/2 = \varepsilon_n$. Now, if there are infinitely many numbers n such that $g(x)(n)\ F_n\ g(y)(n)$, we conclude that $f(x) =_J f(y)$ and $x\ E\ y$.

Case 3b Suppose second that n is such that $t = g(x)(n)\ F_n\ g(y) = s$ fails. As in the previous subcase, write $a = \{m \in n : x(m) = y(m)\}$ and $b = \{m \in n : t(m) = s(m)\}$, and let $z'_n \in [q]$ be the sequence obtained from x by replacing its entries in the set $b \setminus a$ with the corresponding entries of y. Since $\neg t\ F_n\ s$ holds, item I3 shows that $\limsup_k \phi((f(z'_n)\Delta f(z_n) \setminus k) > \varepsilon_n$. However, just as in the previous subcase, we have $\phi(f(x)\Delta f(z'_n)) < \varepsilon_n/4$ and $\phi(f(y)\Delta f(z_n)) < \varepsilon_n/4$ as well. As ϕ is a submeasure, we conclude that $\limsup_k \phi((f(x)\Delta f(y) \setminus k) > \varepsilon_n/2$, in particular $\neg x\ E\ y$. □

Proof of Theorem 9.28 To see that E_1 is indeed in the spectrum, if $B \subset X$ is an I-positive set, use Fact 9.25 to find a sequence $p \in P$ of perfect trees such that $[p] \subset B$ and let $\pi_n : 2^\omega \to [p(n)]$ be a homeomorphism for every number n. Then $\pi : x \mapsto \langle \pi_n(x(n)) : n \in \omega \rangle$ is a Borel reduction of E_1 to $E_1 \restriction B$.

To prove that $P_I^{E_1}$ is a regular $\aleph_0$-distributive subposet of P_I, we will argue that every set of the form $[p]$ for $p \in P$ is an E_1-kernel and then apply Theorem 3.19. Suppose that $B \subset [p]$ is an analytic set such that $[B]_{E_1} \notin I$; we must conclude that $[B]_{E_1} \cap [p] \notin I$. Just find a condition $q \in P$ so that $[q] \subset [B]_{E_1}$, and use the σ-additivity of the σ-ideal I to thin out the trees on the sequence q if necessary and find a number n such that $\forall x \in [q]\ \exists y \in B\ x \restriction (n, \omega) = y \restriction (n, \omega)$. Then, for every number $i > n$ it must be the case that $q(i) \subset p(i)$. Thus, the set $[B]_{E_1} \cap [p]$ contains the product $\Pi_{i\leq n} p(i) \times \Pi_{i>n} q(i)$, and it is I-positive.

To show that the poset $P_I^{E_1}$ yields the model $V[\dot{x}_{\text{gen}}]_{E_1}$, we must show that for analytic I-positive sets $B, C \subset X$, either there is I-positive Borel $B' \subset B$ such that $[B']_{E_1} \cap C \in I$, or there is an I-positive Borel $B' \subset B$ and an I-preserving Borel injection $f : B' \to C$ with $f \subset E_1$. After that, Theorem 3.25 completes the proof.

Thus, fix the sets B, C and find a condition $p \in P$ with $[p] \subset B$. If $[p]_{E_1} \cap C \in I$, then we are done, otherwise find a condition $q \in P$ with $[q] \subset [p]_{E_1} \cap C$. There must be a number $n_0 \in \omega$ such that $\forall n \geq n_0\ q(n) \subset p(n)$. Consider the set $B' = \Pi_{n \in n_0}[p(n)] \times \Pi_{n \geq n_0}[q(n)]$, fix homeomorphisms $\pi_n : [p(n)] \to [q(n)]$ for every $n \in n_0$, and consider the map $f : B' \to C$ defined by $f(x)(n) = \pi_n(x)(n)$ if $n \in n_0$ and $f(x)(n) = x(n)$ for $n \geq n_0$. It is not difficult to see that the function f is a Borel injection with the required properties. □

Proof of Theorem 9.29 Let J be a Borel ideal on ω, then the equivalence relation E_J, the modulo J equality of sequences in $(2^\omega)^\omega$ is a Borel equivalence, it is in the spectrum of the ideal I for the same reason as E_1, and the equalities of this type are cofinal in $\leq_B$ by Fact 2.21. There is an F_σ ideal J such that equality modulo J is bireducible with E_{K_σ}, and so E_{K_σ} is in the spectrum of I as well. □

9.4 Products of superperfect sets

If the ideals in the product do not satisfy the covering property, one cannot expect neat canonization properties of the product ideal. In this section, we will analyze the simplest case: the two dimensional product of the σ-ideal I on ω^ω σ-generated by compact sets. Spinas (2001a) computed the associated σ-ideal and proved a rather complex canonization theorem for smooth equivalence relations. We show that E_0 and E_{K_σ} are in the spectrum of the associated σ-ideal.

Let J be the collection of Borel subsets B of the space $X = \omega^\omega \times \omega^\omega$ such that B contains no rectangle $C \times D$, where $C, D \subset \omega^\omega$ are Borel I-positive sets.

Theorem 9.35 (Jossen and Spinas 2005)

(i) The collection J is a σ-ideal of Borel sets.

(ii) The quotient P_J is proper and adds a dominating real.

(iii) The σ-ideal J is not $\mathbf{\Pi}^1_1$ on $\mathbf{\Sigma}^1_1$.

Proof Our proof of (*i*) is based on an idea of Todorcevic. Define a function $\rho : (\omega)^\omega \to \omega^\omega \times \omega^\omega$ by $\rho(a) = \langle x, y\rangle$ where x is the increasing enumeration of the set $\bigcup_i a(2i)$ and y is the increasing enumeration of the set $\bigcup_i a(2i+1)$. The following easy claim is the heart of our proofs:

Claim 9.36 *Whenever $B \subset (\omega)^\omega$ is a Milliken cube, then there are superperfect trees $S, T \subset \omega^{<\omega}$ such that $[S] \times [T] \subset \rho'' B$;*

Proof For simplicity of notation assume that B is a pure Milliken cube, $B = (a)^\omega$. Let $S \subset \omega^{<\omega}$ be a superperfect tree such that every $x \in [S]$ is an increasing enumeration of the union of some subcollection of the sequence a consisting of sets with even index on a. Let $T \subset \omega^{<\omega}$ be a superperfect tree such that every $x \in [T]$ is an increasing enumeration of the union of some subcollection of the sequence a consisting of sets with odd index on a. We claim that S, T work as required.

Let y, z be some branches of S and T respectively. By induction on $j \in \omega$ define sets $b(2j) = (\mathrm{rng}(y) \setminus \bigcup_{i<j} b(2i)) \cap \min(\mathrm{rng}(z) \setminus \bigcup_{i<j} b(2i+1))$, and $b(2j+1) = (\mathrm{rng}(z) \setminus \bigcup_{i<j} b(2i+1)) \cap \min(\mathrm{rng}(z) \setminus \bigcup_{i\leq j} b(2i))$. It is not difficult to see that the sequence b belongs to $(a)^\omega$ and $g(b) = \langle y, z\rangle$. This concludes the proof. □

The first item of the theorem is now easy. Let S, T be superperfect trees and $[S] \times [T] = \bigcup_i B_i$ be a partition into countably many Borel sets. We must inscribe a product of superperfect trees into one of the Borel sets in the partition. Pulling back through the natural homeomorphisms of $[S]$, $[T]$, and ω^ω, we may assume that in fact $S = T = \omega^{<\omega}$. Now use Theorem 8.29 to find a number $i \in \omega$ and a Milliken cube C such that $\rho'' C \subset B_i$. Claim 9.36 shows that that B_i contains a product of superperfect trees as desired.

The properness of the product was proved in Spinas (2009). To prove that a dominating real is added by the product, let $\max(x, y)$ for $x, y \in \omega^\omega$ be the smallest nondecreasing function dominating both x, y at each entry. The dominating real is then obtained as $\max(x_{\mathrm{lgen}}, x_{\mathrm{rgen}})$, where the arguments denote the left generic and the right generic. This uses a simple forcing-free claim:

Claim 9.37 *For any two superperfect trees $S, T \subset \omega^{<\omega}$ and every $z \in 2^\omega$ there are $x \in [S]$ and $y \in [T]$ such that* $\max(x, y)$ *is on all but finitely many entries bigger than z.*

Proof Let s be the stem of S and let t be the stem of T; without loss of generality $|s| < |t|$. We will find $x \in [S]$ and $y \in [T]$ such that for every $n \geq |s|$, $\max(x, y)(n) > z(n)$. By induction on $i \in \omega$ find splitnodes $s_i \in S$ and $t_i \in T$ so that:

I1. $s = s_0, t = t_0, s_0 \subset s_1 \subset \cdots$ and $t_0 \subset t_1 \subset \cdots$, and $|s_i| < |t_i| < |s_{i+1}|$;
I2. for every $n \in \mathrm{dom}(t_i) \setminus \mathrm{dom}(s_i)$, $s_{i+1}(|s_i|) > z(n)$;
I3. for every $n \in \mathrm{dom}(s_{i+1}) \setminus \mathrm{dom}(t_i)$, $t_{i+1}(|t_i|) > z(n)$.

It is quite clear from the construction that writing $x = \bigcup_i s_i$ and $y = \bigcup_i t_i$, for every $n \geq |s_0|$, it is the case that $\max(x, y)(n) > z(n)$. □

Now suppose for contradiction that S, T are superperfect trees that force in the product that $\max(x_{\mathrm{lgen}}, x_{\mathrm{rgen}})$ does not modulo finite dominate some fixed $z \in \omega^\omega$ in the ground model. The set $\{\langle x, y\rangle \in [S] \times [T] : \max(x, y)$ does not modulo finite dominate $z\}$ is in the σ-ideal J by the previous claim, and so, thinning the trees S, T if necessary, we may assume that it is empty. But then, for every $x, y \in [S] \times [T]$, $\max(x, y)$ modulo finite dominates z, this is a coanalytic formula which persists into the generic extension, and so in the extension $\max(\dot{x}_{\mathrm{gen}}, \dot{y}_{\mathrm{gen}})$ modulo finite dominates z, contradicting the assumed properties of the trees S and T.

The complexity estimate in the third item follows from Zapletal (2008, proposition 3.8.15) – if the quotient poset adds a dominating real then the σ-ideal cannot be $\mathbf{\Pi}^1_1$ on $\mathbf{\Sigma}^1_1$. □

Theorem 9.38 *E_0 is in the spectrum of J.*

Proof For any two increasing functions $x, y \in \omega^\omega$ let $f(x, y) = \{n \in \omega :$ there is an element of $\mathrm{rng}(y)$ between $x(n)$ and $x(n+1)\}$ and let E be the equivalence relation connecting $\langle x_0, y_0\rangle$ and $\langle x_1, y_1\rangle$ if their f-images have finite symmetric difference. This is certainly a Borel equivalence relation reducible to E_0; we will show that it remains bireducible with E_0 on every product of superperfect sets.

Let $S, T \subset \omega^{<\omega}$ be superperfect trees consisting of increasing functions. We must find a continuous reduction of E_0 to $E \restriction [S] \times [T]$. By induction on $n \in \omega$ build:

I1. splitnodes s_n of the tree S so that $s_0 \subset s_1 \subset \cdots$;
I2. splitnodes $\{t_v : v \in 2^n\}$ of the tree T so that $u \subset v \to t_u \subset t_v$;
I3. numbers $i(n)$ and $j(n)$ such that $i(0) = \max(\mathrm{dom}(s_0)) < j(0) < i(1) = \max(\mathrm{dom}(s_1)) < j(1) < \cdots$. For every n and every $v \in 2^n$ the sets $\mathrm{rng}(t_{v^\frown 0} \setminus t_v)$, $\mathrm{rng}(t_{v^\frown 1} \setminus t_v)$ are both nonempty and they are subsets of the intervals $(s_{n+1}(i_n), s_{n+1}(i_n+1))$ and $(s_{n+1}(j_n), s_{n+1}(j_n+1))$ respectively.

This is not difficult to do. Let s_0, t_0 be arbitrary splitnodes of the respective trees S, T. Suppose that $\langle s_n, t_v : v \in 2^n\rangle$ have been found. Let $i_n = \max(\mathrm{dom}(s_n))$. For every $v \in 2^n$ find a splitnode $t_{v^\frown 0}$ in the tree T extending t_v such that the number $t_{v^\frown 0}(\mathrm{dom}(t_v))$ is greater than all elements of $\mathrm{rng}(s_n)$.

Find a splitnode s'_n of the tree S extending s_n such that the number $s'_n(i_n + 1)$ is larger than all elements of $\mathrm{rng}(t_{v\frown 0})$ for all $v \in 2^n$. Let $j(n) = \max(\mathrm{dom}(s'_n))$. For every $v \in 2^n$ find a splitnode $t_{v\frown 1}$ in the tree T extending t_v such that the number $t_{v\frown 1}(\mathrm{dom}(t_v))$ is greater than all elements of $\mathrm{rng}(s'_n)$. Find a splitnode s_{n+1} of the tree S extending s'_n such that the number $s_{n+1}(j(n))$ is larger than all elements of $\mathrm{rng}(t_{v\frown 0})$ for all $v \in 2^n$. This completes the induction step.

Now for every $z \in 2^\omega$ let $h(z) = \langle \bigcup_n s_n, \bigcup_n t_{z \restriction n} \rangle$. It is not difficult to see that $f \circ h(z) = \{i_n : z(n) = 0\} \cup \{j_n : z(n) = 1\}$ modulo finitely many exceptions caused by the interplay between s_0 and t_0. Thus, the continuous function h reduces E_0 to $E \restriction [S] \times [T]$ as required. □

Theorem 9.39 *E_{K_σ} is in the spectrum of J, and J has a rectangular coding function.*

Proof This is an immediate consequence of the fact that the poset P_J adds a dominating real and Corollary 6.55. □

10

The countable support iteration ideals

The countable support iteration is a central operation on forcings since the pioneering work of Shelah (1998). It corresponds to a certain operation on σ-ideals, the transfinite Fubini power, as shown in Zapletal (2008, section 5.1). The natural question then arises whether this operation preserves natural canonization properties of the σ-ideals in question.

The question may be natural, but it also seems to be very hard, even for iterations of length two in the general case. We will prove a rather restrictive canonization result for iterations of a certain broad class of ideals, the best possible canonization result for the special case of iterations of Sacks forcing, and an anticanonization theorem for some other ideals. For a σ-ideal I on a Polish space X and a countable ordinal $\alpha \in \omega_1$ write I^α for the transfinite Fubini power of I. This is a σ-ideal on X^α defined in the next section. There are certain obvious obstacles for canonization on I^α. For ordinal $\beta \in \alpha$ let id_β be the equivalence relation on X^α connecting sequences $x, y \in X^\alpha$ if $x \restriction \beta = y \restriction \beta$. For an equivalence relation F on X, write $\mathrm{id}_\beta \times F \times \mathrm{ev}$ for the equivalence relation connecting $x, y \in X^\alpha$ if $x \restriction \beta = y \restriction \beta$, and $x(\beta)\ F\ y(\beta)$. It turns out that these are the only obstacles in a certain context. We will introduce a preservation property, called the *weak Sacks property*, and prove the following:

Theorem 10.1 *Let I be a σ-ideal on a compact metric space X such that I is $\mathbf{\Pi}^1_1$ on $\mathbf{\Sigma}^1_1$, every I-positive analytic set has a Borel I-positive subset, and the quotient poset P_I is provably proper and has the weak Sacks property. Let $\alpha \in \omega_1$ be a countable ordinal. Then, for every Borel I^α-positive Borel set B and every equivalence relation E on B reducible to E_{K_σ} there is an ordinal $\beta \leq \alpha$ and an equivalence F on X reducible to E and a Borel I^α-positive subset $C \subset B$ such that on C, $E = \mathrm{id}_\beta \times F \times \mathrm{ev}$. In particular,*

$$(\leq_B E_{K_\sigma}) \to_{I^\alpha} \{\mathrm{id}_\beta \times F \times \mathrm{ev} : \beta \leq \alpha, F \leq_B E_{K_\sigma}\}.$$

The weak Sacks property may look quite restrictive, but in practice a good number of forcings satisfy it, it is easy to check and some forcings with this property have nontrivial features in the spectrum – the E_0, E_2, or Silver forcings of previous sections do satisfy the weak Sacks property. We will also show that, in a certain more general case, very complex equivalence relations arise at the first infinite stage of the iteration; however, in the countable support iterations of Sacks forcing a much stronger canonization result holds.

Theorem 10.1 can be used to extend various canonization theorems for a single stage of the iteration to the whole iteration. Here, we will limit ourselves to the following:

Corollary 10.2 *Let I be a σ-ideal on a compact metric space X such that I is $\mathbf{\Pi}^1_1$ on $\mathbf{\Sigma}^1_1$, every I-positive analytic set has a Borel I-positive subset, and the quotient poset P_I is provably proper and has the weak Sacks property. Let E be an equivalence relation Borel reducible to E_{K_σ}. If I has total canonization for equivalences reducible to E, then for every countable ordinal $\alpha \in \omega$,*

$$(\leq_B E) \to_{I^\alpha} \{\mathrm{id}_\beta \times \mathrm{ev} : \beta \leq \alpha\}.$$

In particular, if P_I adds a minimal real degree then the real degrees in the iteration of length α are well-ordered in type α, and E_0 is not in the spectrum of the countable support iteration of E_2 forcing.

In the special case of Sacks forcing, we are able to treat even equivalence relations more complex than E_{K_σ}:

Theorem 10.3 *Let I be the σ-ideal of countable subsets of 2^ω. Let $\alpha \in \omega_1$ be a countable ordinal. Then*

$$\text{analytic} \to_{I_\alpha} \{\mathrm{id}_\beta \times \mathrm{ev} : \beta \leq \alpha\}.$$

On the anticanonization side, we prove the following:

Theorem 10.4 *Suppose that I is a $\mathbf{\Pi}^1_1$ on $\mathbf{\Sigma}^1_1$ equivalence relation on a Polish space X such that every analytic I-positive set has a Borel I-positive subset and the quotient P_I is provably proper. Suppose that E is a Borel equivalence relation on X with all equivalence classes in I, nontrivially ergodic with respect to I. Then the iteration ideal I^ω has a rectangular coding function, and its spectrum is cofinal among the analytic equivalence relations.*

10.1 Iteration preliminaries

Let us first recall the analysis of the countable support iteration of definable proper forcing as introduced in Zapletal (2008, section 5.1):

Definition 10.5 (Zapletal 2008, definition 5.1.1) Let X be a Polish space and I a σ-ideal on X. Let $\alpha \in \omega_1$ be a countable ordinal. The space X^α of all α-sequences of elements of X is equipped with the product topology. The α-th *Fubini iteration* I^α of I is the σ-ideal on the space X^α generated by sets A_f, where $f : X^{<\alpha} \to I$ is an arbitrary function and $A_f = \{\vec{x} \in X^\alpha : \exists \beta \in \alpha\ \vec{x}(\beta) \in f(\vec{x} \restriction \beta)\}$.

Definition 10.6 Let I be a σ-ideal on a Polish space X and J a σ-ideal on a Polish space Y. The symbol $I * J$ stands for the Fubini product of I and J: $I * J = \{B \subset X \times Y : \{x \in X : \{y \in Y : \langle x, y \rangle \in B\} \notin J\} \in I\}\}$.

If I, J are $\mathbf{\Pi}^1_1$ on $\mathbf{\Sigma}^1_1$ σ-ideals such that the quotient forcings are proper and every analytic positive set contains a Borel positive subset, then the two-step iteration of posets P_I and P_J is naturally equivalent to the quotient P_{I*J}. This pattern persists to transfinite iterations. We will need a canonical form of I^α-positive Borel set:

Definition 10.7 (Zapletal 2008, definition 5.1.5) An I, α*-tree* is a tree $p \subset X^{<\alpha}$ such that each node extends into an I-positive set of immediate successors and moreover, whenever $t_0 \subset t_1 \subset \dots$ is an increasing sequence of nodes in p and $\bigcup_n t_n \in X^\beta$ for some $\beta \in \alpha$, then $\bigcup_n t_n \in p$.

Fact 10.8 (Zapletal 2008, section 5.1.3) *Suppose that:*

(i) the σ-ideal I is $\mathbf{\Pi}^1_1$ on $\mathbf{\Sigma}^1_1$ and every analytic I-positive set contains an I-positive Borel subset;
(ii) the quotient forcing P_I is provably proper.

Then:

(i) the σ-ideal I^α is $\mathbf{\Pi}^1_1$ on $\mathbf{\Sigma}^1_1$, and every analytic I^α-positive set contains an I^α-positive Borel subset;
(ii) the coutable support iteration P^α_I of P_I of length α is in the forcing sense equivalent to the quotient P_{I^α};
(iii) every I^α-positive analytic set contains all branches of a Borel I, α-tree.

If suitable large cardinals exist, the above fact holds even for σ-ideals with definition more complicated than $\mathbf{\Pi}^1_1$ on $\mathbf{\Sigma}^1_1$, obviously except for the part of the conclusion that evaluates the complexity of the Fubini iteration ideal. This makes it possible to apply the analysis to forcings such as Laver or Mathias, whose associated ideals are more complicated. Large cardinals or other assumptions transcending ZFC are necessary for the argument to go through in the case of Laver forcing. The countable support iterations in which the iterands vary can be handled in a quite similar way, and we will not bother to explicitly state the trivial adjustments necessary.

10.2 The weak Sacks property

In order to state and prove our main result, we must define a rather common preservation property, provide its strategic characterizations, and apply them to prove common preservation theorems.

Definition 10.9 A *slalom* is a function g with domain ω such that for every $n \in \omega$, $g(n)$ is a set of size $\leq 2^n$. A forcing P has the *weak Sacks property* if for every function $f \in \omega^\omega$ in the extension there is a ground model infinite set $b \subset \omega$ and a ground model slalom g such that for every $n \in b$, $f(n) \in g(n)$ holds.

This is an obvious weakening of the *Sacks property* that requires the infinite set a to be actually equal to ω (Bartoszynski and Judah 1995, definition 6.3.37). The weak Sacks property played an important role in Zapletal (2009), where it is shown that that if I is a $\mathbf{\Pi}^1_1$ on $\mathbf{\Sigma}^1_1$ σ-ideal such that the quotient P_I is bounding, then P_I preserves P-points if and only if P_I adds no independent reals and has the weak Sacks property.

It is not difficult to show that a great number of posets have the weak Sacks property. We give only the most rudimentary examples relevant for our book.

Example 10.10 The E_0 forcing has the (full) Sacks property.

Proof Let P be the combinatorial presentation of the forcing as described in Definition 7.3. Suppose $p \in P$ is a condition, and τ is a P-name for a function in ω^ω. A simple fusion argument will yield a condition $q \leq p$ such that for every n and every splitnode t at n-th level of q, $q \restriction t$ decides the value of $\dot{f}(n)$. Let $g(n) = \{m : \text{for some } t \text{ at } n\text{-th splitting level of } q, q \restriction t \Vdash \dot{f}(n) = \check{m}\}$ and observe that $q \Vdash \forall n\ \tau(n) \in \check{g}(n)$ as desired. □

Example 10.11 The E_2 forcing has the (full) Sacks property.

Proof Let I be the E_2-ideal of Definition 7.41; the quotient P_I is proper by Theorem 7.39. Let $B \in P_I$ be a condition and τ a name for a function in ω^ω. Thinning out the condition B if necessary, we may assume that B is compact and there is a continuous function $f : B \to \omega^\omega$ such that $B \Vdash \tau = f(\dot{x}_{\text{gen}})$. Assume for simplicity that both B, f are lightface Π^0_1. Running the proof of Theorem 7.43 for the function $f \subset 2^\omega \times \omega^\omega$, we see that at n-th step of the fusion we may require the sets $\{A_t : t \in 2^n\}$ to be such that the value $f(x)(n)$ is the same for all $x \in A_t$, equal to some $r(t) \in \omega$. Let $C \subset B$ be the E_2-positive result of the fusion, and let g be the slalom defined by $g(n) = \{r(t) : t \in 2^n\}$. It is immediate that for all $x \in C$ and all $n \in \omega$, $f(x)(n) \in g(x)$, and so $C \Vdash \forall n\, \tau(n) \in \check{g}(n)$ as required. □

Example 10.12 Whenever X is a compact metric space and I is a σ-ideal σ-generated by an F_σ (in the hyperspace $K(X)$) set of compact subsets of X, then P_I has the weak Sacks property.

Proof Zapletal (2008, theorem 4.1.8) shows that the poset P_I is bounding, and its proof shows that it preserves P-points; the weak Sacks property formally follows from these two properties by Zapletal (2009). □

Example 10.13 The Silver forcing has the (full) Sacks property.

Proof A fusion argument entirely identical to Example 10.10. □

The weak Sacks property will be used in the proofs below in the following optically stronger form:

Claim 10.14 *Suppose that P is a proper forcing. If P has the weak Sacks property, then for every collection $\{\chi_n : n \in \omega\}$ of names for functions with domain ω there is a stronger condition $q \le p$, a slalom g and an infinite set $b \subset \omega$ such that for every $n \in \omega$ for all but finitely many numbers $m \in b$, we have $q \Vdash \chi_n(m) \in \check{g}(m)$.*

Proof Use the properness of P to strengthen p if necessary and find a countable set c such that $p \Vdash \bigcup_n \operatorname{rng}(\chi_n) \subset \check{c}$. Let τ be the name for a function from ω to $[c]^{<\aleph_0}$ given by $\tau(i) = \{\chi_n(2i) : n \in i\}$. An application of the weak Sacks property finds a condition $q \le p$, an infinite set $a \subset \omega$, and a slalom h such that $q \Vdash \forall i \in a\, \tau(i) \in \check{h}(i)$. Now let g be a slalom such that for all $i \in a$, $g(2i)$ is the union of all sets in $h(i)$ of size $\le i$ (which has size $\le 2^i \cdot i \le 2^{2i}$). It is not difficult to see that writing $b = \{2i : i \in a\}$, the condition q and b, g have the desired property. □

10.3 A game reformulation

In order to state and prove our main result, we cannot avoid a game reformulation of the weak Sacks property. We will use a Boolean game on P, and in the case of a quotient poset, also another game on the underlying Polish space. The Boolean game is instrumental in showing that the weak Sacks property is preserved under the countable support iteration, while the Polish game is necessary for the canonization result.

Definition 10.15 Let P be a partial order. The game $G_0(P)$ between Players I and II proceeds in infinitely many rounds in the following way. Player II first indicates an initial condition $p_{\text{ini}} \in P$. After that, in round n Player I indicates a natural number $i_n \in \omega$, Player II plays a P-name τ_n for a natural number, and Player I answers with a set $a_n \subset \omega$ of size i_n. In the end, Player I wins if the *result* of the play, the Boolean expression $p_{\text{ini}} \wedge \bigwedge_n \|\tau_n \in \check{a}_n\|$, is a nonzero element in the completion of the poset P.

Definition 10.16 Let I be a σ-ideal on a Polish space X. The game $G_1(I)$ between Players I and II proceeds as follows: Player II starts out with a compact I-positive set $K_{\text{ini}} \subset X$. Then, each round $n \in \omega$ consists of three moves: first Player II chooses a set $A_n \in I$, then Player I chooses a finite set $K_n \subset K_{\text{ini}}$, and then Player II chooses an open set $O_n \subset X$. We require that $K_n \subset \bigcap_{m \in n} O_m \setminus A_n$, and $O_n \supset K_n$. Player I wins if the *result* of the play, the set $K_{\text{ini}} \cap \bigcap_n O_n$, is I-positive.

Theorem 10.17 *Let P be a poset. The following are equivalent:*

(i) P has the weak Sacks property;
(ii) Player II has no winning strategy in the game G_0.

Proof (*ii*) implies (*i*). To see that, assume (*ii*) holds and $p \in P$ and $\dot{f}$ is a P-name for a function in ω^ω. Consider the strategy for Player II in which he indicates $p_{\text{ini}} = p$ and at round n, once Player I indicates i_n, Player II answers with the name $\dot{f}(j_n)$ for some number j_n greater than i_n as well as j_m for $m \in n$, and Player I finishes the round with some set $g(n)$ of size i_n. As this is not a winning strategy for Player II, Player I has a winning counterplay, and the result of that counterplay is a condition $q \leq p$ which forces $\forall n\ \dot{f}(j_n) \in \check{g}(n)$ as required in the weak Sacks property.

To see that (*i*) implies (*ii*), suppose that P has the weak Sacks property and σ is a strategy for Player II in the game G_0; we must produce a winning counterplay for Player I. Let $p \in P$ be the initial move of the strategy σ. For every $n \in \omega$ let χ_n be the name for a function with domain ω defined by letting $\chi_n(m)$ to be the set of those τ such that for some play t of length $\leq n$ in which Player I uses only numbers $\leq n$ and subsets of m and Player II uses the strategy σ, if Player I extends t by playing m, then σ answers with τ. It is clear that the values of τ_n are finite sets whose size is bounded by some constant depending only on n. Claim 10.14 now provides a condition $q \leq p$, a slalom g, and an infinite set $b \subset \omega$ such that for all n, for all but finitely many $m \in b$, $q \Vdash \chi_n(m) \subset g(m)$.

By induction on $i \in \omega$ build plays t_i so that $0 = t_0 \subset t_1 \subset \cdots$ in which Player II follows the strategy σ, so that for every round i, $q \Vdash \tau_i \in a_i$; then $\bigcup_i t_i$ is a winning counterplay against the strategy σ, since the condition q is below the result of this counterplay. To perform the induction step, let n be so large that moves of Player II in t_i use only numbers smaller than n. Use the property of the slalom g to find a number m so large that $q \Vdash \tau_n(m) \in \check{g}(m)$, and extend the play t_i by letting Player I make the moves m and $g(m)$. This concludes the induction step and the proof. □

Theorem 10.18 *Suppose that I is a σ-ideal on a Polish space X such that the quotient forcing P_I is proper and bounding. The following are equivalent:*

(i) P_I has the weak Sacks property;
(ii) Player II has no winning strategy in the game G_1.

Proof (*ii*) implies (*i*). Suppose that $K \subset X$ is a Borel I-positive set, and $\dot{f}$ is a P_I-name for a function in ω^ω. Since the poset P_I is bounding, we may thin down K to a compact I-positive set and find a continuous function $g : K \to \omega^\omega$ such that $K \Vdash \dot{f} = \dot{g}(\dot{x}_{\mathrm{gen}})$. Consider the strategy for Player II in which he indicates $K_{\mathrm{ini}} = K$ and at round n he plays $A_n = 0$, and after Player I responds with a finite set K_n, Player II finds a number j_n greater than the size of K_n and all the numbers j_m for $m \in n$, also for each $x \in K_n$ finds a basic open set O_x containing x such that for every $y \in O_x \cap K$ we have $g(y(j_n)) = g(x(j_n))$; then, set $O_n = \bigcup_{x \in K} O_x$. Since this is not a winning strategy, there is a counterplay against it winnng for Player I, and its result is a condition which forces $\forall n\, \dot{f}(j_n) \in h(n)$, where $h(n) = \{g(x)(j_n) : x \in K_n\}$. This verifies the weak Sacks property of the poset P_I.

To see why (*i*) implies (*ii*), suppose that P_I has the weak Sacks property and let σ be a strategy for Player II in the game G_1. We must produce a counterplay against σ winning for Player I. Let K_{ini} be the initial move of the strategy σ. Let M be a countable elementary submodel of a large structure containing all objects mentioned so far, and let $L \subset K_{\mathrm{ini}}$ be a compact I-positive set consisting of M-generic points only, such that all sets in M are relatively clopen in L. Such a set exists by the properness and boundedness assumption (Zapletal 2008, theorem 3.3.2).

For every finite play t ending with an open set move of Player II that follows the strategy σ write $[t]$ for the intersection of all open sets played during the course of t. We let $[0] = X$.

Claim 10.19 *For every finite play $t \in M$ observing the strategy σ and every $m \in \omega$ there is a finite collection $S(t, m)$ of pairwise disjoint open sets in the* *model M such that $[t] \cap L \subset \bigcup S(t, m)$ and for every set $S \subset S(t, m)$ of size $\leq 2^m$ there is a play s extending t such that $\bigcup S \cap L \subset [s]$.*

Proof Let $A \in I$ be the next move of Player II past t dictated by the strategy σ. Since $A \in M$, any 2^m-tuple from the compact set $L \cap [t]$ is a legal move of Player I after t. Let $R = \{[s] : s$ extends t by one more round in which Player I indicates some 2^m-tuple of points in $L \cap [t]$ and II follows the strategy $\sigma\}$ and note that $R \subset M$. The space $(L \cap [t])^{2^m}$ is compact and a subset of the union $\bigcup\{O^{2^m} : O \in R\}$. Thus, there is a finite collection $R(t, m) \subset R$ such that $(L \cap [t])^{2^m} \subset \bigcup\{O^{2^m} : O \in R(t, m)\}$. By the choice of the set L, for any Boolean combination b of sets in $R(t, m)$ there is an open set $O_b \in M$ such that $L \cap [t] \cap b = L \cap O_b$. Let $S(t, m) = \{O_b : b$ is a complete Boolean combination of sets in $R(t, m)$ and $L \cap O_b \neq 0\}$. We claim that this collection works.

First of all, $L \cap [t] \subset \bigcup S(t, m)$ follows from the choice of the set $R(t, m)$. Second, if $S \subset S(t, m)$ is a collection of size $\leq 2^m$ then there is a set $d \subset L \cap [t]$ of size $\leq 2^m$ which meets all sets in S. There must be an open set $O \in R(t, m)$ with $d \subset O$. By the elementarity of the model M, there must be a play s in the model M extending t by one round such that $[s] = O$. This play works as required: for every set $P \in S$, the set $P \cap O \cap L \cap [t]$ is nonempty and so in fact $P \cap L \subset O \cap L \cap [t]$. □

For every finite play $t \in M$ let χ_t be the name for the function defined by $\chi_t(m) =$ the unique element of $S(t, m)$ containing $\dot{x}_{\text{gen}}$ if such exists, otherwise $\chi_t(m) =$ trash. Use Claim 10.14 to find a compact I-positive set $L' \subset L$, an infinite set $b \subset \omega$ and a slalom g such that for every $t \in M$, for all but finitely many numbers $m \in \omega$, $L' \Vdash \chi_t(m) \in g(m)$. Removing a small subset of L' if necessary, we may assume that intersections of L' with open sets are either empty or I-positive.

Now, by induction on $i \in \omega$ build plays $t_i \in M$ observing the strategy σ such that $0 = t_0 \subset t_1 \subset \cdots$ and $L' \subset [t_i]$. Once this is performed, the play $\bigcup_i t_i$ is obviously winning for Player II since its result contains the I-positive compact set L'. The basis $t_0 = 0$ satisfies the induction hypothesis. Now suppose that t_i has been found. Find a number $m \in b$ such that $L' \Vdash \tau_{t_i}(m) \in g(m)$. Since g is a slalom, the set $g(m)$ contains at most 2^m many elements of $S(t_i, m)$. We will now argue that $L' \subset \bigcup(g(m) \cap S(t_i, m))$. For every $x \in L'$, since $x \in [t_i]$, x must belong to a unique element $O \in S(t_i, m)$, and as $O \cap S(t_i, m) \Vdash \chi_{t_i}(m) = O$, it must be the case that $O \in S(t_i, m) \cap g(m)$ and so $x \in \bigcup(g(m) \cap S(t_i, m))$. By the choice of the collection $S(t_i, m)$, there is an extension t_{i+1} such that $[t_{i+1}] \supset \bigcup(g(m) \cap S(t_i, m))$. This concludes the induction step and the proof of the theorem. □

Theorem 10.20 *Suppose that I is a $\mathbf{\Pi}^1_1$ on $\mathbf{\Sigma}^1_1$ σ-ideal on a Polish space X such that the quotient forcing P_I is proper. Then the games $G_0(P_I)$, $G_1(P_I)$ are both determined.*

The proof uses the unraveled versions of both games which we now proceed to define. Suppose that I is $\mathbf{\Pi}^1_1$ on $\mathbf{\Sigma}^1_1$ ideal on a Polish space X.

Definition 10.21 Let $A \subset 2^\omega \times X$ be an analytic set universal for analytic subsets of X, and let $D \subset \langle(2^\omega)^{\omega\times\omega} \times \omega^\omega$ be a closed set projecting into the analytic set $\{y \in (2^\omega)^{\omega\times\omega} : \bigcap_n \bigcup_m A_{y(n,m)} \notin I\}$. The *unraveled version of* G_0 is the game played as follows: Player II starts with an initial I-positive Borel set $B_{\text{ini}} \subset X$. Then, in each round n Player I starts with a number i_n, Player II answers with a name τ_n for a natural number, Player I answers with a finite set $a_n \subset \omega$ of size i_n. This concludes round n except for the following. At certain rounds of his choice, Player I may also indicate points $y(m,k) \in 2^\omega$ for some m, k as well as a natural number j_l for some $l \in \omega$. It must be the case that $A_{y(m,k)}$ is a Borel I-positive subset of B_{ini} and it forces $\tau_m \in \check{a}_m$. In the end, Player I wins if he chose the points $y(m,k)$ for all $m, n \in \omega$, the numbers j_l for all $l \in \omega$, and the point $\langle\langle y(m,k) : m, k \in \omega\rangle, \langle j_l : l \in \omega\rangle\rangle$ belongs to the closed set D.

Definition 10.22 Let $\mathcal{O}$ be any countable basis of the space X closed under finite unions and intersections. Let D be any closed subset of $K(X) \times \mathcal{O}^\omega \times \omega^\omega$ projecting into the analytic set $\{\langle K, O_n : n \in \omega\rangle : K \cap \bigcap_n O_n \notin I\}$. The *unraveled version of* G_1 is the game played as follows. Let I be a σ-ideal on a Polish space X. The game $G_1(I)$ between Players I and II proceeds as follows: Player II starts out with a compact I-positive set K_{ini}. Then, each round $n \in \omega$ consists of four moves: first Player II chooses a set $A_n \in I$, then Player I chooses a finite set $K_n \subset K_{\text{ini}}$, then Player II chooses an open set $O_n \in \mathcal{O}$ and Player I finishes with an open set $P_n \in \mathcal{O}$. We require that $K_n \subset \bigcap_{m \in n} P_m \setminus A_n$, $O_n \supset K_n$, and $K_n \subset P_n \subset O_n$. At certain rounds of his choice, Player I may also indicate a number $j_l \in \omega$ for some $l \in \omega$. Player I wins if he chose j_l for all $l \in \omega$ and $\langle K_{\text{ini}}, \langle P_n : n \in \omega\rangle, \langle j_l : l \in \omega\rangle\rangle \in D$.

Proof of Theorem 10.20 The argument follows the lines of the proofs in Zapletal (2008, section 3.10). We will start with the case of G_0. The unraveled version of G_0 is certainly determined by Borel determinacy (Martin 1985), as the payoff set is G_δ for Player I in the wide tree of all possible plays. It is clear that the unraveled version is more difficult for Player I than the original version, so it will be enough to show that if P_I has the weak Sacks property, then Player II has no winning strategy in the unraveled version. Suppose that σ is a strategy for Player II; we must find a winning counterplay.

For every finite set $a \subset 2^\omega$ and every $k \in \omega$ define $\tau(a, k)$ to be the name for a function with domain ω. The value $\tau(a, k)(m)$ is defined as the set of all τ such that for some play p of not more than k many rounds in which Player I uses only numbers smaller than k, sets below k, and points in the set a and Player II uses the strategy σ, if Player I extends the play p with the number m then the strategy σ responds with τ. Note that the values of the function $\tau(a, k)$ are finite subsets of ω whose cardinality is bounded by some fixed number depending on a, k only.

Let M be a countable elementary submodel of a large enough structure containing σ. Let $B \in P_I \cap M$ be the initial move of the strategy σ. Use properness of P_I to conclude that the set C of all P_I-generic points of B over M is I-positive.

Claim 10.23 *There is a Borel I-positive set $C' \subset C$, an infinite set $b \subset \omega$ and a slalom g such that for every finite set $a \subset 2^\omega \cap M$ and every number $k \in \omega$, for all but finitely many numbers $m \in b$ it is the case that for every $x \in C'$, $\tau(a, k)(m)/x \subset g(m)$ holds.*

Proof First use Claim 10.14 to find an a Borel I-positive set $C' \subset C$, an infinite set $b \subset \omega$ and a slalom g such that for every finite set $a \subset 2^\omega \cap M$ and every number $k \in \omega$, for all but finitely many numbers $m \in b$ it is the case that $C' \Vdash \tau(a, k)(m) \subset g(m)$. Removing an I-small set from C', we get a set with the properties required in the claim. □

Choose an enumeration $\{y_i : i \in \omega\}$ of the countable set $M \cap 2^\omega$. By induction on $m \in \omega$ find increasing numbers $i_m \in b$ and sets a_m such that:

I1. whenever $k_m = \max(\bigcup\{a_{m'} : m' \in m\})$, for every $x \in C'$ we have $\tau(\{y'_m : m \in m\}, k_m)(i_m)/x \subset g(i_m)$;

I2. $a_m = g(i_m)$.

This is easy to do using the properties of the set $C' \subset C$. Let $y \in (2^\omega)^{\omega\times\omega}$ be a function such that $y(\cdot, m)$ enumerates all the points $z \in M$ such that $A_z \Vdash \phi_m$ and $A_z \subset B$. Observe that $C' \supset \bigcap_n \bigcup_m A_{y(n,m)}$ since every element $x \in C'$ is P_I-generic over M and so by the forcing theorem, it must meet some set $A \in P_I \cap M$ forcing the sentence $\tau(\{y'_m : m \in m\}, k_m)(i_m) \subset g(i_m)$. Thus, the set $\bigcap_n \bigcup_m A_{y(n,m)}$ is I-positive, and therefore there must be a point $z \in \omega^\omega$ such that $\langle y, z\rangle \in D$. Finally, we are ready to construct a winning counterplay against the strategy σ. Player I simply produces the numbers i_m, sets a_m, and the points y and z with the following additional requirement: he places the point $y(n, m)$ to y only at some round m' such that $y(n, m) = y_i$ for some $i \in m'$, and he places new entries to the sequence z only in rounds where they are smaller than the numbers i_m. This completes the proof that Player II has no winning strategy in the determined unraveled version of G_0, therefore Player I has it and so Player I has a winning strategy even in the more difficult game G_0.

For the proof of the determinacy of G_1, note that the unraveled version of G_1 has a Borel payoff, and therefore it is determined by Martin (1985). Since the unraveled version of G_1 is clearly more difficult than G_1 itself for Player I, it is enough to show that a winning strategy σ for Player II in the unraveled version can be transformed into a winning strategy τ for Player II in G_1 itself. To describe the strategy τ, it starts with the same compact set K_{ini} as σ. The play of G_1 progresses with some moves K_n, O_n for $n \in \omega$. At round n, for all functions $v : n \to n+1 \cup \{\text{pass}\}$ bounded by the identity function on ω, Player II will on the side create plays $p(v)$ of the unraveled version of G_1 so that:

- $p(v)$ is a play of n many rounds in which Player II followed the strategy σ, Player I played the sets K_m for $m \in n$ and the sets $P_m = O_m$ for $m \in n$, and the sequence v;
- $u \subset v$ implies $p(u) \subset p(v)$.

This is not difficult to do as follows. To pass from round n to round $n+1$, in the play of the unraveled game Player II will make the move $A_{n+1} \in I$ equal to the union of all sets that the strategy σ asks Player II to play after the plays $p(v)$. Player I responds with a finite set K_{n+1} which is appended to all games $p(v)$. The answer O_{n+1} of Player II in the unraveled game is now the intersection of all the moves that the strategy σ advises Player II to play after $p(v)$ – this is an open set which contains K_{n+1}, and therefore a legal next move for Player II in the play of G_1 under construction. The games $p(v)$ are extended by letting Player I make the move O_{n+1}, and either playing some number $\leq n+2$ or passing. This completes the description of the strategy τ.

To see that τ is indeed a winning strategy, suppose that Player II lost a play in which he used it. Then, the result of the play would be an I-positive set, and there would be a branch $z \in \omega^\omega$ such that $\langle K_{\text{ini}}, \langle O_n : n \in \omega \rangle, z \rangle \in D$. Find a function $v : \omega \to \omega \cup \{\text{pass}\}$ whose numerical values stay below the identity and form the sequence z. Then the play $\bigcup_n p(v \restriction n)$ would have to be a counterplay in the unraveled game against the strategy σ that Player II lost, contradicting the assumption that σ was winning. □

Theorem 10.24 *Let I be a $\mathbf{\Pi}^1_1$ on $\mathbf{\Sigma}^1_1$ σ-ideal on a Polish space X such that the quotient poset is provably proper and has the weak Sacks property. Then the countable support iterations of P_I of arbitrary length has the weak Sacks property as well.*

Proof This is a strategy diagonalization argument for the winning strategies for Player I in the game $G_0(I)$. It is completely parallel to the proofs in Zapletal (2008, section 6.3.5) and we omit it. □

10.4 The canonization result

This section contains the proof of Theorem 10.1. To establish a suitable terminology for the equivalence relation E_{K_σ}, define an extended-value metric d on ω^ω by setting $d(x, y) = \sup\{|y(n) - z(n)| : n \in \omega\}$. The distance can be infinite, and E_{K_σ} connects those points in ω^ω with finite distance. The proof of the theorem proceeds by induction on α, and starts with the investigation of the successor step, where in fact most of the difficulty lies.

Lemma 10.25 *Let I, J be $\mathbf{\Pi}^1_1$ on $\mathbf{\Sigma}^1_1$ σ-ideals on Polish spaces X, Y such that provably in ZFC, every analytic I (or J)-positive set contains a Borel I (or J)-positive subset and the quotients P_I, P_J are proper and have the weak Sacks property. Let E be an equivalence relation on $X \times Y$ reducible to E_{K_σ}, and let $B \subset X \times Y$ be an $I * J$-positive Borel set. There is a Borel $I * J$-positive Borel set $C \subset B$ such that:*

(i) either there is an equivalence relation $F \leq_B E$ on X such that $E \restriction C = F \times \mathrm{ev}$;
(ii) or there is an equivalence relation $G \leq_B E$ on Y such that $E \restriction C = \mathrm{id} \times G$.

Proof To simplify the notation, assume that both σ-ideals I, J live on the Cantor space 2^ω. Let $B \subset 2^\omega \times 2^\omega$ be a Borel $I * J$-positive set, and let E be a Borel equivalence relation on B reducible to E_{K_σ}. Let $h : B \to \omega^\omega$ be a Borel function reducing E to E_{K_σ}. Since the iteration $P_I * P_J$ is a bounding forcing, by Fact 2.52 we can thin out B if necessary to a compact $I * J$-positive set on which the reduction h is continuous. Whenever $x \in 2^\omega$ is a point, write E_x for the equivalence relation on B_x given by $y \; E_x \; z$ if and only if $\langle x, y\rangle \; E \; \langle x, z\rangle$. Consider the set $A = \{x \in 2^\omega : E_x$ has a J-positive class$\}$; this is an analytic set as J is $\mathbf{\Pi}^1_1$ on $\mathbf{\Sigma}^1_1$. The treatment divides into two cases.

Case 1 Suppose first that $A \notin I$. The assumptions show that there is an Borel I-positive subset $A' \subset A$. Let $B' = \{\langle x, y\rangle \in B : x \in A' \wedge [y]_{E_x} \notin J$. This is an analytic set such that for every $x \in A'$, the section B'_x is nonempty. By Fact 2.51(*ii*) we may thin out A' if necessary and find a Borel function $g : A' \to 2^\omega$ such that $g \subset B'$. Let $C \subset B$ be the set $\{\langle x, y\rangle \in B : x \in A' \wedge \langle x, y\rangle \; E \; \langle x, g(x)\rangle\}$. This is clearly a Borel K-positive set and if $\langle x_0, y_0\rangle, \langle x_1, y_1\rangle$ are two points in C then they are E-related just in case $\langle x_0, g(x_0)\rangle, \langle x_1, g(x_1)\rangle$ are. Item (*i*) of the claim holds with the equivalence relation F connecting points x_0, x_1 if $\langle x_0, g(x_0)\rangle\ E\ \langle x_1, g(x_1)\rangle$.

Case 2 Suppose now that $A \in I$. In this case, we will find a compact $I * J$-positive set $C \subset B$ such that its vertical sections are pairwise E-unrelated. This will yield (*ii*) of the lemma. Just let G be the equivalence relation on 2^ω connecting points $y, z \in 2^\omega$ if they belong to the same vertical section C_x of C and $y\ E_x\ z$. It is easy to see that the equivalence relation G is reducible E, and $\mathrm{id} \times G = E \restriction C$.

The set $A' = \{x \in 2^\omega : B_x \notin J\}$ is analytic and I-positive, as $B \notin I * J$. By the assumptions, thinning out the set A' if necessary we may assume that it is Borel. The set $A \in I$ is analytic, and as the σ-ideal is $\mathbf{\Pi}^1_1$ on $\mathbf{\Sigma}^1_1$, by the first reflection theorem A has a Borel superset in I. Subtracting this superset from A' if necessary, we may assume that $A' \cap A = 0$. Let $\dot{\tau}$ be a P_I-name for a winning strategy for Player I in the unraveled version of the game $G_1(J)$. Let M be a countable elementary submodel of a large enough structure. First, an easy claim.

Claim 10.26 *If x is P_I-generic over M then $M[x] \models$ all equivalence classes of E_x are in the ideal J. In addition, if q is any infinite play against the strategy σ/x with moves in the model M, the result of q is a J-positive subset of B_x.*

Proof The statement "all equivalence classes of E_x are in J" is coanalytic, and therefore absolute between $M[x]$ and V. The statement "σ/x is a winning strategy for Player I" is a wellfoundedness statement about the tree of all possible counterplays, therefore absolute between the wellfounded model $M[x]$ and V. □

Since the poset P_I is bounding, by Fact 2.52 there is an I-positive set $A'' \subset A'$ consisting of P_I-generic points over M, and such that all Borel subsets of X in M are relatively clopen in A''. Towards the construction of the set C, by induction on $n \in \omega$ build plays p_n against the strategy σ, names $\dot{q}_n \in M$ for plays against the strategy $\dot{\tau}$, and clopen sets $O_n \subset 2^\omega \times 2^\omega$ such that:

I1. $p_0 \subset p_1 \subset \cdots$ and the initial move of p_0 is the set A''. Moreover, the closure of O_{n+1} is a subset of O_n for each $n \in \omega$;
I2. for every $x \in [p_n] \cap A$, $q_{n+1}/x \supset q_n/x$. The initial move of the play q_n/x is some J-positive compact subset of B_x;
I3. for every $x \in [p_n] \cap A$, $\{x\} \times [\dot{q}_n/x] \subset O_n$;
I4. for every $\langle x_0, y_0\rangle$ and $\langle x_1, y_1\rangle$ in $O_n \cap B$, if $x_0 \restriction n \neq x_1 \restriction n$ then $d(f(x_0, y_0), f(x_1, y_1) > n$.

Once the induction is performed, let $C = B \cap \bigcap_n O_n$. In view of I4, it must be the case that the vertical sections of C are pairwise E-unrelated. It is also the case that C is $I * J$-positive. To verify the positivity of the set C, note that the result of the play $\bigcup_n p_n$ is I-positive, and for every point x in the result, the vertical section C_x is a J-positive subset of the section B_x as it includes the result of the play $\bigcup_n \dot{q}_n/x$, which is J-positive by Claim 10.26. This will complete the proof of Case 2 by the first paragraph of Case 2.

The induction is elementary except for arranging the last item. Start with $p_0 = 0, \dot{q}_0 = 0$ and $O_0 = 2^\omega \times 2^\omega$. Suppose that p_n, O_n, q_n have been obtained. Consider the extension of p_n in which Player II first plays the empty set and the strategy σ responds with a finite set $K \subset 2^\omega$. For each $x \in K$, in the model $M[x]$ consider a tentative infinite extension q_x of the play $\dot{q}_n/x$ against the strategy τ/x in which the strategy indicates finite sets $L(m, x)$ for $m \in \omega$. We can pick such an extension that in addition satisfies that the finite sets $L(m, x)$ are pairwise E_x-unrelated: Since the classes of E_x are J-small, Player I can always start the round m with the E_x-saturation of the set $\bigcup_{k \in m} L(k, x)$, and this forces the strategy τ/x to answer with a set $L(m, x)$ that is E_x-unrelated to all the sets played previously.

The next step is to find numbers $m_x \in \omega$ for $x \in K$ such that the sets $\{x\} \times L(m_x, x)$ for $x \in K$ are pairwise E-unrelated. To do this, just enumerate $K = \{x_i : i \in j\}$ and choose the numbers m_{x_i} by induction on i. The induction step is easy: the finite set $\{x_k\} \times L(m_{x_k}, x_k)$ can be E-connected with only finitely many of the pairwise E-unrelated sets $\{x_i\} \times L(m, i)$ for every $k \in i$, and so for sufficiently large number m_{x_i}, the set $\{x_i\} \times L(m_{x_i}, x_i)$ must be E-unrelated to $\{x_k\} \times L(m_{x_k}, x_k)$ for all $k \in i$.

Now, find $k \in \omega$ large enough such that for distinct points $x, x' \in K$ and $y \in L(m_x, x)$, $y' \in L(m_{x'}, x')$ we have $d(f(x, y) \restriction k, f(x', y') \restriction k) > n + 1$; write $S(x) = \{f(x, y) \restriction k : y \in L(m_x, x)\}$. For each $x \in K$, use the continuity of the function f to find a clopen set $O_x \subset 2^\omega$ such that such that $L_x^{m_x} \subset O_x$ and $f''(B_x \cap O_x) \restriction k = S(x)$. Consider the finite play r_x which follows the play q_x up to the point where the strategy τ/x indicates the set $L(m_x, x)$, after which Player II responds with O_x which is the final move of the play $r_n \in M_x$. By the forcing theorem, there must be a P_I-name $\dot{q}^x_{n+1}$ such that $r_x = \dot{q}^x_{n+1}/x$ and a condition $A_x \in P_I$ such that $x \in A_x$ and $A_x \Vdash [\dot{q}^x_{n+1}] \subset O_x$. By the choice of A'' there must be a clopen set $P_x \subset 2^\omega$ so that $A \cap O_x \subset A_x$ and $f''(P_x \times O_x) \restriction k = S(x)$.

The induction step is now completed by letting $p_{n+1} = p_n{}^\frown 0^\frown K^\frown \bigcup_{x \in K_n} P_x$, q_{n+1} equal to any name in M such that $P_x \Vdash q_{n+1} = q^x_{n+1}$ for all $x \in K$, and $O_{n+1} = \bigcup_{x \in K} P_x \times O_x$. □

The theorem is now fairly easy to prove by induction on α. The successor step is immediately handled by the lemma. For the limit step, we will consider the iterations of length ω, and the rest will follow by the usual bootstrapping argument. For every number $n \in \omega$ let I^n be the n-fold Fubini power of the σ-ideal I on the space X^n, and let $I^{\geq n}$ be the iterated Fubini power of the σ-ideal I on the space $X^{\geq n} = X^{\omega \setminus n}$. The usual forcing factorization theorems for iteration show that $P_{I^\omega} = P_{I^n} * P_{I^{\geq n}}$.

Now assume that $B \subset X^\omega$ is a Borel I^ω-positive set and E is an equivalence relation on it reducible to E_{K_σ}. For every $n \in \omega$, let $A_n = \{\vec{x} \in X^n :$ there is an $I^{\geq n}$-positive set $D \subset X^{\geq n}$ such that the product $\{\vec{x}\} \times D$ is a subset of the set B and of a single E-equivalence class$\}$. As the ideal $I^{\omega \setminus n}$ is $\mathbf{\Pi}^1_1$ on $\mathbf{\Sigma}^1_1$, the set $A_n \subset X^n$ is analytic. There is now a split into two cases:

Case 1 Suppose first that there is $n \in \omega$ such that the set $A_n \subset X^n$ is I^n-positive. By Case 1 of Lemma 10.25 applied to the σ-ideals I^n and $I^{\geq n}$, there is a Borel set $B' \subset B$ and an equivalence relation $F \leq_B E$ on X^n such that $E \restriction B'$ is equal to $F \times \text{ev}$. Then, it is possible to use the induction hypothesis for the iteration of length n to get the desired simplification of the equivalence relation E on some I^ω-positive set $C \subset B$.

Case 2 Suppose now that the sets $\{A_n : n \in \omega\}$ are I^n-small for every $n \in \omega$. The set $A = \bigcup_n (A_n \times X^{\geq n}) \subset X^\omega$ is analytic and I^ω-small. Since the σ-ideal I^ω is $\mathbf{\Pi}^1_1$ on $\mathbf{\Sigma}^1_1$ by Fact 10.8, the first reflection theorem implies that A has an I^ω-small Borel superset A'. Look at the set $B' = B \setminus A'$; it is I^ω-positive, and so contains an I^ω-positive compact subset $B'' \subset B'$. We will find a Borel I-positive set $C \subset B''$ such that $E \restriction C = \text{id}$.

Choose a condition $\langle p_n : n \in \omega \rangle$ in the iteration of length ω which forces the generic to belong to the set B'', let $n \in \omega$, and consider what happens after the first n stages of the iteration. The remainder of the iteration is forced to be equal to the two-step iteration of P_I and $P_{I^{>n}}$ under some condition, and that condition forces the tail of the generic sequence to belong to some compact subset of $X \times X^{>n}$ such that the E-equivalence classes in its vertical sections are $I^{>n}$-small (recall the definitions of the sets A_n here). By Case 2 of Lemma 10.25 applied to I and $I^{>n}$ in the P_{I^n}-extension, the compact subset can be thinned out so that points in its distinct vertical sections are E-unrelated. By induction on $n \in \omega$, one can then strengthen the condition $\langle p_n : n \in \omega \rangle$ so that for every $n \in \omega$, $p \restriction n$ forces that there is a compact subset of $X \times X^{>n}$ such that points in its distinct vertical sections are E-unrelated, and the remainder of the condition forces the tail of the generic sequence to belong to this compact set. Now let M be a countable elementary submodel of a large structure containing the condition $\langle p_n : n \in \omega \rangle$, and let

$C \subset X^\omega$ be the Borel I^ω-positive set of all M-generic sequences meeting this condition. By an absoluteness argument, $C \subset B''$. Also, for any two distinct sequences $\vec{x}, \vec{y} \in C$, if n is the largest number such that $\vec{x} \restriction n = \vec{y} \restriction n$, the forcing theorem shows that there is a compact subset of $X \times X^{>n}$ in the model $M[\vec{x} \restriction n]$ in which distinct vertical sections contain no E-related points, and the tails of the sequences belong to this compact set, hence $\vec{x}, \vec{y}$ must be E-unrelated. Thus, $E \restriction C = \mathrm{id}$ as required.

This completes the proof of Theorem 10.1.

Proof of Corollary 10.2 Let I be the σ-ideal on X, let $\alpha \in \omega_1$ be a countable ordinal, let $B \subset X^\alpha$ be a Borel I^α-positive set, and let F be an equivalence relation on B Borel reducible to E. Let $h : B \to \omega^\omega$ be a function reducing F to E_{K_σ}. Thinning out the set B if necessary, we may assume that B is compact, $h \restriction B$ is continuous, and there is an ordinal $\beta \leq \alpha$ and an equivalence relation G on X reducible to E such that $F \restriction B = \mathrm{id}_\beta \times G \times \mathrm{ev}$. Thinning out further if necessary, we may adjust the function h so that $h(z)$ depends only on $z \restriction \beta+1$ for $z \in B$.

For every $x \in X^\beta$ write $B_x = \{y \in X : \{z \in X^{\alpha \setminus \beta+1} : x^\frown y^\frown z \in B\} \notin I^\alpha \setminus \beta + 1\}$. The set $\{\langle x, y\rangle \in X^\beta \times X : y \in B_x\}$ is analytic since the ideal $I^{\alpha\setminus\beta+1}$ is $\mathbf{\Pi}^1_1$ on $\mathbf{\Sigma}^1_1$. Choose a closed set $D \subset X^\beta \times X \times \omega^\omega$ projecting into it. Consider the following sets:

- $A = \{x \in X^\beta : B_x \notin I\}$;
- for every $n \in \omega$, $A_{\mathrm{ev},n} = \{\langle x, Y, g\rangle : x \in X^\beta,\ Y \subset X$ is I-positive compact, $g : Y \to \omega^\omega$ is continuous, $\forall y \in Y\ \langle x, y, g(y)\rangle \in D$ and $\forall y_0, y_1 \in C\ d(h(x^\frown y_0), h(x^\frown y_1)) \leq n\}$;
- $A_{\mathrm{id}} = \{\langle x, Y, g\rangle : Y \subset X$ is I-positive compact, $g : Y \to \omega^\omega$ is continuous, $\forall y \in Y\ \langle x, y, g(y)\rangle \in D$ and $\forall y_0, y_1 \in Y\ d(h(x^\frown y_0), h(x^\frown y_1)) = \infty\}$.

First observe that all these sets are analytic, since I is $\mathbf{\Pi}^1_1$ on $\mathbf{\Sigma}^1_1$ and the universal quantifiers range over compact sets only. Second, we claim that A is the projection of the set $\bigcup_n A_{\mathrm{ev},n} \cup A_{\mathrm{id}}$. Whenever $x \in A$, then by the canonization assumption the I-positive set B_x can be thinned down to an I-positive compact set Y on which G is either id or ev; if it is equal to ev, then by σ-additivity of I the set Y can be further thinned down so that the values $h(x, y)$ for $y \in Y$ are all within some fixed d-distance n from some point in ω^ω. Finally, using uniformization (Fact 2.51), Y can be thinned down further to find the continuous function $g : Y \to \omega^\omega$ such that for all $y \in Y$, $\langle x, y, g(y)$ holds. Then the triple $\langle x, Y, g\rangle$ belongs to one of the sets A_{id} or $A_{\mathrm{ev},n}$ for some $n \in \omega$.

The set A is I^β-positive. By the σ-additivity of the ideal I^β, the projection of one of the sets A_{id}, $A_{\mathrm{ev},n}$ must be I^β-positive. For definiteness assume that $p(A_{\mathrm{id}}) \notin I_\beta$. Use the I^β-uniformization (Fact 2.51) to thin down $p(A_{\mathrm{id}})$ to a compact I^β-positive set A' and continuous functions k, l on A' such that for every $x \in A'$, $\langle x, k(x), l(x)\rangle \in A_{\mathrm{id}}$. Let $C = \{z \in B : z \restriction \beta \in A'$ and $z(\beta) \in k(z \restriction \beta)\}$. This is an I^α-positive compact set on which $F \restriction C = \mathrm{id}_\beta$. This completes the proof of the corollary. □

10.5 The iteration of Sacks forcing

In this section, we will prove Theorem 10.3. To simplify the notation, write $X = 2^\omega$, $X^\alpha = (2^\omega)^\alpha$, let I be the σ-ideal of countable sets on X, and let P_α to be the poset of I^α-positive analytic subsets of X^α ordered by inclusion. Note that every analytic I^α-positive set has a Borel I^α-positive subset by Fact 10.8, and the Borel I^α-positive set has in turn a compact I^α-positive subset as the iteration is bounding by Fact 2.52. For a condition $p \in P_\alpha$ and an ordinal $\beta \in \alpha$ write $p \restriction \beta = \{\vec{x} \in X^\beta : \{\vec{y} \in X^{\alpha\setminus\beta} : \vec{x}^\frown\vec{y} \in p\} \notin I^{\alpha\setminus\beta}\}$. The σ-ideal $I^{\alpha\setminus\beta}$ is $\mathbf{\Pi}^1_1$ on $\mathbf{\Sigma}^1_1$ by Fact 10.8. Thus, $p \restriction \beta$ is an I^β-positive analytic set and as such a condition in the poset P_β.

We will need use a certain reduced product of the poset P_α. For an ordinal $\beta \in \alpha$, consider the *reduced product* P^α_β consisting of pairs $\langle p, q\rangle \in P_\alpha \times P_\alpha$ for which $p \restriction \beta = q \restriction \beta$. The reduced product adds sequences $\vec{x}_{\mathrm{lgen}}, \vec{x}_{\mathrm{rgen}} \in (2^\omega)^\alpha$ which coincide on their first β many coordinates. If M is a countable elementary submodel of a large structure, we call sequences $\vec{x}, \vec{y} \in (2^\omega)^\alpha$ *reduced product generic* over M if the set $\beta = \{\gamma \in \alpha : \vec{x}(\gamma) = \vec{y}(\gamma)\}$ is an ordinal and the sequences are P^α_β-generic for the model M. As in the product case, there is a key lemma:

Lemma 10.27 *Let M be a countable elementary submodel of a large enough structure, and $B \in P_{I_\alpha} \cap M$ is a condition. There is a Borel I_α-positive set $C \subset B$ consisting of pairwise reduced product generic sequences over M.*

Proof Consider a fusion-type poset Q for P_{I_α}. A condition $p \in Q$ is a triple $p = \langle a_p, n_p, r_p\rangle$ where $a_p \subset \alpha$ is a finite set, $n_p \in \omega$, and r_p is a function from $(2^{n_p})^{a_p}$ to $P_{I_\alpha} \restriction B$ such that for every $s, t \in \mathrm{dom}(r_p)$ and every ordinal $\beta \in a_p$, if $s \restriction \beta = t \restriction \beta$ then $r_p(t) \restriction \beta = r_p(s) \restriction \beta$. If M is a countable elementary submodel of a large structure and $G \subset Q$ is a filter generic over M, then one can form the set $C \subset (2^\omega)^\alpha = \bigcap_{p\in G} \bigcup \mathrm{rng}(r_p)$. This is the desired set. Checking its required properties is easy and left to the reader; we only state the central claim:

Claim 10.28 *Let $\beta \in \alpha$, $p \in Q$ and let $D \subset P^\alpha_\beta$ be an open dense set. There is a condition $q \leq p$ such that $\beta \in a_q$ and for every pair $s, t \in \text{dom}(r_q)$ such that β is the smallest ordinal on which they differ, $\langle r_q(s), r_q(t)\rangle \in D$.*

□

Now, suppose that E is an analytic equivalence relation on a Borel I_α-positive set $B \subset (2^\omega)^\alpha$. Let β be the least ordinal such that there is a condition $\langle B_0, C_0\rangle \in P^\alpha_\beta$ such that $\langle B_0, C_0\rangle \Vdash \vec{x}_{\text{lgen}}\ E\ \vec{x}_{\text{rgen}}$. Observe that then $\langle B_0, B_0\rangle \Vdash \vec{x}_{\text{lgen}}\ E\ \vec{x}_{\text{rgen}}$: given a P^α_β-generic pair $\langle \vec{x}_0, \vec{x}_1\rangle$ below the condition $\langle B_0, B_0\rangle$, a simple mutual genericity argument will produce a sequence $\vec{x}_2$ such that both pairs $\langle \vec{x}_0, \vec{x}_2\rangle$ and $\langle \vec{x}_1, \vec{x}_2\rangle$ are P^α_β-generic below the condition $\langle B_0, C_0\rangle$. By the forcing theorem then, $\vec{x}_0\ E\ \vec{x}_2\ E\ \vec{x}_1$, and by the transitivity of the relation E, $\vec{x}_0\ E\ \vec{x}_1$.

Now let M be a countable elementary submodel of a large structure containing E and B_0 and use the lemma to produce a Borel I_α-positive set $C \subset B_0$ consisting of pairwise reduced product generic sequences. For any two sequences $\vec{x}, \vec{y} \in C$ write $\gamma(\vec{x}, \vec{y})$ to denote the set of all ordinals on which their values agree. If $\gamma < \beta$ then $M[\vec{x}, \vec{y}] \models \neg\vec{x}\ E\ \vec{y}$ by the minimality of the ordinal β and the forcing theorem; the Mostowski absoluteness between $M[\vec{x}, \vec{y}]$ and V then shows that $\neg\vec{x}\ E\ \vec{y}$ holds in V. If $\gamma = \beta$ then by a similar argument $\vec{x}\ E\ \vec{y}$. If $\gamma > \beta$ then we must use the transitivity of the equivalence relation E again. Use a mutual genericity argument to produce a sequence $\vec{z} \in B$ such that both pairs $\langle \vec{x}, \vec{z}\rangle$ and $\langle \vec{y}, \vec{z}\rangle$ are P^α_β-generic over M, use the forcing theorem and absoluteness to argue that $\vec{x}\ E\ \vec{z}\ E\ \vec{y}$, and by transitivity $\vec{x}\ E\ \vec{y}$ again. In conclusion, $E \restriction C = \text{id}_\beta \times \text{ev}$ as desired.

10.6 Anticanonization results

Proof of Theorem 10.4 Let I be a $\mathbf{\Pi}^1_1$ on $\mathbf{\Sigma}^1_1$ σ-ideal on a Polish space X, and let E be a Borel nontrivially I-ergodic equivalence relation on X. Let $f(\vec{x}, \vec{y})(n) = 1$ if $\vec{x}(n)\ E\ \vec{y}(n)$. We claim that this is a rectangular coding function for I^ω. Suppose that $B, C \subset X^\omega$, are Borel I^ω-positive sets. By Fact 10.8, it contains all branches of a Borel I-branching trees $S, T \subset X^{<\omega}$ respectively. Let $z \in 2^\omega$ be an arbitrary point. By induction on $n \in \omega$, build nodes $s_n \in S$, $t_n \in T$ of length n such that $s_{n+1}(n)\ E\ t_{n+1}(n)$ iff $z(n) = 1$. The induction step uses the ergodicity of the equivalence relation E: the sets $\{x \in X : s_n{}^\frown x \in T\}$ and $\{x \in X : t_n{}^\frown x \in T\}$ are I-positive, and the nontrivial ergodicity of E implies the existence of both an E-related and E-unrelated pair whose components come from the respective two sets. In the end, the sequences $\vec{x} = \bigcup_n s_n$ and $\vec{y} = \bigcup_n y_n$ satisfy $f(\vec{x}, \vec{y}) = z$ as required.

For the spectrum, suppose that J is an analytic ideal on ω and let E_J be the equivalence relation on X^ω defined by $\vec{x}\ E_J\ \vec{y}$ if the set $\{n \in \omega : \neg\vec{x}(n)\ E\ \vec{y}(n)\}$ belongs to the ideal J. The argument from the previous paragraph shows that $=_J$ is reducible to E_J restricted to any Borel I-positive subset of X^ω. Thus the complexity of equivalence relations in the iteration grows to encompass a cofinal set of Borel equivalence relations as well as a complete analytic equivalence relation. □

We do not know anything tangible about the canonization properties of iterations of posets that add unbounded reals. The following theorem shows that the mutual genericity arguments from the iteration of Sacks forcings above cannot work: there is a rectangular coding function. The argument is a slight upgrade of the proof of Velickovic–Woodin coding on triples (Velickovic and Woodin 1998).

Theorem 10.29 *Let I be the σ-ideal of σ-compact subsets of ω^ω. There is a Borel function $f : \omega^\omega \times (\omega^\omega \times \omega^\omega) \to 2^\omega$ such that for every I-positive Borel set $B \subset \omega^\omega$ and every $I * I$-positive Borel set $C \subset \omega^\omega \times \omega^\omega$ the image $f''B \times C$ contains a nonempty open set.*

Corollary 10.30 *Let J be a $\mathbf{\Pi}^1_1$ on $\mathbf{\Sigma}^1_1$ σ-ideal on a Polish space X, such that every analytic J-positive set contains a Borel J-positive subset. If the poset P_J is proper and adds an unbounded real then the two-step iteration ideal J has a rectangular coding function.*

Proof of Corollary 10.30 Let $\dot{z}$ be a P_J-name for an unbounded real, and use Fact 2.51 to find a Borel J-positive set $B \subset X$ and a Borel function $g : B \to \omega^\omega$ such that $B \Vdash \dot{z} = \dot{g}(\dot{x}_{\text{gen}})$. It must be the case that the preimages of I-small sets are J-small. Consider the function $h : B \times B \to 2^\omega$ given by $h(\langle y_0, y_1\rangle, \langle x_0, x_1\rangle) = f(g(y_0), g(x_0), g(x_1))$. The h-image of any Borel $J * J$-positive rectangle has nonempty interior, and so Proposition 2.67 yields a coding function on the $J * J$-positive set $B \times B$. □

Proof of Theorem 10.29 To define the functional value $f(y, x_0, x_1)$, let $\langle n_i : i \in \omega\rangle$ enumerate those numbers n such that $x_0(m) > y(m)$ for some m between $x_1(n)$ and $x_1(n+1)$. Let $f(y, x_0, x_1)(i) = 0$ if in the interval $[x_1(n_i), x_1(n_i + 1))$ the values of x, y oscillate fewer than four times, and $f(y, x_0, x_1)(i) = 1$ otherwise. If some of the numbers n_i are undefined, let $f(y, x_0, x_1) = \text{trash}$. This function works.

Suppose that $B \subset \omega^\omega$ and $C \subset \omega^\omega \times \omega^\omega$ are I-positive and $I * I$-positive Borel sets respectively. Let T be a superperfect tree such that $[T] \subset C$. For simplicity, assume that the trunk of T is 0; we will show that $f''B \times C = 2^\omega$.

Let $z \in 2^\omega$ be a prescribed value; we must find points $y \in B$ and $\langle x_0, x_1\rangle \in C$ such that $f(y, x_0, x_1) = z$.

For the iteration of Miller forcing of length 2, write $\dot{x}_{0\text{gen}}$ for the first generic point and $\dot{x}_{1\text{gen}}$ for the second. Find a condition p in the iteration which forces the pair $\langle \dot{x}_{0\text{gen}}, \dot{x}_{1\text{gen}}\rangle$ into $\dot{C}$. Find a countable elementary submodel M of a large enough structure and by induction on $i \in \omega$ build finite integer sequences s_i, t_i, u_i, numbers n_i and conditions p_i in the iteration such that:

I1. the sequences s_i extend as i increases, the same for t_i and u_i. The length $|s_i|$ is smaller than the last entry on the sequence u_i, the length of t_i is larger than the last entry of u_i, and $n_i = |u_i| - 1$;

I2. s_i is a splitnode of the tree S, and p_i is a condition of the form $\langle T_i, \dot{U}_i\rangle$ such that t_i is a splitnode of the tree T_i and p_i forces u_i to be a splitnode of the tree $\dot{U}_i$;

I3. $f(s_i, t_i, u_i) = z \restriction i$;

I4. the conditions p_i decrease as i increases and $p_i \in M$ meets the i-th open dense subset of the iteration in the model M in some fixed enumeration.

If this induction succeeds, then in the end consider the points $y = \bigcup_i s_i$, $x_0 = \bigcup_i t_i$ and $x_1 = \bigcup_i u_i$. Clearly, $f(y, x_0, x_1) = z$ by item I3. The point $y \in \omega^\omega$ is a branch through the tree S and therefore an element of the set B. Writing g for the filter generated by the conditions $\{p_i : i \in \omega\}$ in the two-step iteration, it is obvious that g is M-generic and $x_0 = \dot{x}_{0\text{gen}}/g$, $x_1 = \dot{x}_{1\text{gen}}/g$ and so $\langle x_0, x_1\rangle \in C$ by the forcing theorem. This will conclude the proof.

To start the induction, just find a condition $p_0 \leq p$ in the model M such that $p_0 = \langle T_0, \dot{U}_0\rangle$, t_0 is the trunk of T_0 and T_0 forces some specific sequence u_0 to be the trunk of $\dot{U}_0$; moreover, let $s_0 = 0$. Now suppose s_i, t_i, u_i are given, and assume that $z(i) = 0$ (the case $z(i) = 1$ is similar). Use the superperfect property of the trees S, T_i to choose splitnodes $s_{i+1} \in S$ and $t' \in T_i$ such that $|s_{i+1}| < |t'|$, $s_i \subset s_{i+1}$, $t_i \subset t'$, and $s_{i+1}(|s_i|) > t_i(n_i)$, and moreover, the sequences s_{i+1}, t' oscillate only twice on $\text{dom}(s_{i+1}) \setminus n_i$. Find some condition $T' \leq T_i \restriction t'$ in the Miller forcing in the model M indicating a number $m > \text{dom}(s_{i+1})$ such that $u' = u_i^\frown m \in \dot{U}_i$. Find some condition $p_{i+1} \leq \langle T', \dot{U}_i \restriction u'\rangle$ in the model M in the $i+1$-th open dense set and strengthen the first coordinate if necessary to decide the first splitnode of $\dot{U}_{i+1}$ to be some definite u_{i+1}. Further, we can strengthen the first coordinate if necessary to make sure that its trunk t_{i+1} is longer than the last entry on u_{i+1}. This concludes the induction step. □

References

Bagaria, J. and DiPrisco, C. A. (2009) Parameterized partition relations on the real numbers, *Archive for Mathematical Logic* **48**, 201–226.

Balcar, B. and Simon, P. (1989) Disjoint refinement. In *Handbook of Boolean Algebras*. Amsterdam: North Holland, pp. 333–388.

Balcar, B., Jech, T. and Pazák, T. (2005) Complete CCC Boolean algebras, the order sequential topology, and a problem of von Neumann, *Bulletin of the London Mathematical Society* **37**, 885–898.

Bartoszynski, T. and Judah, H. (1995) *Set Theory: On the Structure of the Real Line*. Wellesley, MA: A K Peters.

Becker, H. and Kechris, A. (1996) *The Descriptive Set Theory of Polish Group Actions*. London Mathematical Society Lecture Notes Series 232. Cambridge: Cambridge University Press.

Bell, J. L. (2005) *Set Theory*. Oxford Logic Guides 47. Oxford: Clarendon Press.

Blass, A. (1989) Applications of superperfect forcing and its relatives. In *Set Theory and its Applications, Toronto 1987*. Lecture Notes in Mathematics 1401. Berlin: Springer Verlag, pp. 18–40.

Brendle, J., Hjorth, G. and Spinas, O. (1995) Regularity properties for dominating projective sets, *Annals of Pure and Applied Logic* **72**, 291–307.

Cichon, J., Morayne, M., Pawlikowski, J. and Solecki, S. (1991) Decomposing Baire functions, *Journal of Symbolic Logic* **56**, 1273–1283.

Clemens, J. and Zapletal, J. (2012) Dichotomies for ideals, in preparation.

Connes, A., Feldman, J. and Weiss, B. (1981) An amenable equivalence relation is generated by a single transformation, *Ergodic Theory and Dynamical Systems* **1**, 431–450.

Debs, G. (1995) Polar σ-ideals of compact sets, *Transactions of the American Mathematical Society* **347**, 317–338.

Debs, G. and Saint-Raymond, J. (1987) Ensembles boréliens d'unicité et d'unicité au sens large, *Annals de l'Institut Fourier (Grenoble)* **37**, 217–239.

Doucha, M. (2013) Ph.D. thesis, Charles University.

Erdős, P. and Rado, R. (1950) A combinatorial theorem, *Journal of the London Mathematical Society* **25**, 249–255.

Farah, I. and Zapletal, J. (2004) Between Maharam's and von Neumann's problems, *Mathematical Research Letters* **11**, 673–684.

Feldman, J. and Moore, C. C. (1977) Ergodic equivalence relations, cohomology and von Neumann algebras, *Transactions of the American Mathematical Society* **234**, 289–324.

Foreman, M. and Magidor, M. (1995) Large cardinals and definable counterexamples to the continuum hypothesis, *Annals of Pure and Applied Logic* **76**, 47–97.

Galvin, F. and Prikry, K. (1973) Borel sets and Ramsey's theorem, *Journal of Symbolic Logic* **38**, 193–198.

Gao, S. (2009) *Invariant Descriptive Set Theory*. Boca Raton, CA: CRC Press.

Grigorieff, S. (1971) Combinatorics on ideals and forcing, *Annals of Mathematical Logic* **3**, 363–394.

Groszek, M. (1987) Combinatorics of ideals and forcing with trees, *Journal of Symbolic Logic* **52**, 582–593.

Groszek, M. (1994) ω_1^* as an initial segment of the *c* degrees, *Journal of Symbolic Logic* **59**, 956–976.

Harrington, L., Kechris, A. S. and Louveau, A. (1990) A Glimm–Effros dichotomy for Borel equivalence relations, *Journal of the American Mathematical Society* **3**, 903–928.

Harrington, L., Marker, D. and Shelah, S. (1988) Borel orderings, *Transactions of the American Mathematical Society* **310**, 293–302.

Hjorth, G. (2000) Actions by the classical Banach spaces, *Journal of Symbolic Logic* **65**, 392–420.

Humke, P. D. and Preiss, D. (1985) Measures for which σ-porous sets are null, *Journal of the London Mathematical Society* **32**, 236–244.

Ishiu, T. (2005) α-properness and Axiom A, *Fundamenta Mathematicae* **186**, 25–37.

Jech, T. (2002) *Set Theory*. New York: Springer Verlag.

Jech, T. and Balcar, B. (2006) Weak distributivity, a problem of von Neumann, and the mystery of measurability, *Bulletin of Symbolic Logic* **12**, 241–265.

Jossen, S. and Spinas, O. (2005) A two-dimensional tree ideal. In *Logic Colloquium 2000*. Lecture Notes in Logic 19. Urbana, IL: Association for Symbolic Logic, pp. 294–322.

Judah, H., Roslanowski, A. and Shelah, S. (1994) Examples for souslin forcing, *Fundamenta Mathematicae* **144**, 23–42. Math.LO/9310224.

Kanamori, A. (1994) *The Higher Infinite*. New York: Springer-Verlag.

Kanovei, V. (1999) On non-well-founded iterations of the perfect set forcing, *Journal of Symbolic Logic* **64**, 191–207.

Kanovei, V. (2008) *Borel Equivalence Relations*. University Lecture Series 44. Providence, RI: American Mathematical Society.

Kanovei, V. and Reeken, M. (2000) New Radon-Nikodym ideals, *Mathematika* **47**, 219–227.

Kechris, A. and Louveau, A. (1989) *Descriptive Set Theory and the Structure of Sets of Uniqueness*. Cambridge: Cambridge University Press.

Kechris, A. and Louveau, A. (1997) The structure of hypersmooth Borel equivalence relations, *Journal of American Mathematical Society* **10**, 215–242.

Kechris, A. and Miller, B. (2004) *Topics in Orbit Equivalence*. Lecture Notes in Mathematics 1852. New York: Springer Verlag.

Kechris, A., Louveau, A. and Woodin, H. (1987) The structure of σ-ideals of compact sets, *Transactions of the American Mathematical Society* **301**, 263–288.

Kechris, A. S. (1977) On a notion of smallness for subsets of the Baire space, *Transactions of the American Mathematical Society* **229**, 191–207.

Kechris, A. S. (1994) *Classical Descriptive Set Theory*. New York: Springer Verlag.

Kellner, J. and Shelah, S. (2011) Saccharinity, *Journal of Symbolic Logic* **74**, 1153–1183. Math.LO/0511330.

Klein, O. and Spinas, O. (2005) Canonical forms of Borel functions on the Milliken space, *Transactions of the American Mathematical Society* **357**, 4739–4769.

Lindenstrauss, J., Preiss, D. and Tišer, J. (2012) *Fréchet Differentiability of Lipschitz Functions and Porous Sets in Banach Spaces*. Annals of Mathematics Studies 179. Princeton, NJ: Princeton University Press.

Martin, D. A. (1985) A purely inductive proof of Borel determinacy. In A. Nerode and R. A. Shore (eds), *Recursion Theory*. Proceedings of Symposia in Pure Mathematics 42. Providence, RI: American Mathematical Society, pp. 303–308.

Matet, P. (1988) Some filters of partitions, *Journal of Symbolic Logic* **53**, 540–553.

Mathias, A. R. (1977) Happy families, *Annals of Mathematical Logic* **12**, 59–111.

Mazur, K. (1991) F_σ-ideals and $\omega_1\omega_1*$ gaps in the Boolean algebra $P(\omega)/I$, *Fundamenta Mathematicae* **138**, 103–111.

Miller, A. W. (1984) Rational perfect set forcing. In *Axiomatic Set Theory*. Contemporary Mathematics 31. Providence, RI: American Mathematical Society, pp. 143–159.

Mycielski, J. (1964) Independent sets in topological algebras, *Fundamenta Mathematicae* **55**, 139–147.

Nešetřil, J. (2005) Ramsey classes and homogeneous structures, *Combinatorics, Probability and Computing* **14**, 171–189.

Pelant, J. and Zelený, M. (2004) The structure of the σ-ideal of σ-porous sets, *Commentationes Mathematicae Universitatis Carolinae* **45**, 37–72.

Pestov, V. (2006) *Dynamics of Infinite-Dimensional Groups*. University Lecture Series 40. Providence, RI: American Mathematical Society.

Pol, R. and Zakrzewski, P. (2012) On Borel mappings and σ-ideals generated by closed sets, *Advances in Mathematics* **231**, 651–663.

Prömel, H. J. and Voigt, B. (1985) Canonical forms of Borel-measurable mappings $\Delta : [\omega]^\omega \to R$, *Journal of Combinatorial Theory Series A* **40**, 409–417.

Rajchman, A. (1922) Sur l'unicité du développement trigonométrique, *Fundamenta Mathematicae* **3**, 287–302.

Rojas, D. (2006) Using determinacy to inscribe compact non-σ-porous sets into projective non-σ-porous sets, *Real Analysis Exchange* **32**, 55–66.

Rosendal, C. (2005) Cofinal families of Borel equivalence relations and quasiorders, *Journal of Symbolic Logic* **70**, 1325–1340.

Rosłanowski, A. (2006) n-localization property, *Journal of Symbolic Logic* **71**, 881–902.

Sabok, M. (2009) σ-continuity and related forcings, *Archive for Mathematical Logic* **48**, 449–464.

Sabok, M. (2012a) Forcing, games, and families of closed sets, *Transactions of the American Mathematical Society* **364**, 4011–4039.

Sabok, M. (2012b) Complexity of Ramsey null sets, *Advances in Mathematics* **230**, 1184–1195.

Sabok, M. and Zapletal, J. (2011) Forcing properties of ideals of closed sets, *Journal of Symbolic Logic* **76**, 1075–1095.

Shelah, S. (1998) *Proper and Improper Forcing*. 2nd edn. New York: Springer Verlag.

Shelah, S. and Zapletal, J. (2011) Ramsey theorems for product of finite sets with submeasures, *Combinatorica* **31**, 225–244.

Silver, J. (1970) Every analytic set is Ramsey, *Journal of Symbolic Logic* **35**, 60–64.

Silver, J. (1980) Counting the number of equivalence classes of Borel and coanalytic equivalence relations, *Annals of Mathematical Logic* **18**, 1–28.

Solecki, S. (1994) Covering analytic sets by families of closed sets, *Journal of Symbolic Logic* **59**, 1022–1031.

Solecki, S. (1999) Analytic ideals and their applications, *Annals of Pure and Applied Logic* **99**, 51–72.

Solecki, S. and Spinas, O. (1999) Dominating and unbounded free sets, *Journal of Symbolic Logic* **64**, 75–80.

Spinas, O. (2001a) Canonical behavior of Borel functions on superperfect rectangles, *Journal of Mathematical Logic* **1**, 173–220.

Spinas, O. (2001b) Ramsey and freeness properties of Polish planes, *Proceedings of the London Mathematical Society* **82**, 31–63.

Spinas, O. (2009) Proper products, *Proceedings of the American Mathematical Society* **137**, 2767–2772.

Steprans, J. (1993) A very discontinuous Borel function, *Journal of Symbolic Logic* **58**, 1268–1283.

Stern, J. (1984) On Lusin's restricted continuum problem, *Annals of Mathematics* **120**, 7–37.

Taylor, A. D. (1976) A canonical partition relation for finite subsets of ω, *Journal of Combinatorial Theory Series A* **21**, 137–146.

Todorcevic, S. (2010) *Introduction to Ramsey Spaces*. Annals of Mathematics Studies 174. Princeton, NJ: Princeton University Press.

Velickovic, B. and Woodin, W. H. (1998) Complexity of the reals in inner models of set theory, *Annals of Pure and Applied Logic* **92**, 283–295.

Zajíček, L. (1987/88) Porosity and σ-porosity, *Real Analysis Exchange* **13**, 314–350.

Zajíček, L. and Zelený, M. (2005) Inscribing compact non-σ-porous sets into analytic non-σ-porous sets, *Fundamenta Mathematicae* **185**, 19–39.

Zapletal, J. (2003) Isolating cardinal invariants, *Journal of Mathematical Logic* **3**, 143–162.

Zapletal, J. (2004) *Descriptive Set Theory and Definable Forcing*. Providence, RI: American Mathematical Society.

Zapletal, J. (2008) *Forcing Idealized*. Cambridge Tracts in Mathematics 174. Cambridge: Cambridge University Press.

Zapletal, J. (2009) Preservation of P-points and definable forcing, *Fundamenta Mathematicae* **204**, 145–154.

Zapletal, J. (2011) Overspill and forcing, preprint.

Index

Page numbers in italics indicate where the concept is defined.

$\geq_K$, *43*, 106, 171, 186
$\leq_B$, *16*
μ-ergodic, *8*
ω_1 inaccessible to the reals, *13*, 87, 97, 134
$\mathbf{\Delta}^1_2$ on $\mathbf{\Sigma}^1_1$, *86*, 98, 135
$\mathbf{\Pi}^1_1$ on $\mathbf{\Sigma}^1_1$, 10, *14*, 38, 95, 97, 103, 121, 153, 172, 190, 208, 232, 252

absoluteness
  Mostowski, *23*
  Shoenfield, *23*, 71, 79, 89, 91, 116, 210
Axiom of Determinacy, 86

capacity, 96
c.c.c., 10
coding function, *6*, 42, 106, 113, 170, 185, 217, 262
concentration of measure, *42*, 108, 138
continuous reading of names, *35*, 100, 177, 208, 218, 225
countable chain condition, *see* c.c.c.
cube
  s-cube, *152*
  Ellentuck, *196*
  Hamming, *42*, 108
  Matet, *207*
  Milliken, *202*
  pure Matet, *207*
  pure Milliken, *202*
  reverse, *152*, 158
  Silver, *190*

E_0-box, *223*
equivalence
  $=_J$, *16*, 121, 129, 172, 207, 232
  E_0, *16*, 217
  E_1, *16*, 76, 150, 233
  E_2, *16*, 172, 184
  E_{K_σ}, *16*, 191, 200, 207
  E_{tail}, 203
  μ-ergodic, *18*, 140
  classifiable by countable structures, *16*, 68, 190, 207, 232
  countable, 63, 203
  ergodic, *10*, 45, 49
  essentially countable, 9, *16*, 67, 230
  generically ergodic, *18*, 115
  hypersmooth, *16*, 76, 157
  nontrivially ergodic, *10*, 136, 140
  orbit, *17*, 21, 56, 64, 68, 69, 78, 136, 150, 167, 233
  smooth, *16*, 62, 159, 224
  trim, *71*
ev, 16

forcing
  $< \omega_1$-proper, *92*
  E_0, *147*
  E_2, *172*
  P_I, *33*
  P^E_I, *56*, 184, 191, 199, 233
  Coll(ω, κ), *19*, *45*, *54*
  bounding, *20*, 34, 69, 137, 190
  Cohen, *115*
  Gandy–Harrington, *15*, 174, 180, 183
  Laver, *134*
  Matet, *208*
  nowhere c.c.c., *20*, 64, 68, 92, 102, 135
  proper, 10, *20*, 34, 208

reduced product, 148, 187
weakly homogeneous, *18*, 64
fusion sequence, 158, 191, 208, 233

game
Boolean, 125, 219, 249
integer, 80
refinement, 125
generic ultrapower, *31*, 50, 52
generically ergodic, *8*

id, 16
ideal
σ-continuity, 96, 123, 133
σ-finite Hausdorff mass, 57, 69, 84, 96
box product, 218
c.c.c., 38
calibrated, *103*
game, *81*, 121, 137
generated by closed sets, 124
Laver, 85, *134*
meager, 115
metric porosity, 123, 133
porosity, *120*
Ramsey null, 40
interpretation, *24*, 28, 32, 50

Katětov order, *43*, 102, 106, 120, 136, 146, 171
kernel, *58*, 195, 239

large enough structure, *21*

Maharam algebra, 38
model
$V[[x]]_E$, *54*, 74, 184, 233
$V[x]_E$, *45*, *60*, 184, 191, *217*, 233

property
covering, *103*, 106, *218*, 219
free set, *5*, 36, 103, 111, 113, 133
Fubini, *35*
Laver, *20*, 122
mutual generics, 36, 188
rectangular Ramsey, *35*, 40, 95, 96, 133, 232
selection, *90*, 100, 189, 190, 208
separation, *74*, 235
Silver, *4*, *97*, 103, 110, 111
weak Laver, 111
weak Sacks, 12, *247*

real
dominating, *20*, 134
independent, *20*, 40, 69, 137, 190, 208, 247
infinitely equal, *122*, 129
rew, *12*

scrambling tool, *81*
slalom, *247*
spectrum, *6*, 184, 217, 233

total canonization, 3, 64, 68, 74, 91, 95, 97, 102, 103, 121, 129, 134, 137
V-generic, 13